OPTIMIZATION OF FINITE DIMENSIONAL STRUCTURES

OPTIMIZATION OF FINITE DIMENSIONAL STRUCTURES

MAKOTO OHSAKI

HIROSHIMA UNIVERSITY
HIGASHI-HIROSHIMA, JAPAN

CRC Press
Taylor & Francis Group
Boca Raton   London   New York

CRC Press is an imprint of the
Taylor & Francis Group, an **informa** business

CRC Press
Taylor & Francis Group
6000 Broken Sound Parkway NW, Suite 300
Boca Raton, FL 33487-2742

First issued in paperback 2017

ISBN 13: 978-1-138-11365-7 (pbk)
ISBN 13: 978-1-4398-2003-2 (hbk)

Library of Congress Cataloging-in-Publication Data

Ohsaki, Makoto, 1960-
   Optimization of finite dimensional structures / author, Makoto Ohsaki.
      p. cm.
   "A CRC title."
   Includes bibliographical references and index.
   ISBN 978-1-4398-2003-2 (hardcover : alk. paper)
   1. Structural optimization. 2. Finite element method. I. Title.

TA658.8.O678 2011
624.1'7713--dc22

2010018870

Visit the Taylor & Francis Web site at
http://www.taylorandfrancis.com

and the CRC Press Web site at
http://www.crcpress.com

Preface

The attempt to find mechanically efficient structural designs and shapes was initiated mainly in the fields of mechanical engineering and aeronautical engineering, which established the field known as *structural optimization*. Many practically acceptable results have been developed for application to automobiles and aircraft. Some examples are structural components, including the wings of aircraft and engine mounts of automobiles, which can be fully optimized using efficient shape optimization techniques.

In contrast, regarding civil engineering and architectural engineering, structural optimization is difficult to apply because structures in these fields are not mass products: structures are designed in accordance with their specific design requirements. Furthermore, the structure's shape and geometry are determined by a designer or an architect in view of nonstructural performance, including the aesthetic perspective. Therefore, the main role of structural engineers is often limited to selection of materials, determination of member sizes through structural analyses, planning details of the construction process, and so on. However, for special structures, such as shells, membrane structures, spatial long-span frames, and highrise buildings, the structural shape should be determined in view of the responses against static and dynamic loads. In truth, the beauty of the structural form is related closely to the mechanical performance of the structure. Therefore, cooperation between designers and structural engineers is very important in designing such structures.

Even for building frames, because of the recent trend of *performance-based design*, optimization has been identified as a powerful tool for designing structures under constraints imposed on practical performance measures, including elastic/plastic stresses and displacements under static/dynamic design loads. Furthermore, recent rapid advancements in the areas of computer hardware and software enabled us to carry out structural analysis many times to obtain optimal or approximately optimal designs. Optimization of real-world structures with realistic objective function and constraints is possible through quantitative evaluation of nonstructural performance criteria, e.g., aesthetic properties, and life-cycle costs, including costs of construction, fabrication, and maintenance.

Many books describing structural optimization have been published since the 1960s; e.g., Hemp (1973), Rozvany (1976), Haug and Cea (1981), Haftka, Gürdal, and Kamat (1990), Papalambos and Wilde (2000), Bendsøe and Sigmund (2003), Arora (2004), etc. These books are mainly classifiable into the following three categories:

1. Basic theories and methodologies for optimization with examples of small structural optimization problems.

2. Continuum-based approaches for application to mechanical and aeronautical structures.

3. Theoretical and analytical results of structural optimization in earlier times without the assistance of computer technology.

Using books of the first category, readers can learn only the concepts and some difficult theories of structural optimization without application to large-scale structures. On the other hand, for the books of the second category, a good background in applied mathematics and continuum mechanics is needed to fully understand the basic concepts and methods. Unfortunately, most researchers, practicing engineers, and graduate students in the field of civil engineering have no such background and are not strongly interested in the basic theories or methods of structural optimization. Also, in mechanical engineering, the finite element approach is used for practical applications, and complex practical design problems are solved in a finite dimensional formulation.

The derivatives of objective and constraint functions, called design sensitivity coefficients, should be computed if a gradient-based approach is used for structural optimization. However, most methods of *design sensitivity analysis* are developed mainly for a continuum utilizing variational principles, for which sensitivity coefficients are to be computed for a functional, such as compliance that can be formulated in an integral of a bilinear form of response. For finite dimensional structures, including trusses and frames, variational formulations are not needed, and sensitivity coefficients can be found simply by differentiating the governing equations in a matrix-vector form.

Another important aspect of optimization in civil engineering is that the design variables often have discrete values: the frame members are usually selected from a pre-assigned list or catalog of available sections. Furthermore, some traditional layouts are often used for plane and spatial trusses and for latticed domes. Therefore, the optimization problem often turns out to be a combinatorial problem, a fact that is not fully introduced into most books addressing the study of structural optimization.

This book introduces methodologies and applications that are closely related to design problems of *finite dimensional structures*, to serve thereby as a bridge between the communities of structural optimization in mechanical engineering and the researchers and engineers in civil engineering. The book provides readers with the basics of optimization of frame structures, such as trusses, building frames, and long-span structures, with descriptions of various applications to real-world problems.

Recently, many efficient techniques of optimization have been developed for convex programming problems, e.g., semidefinite programming and interior point algorithms, which are extensions of the approaches used for linear and quadratic programming problems. The book introduces application of these methods to optimization of finite-dimensional structures. Approximate methods resembling the conventional optimality criteria approaches have also been developed with no reference to the pioneering papers in the 1960s and 1970s. Therefore, it is extremely important to describe their development history to young researchers so that similar methods are not re-developed with no knowledge related to conventional approaches. For that reason, another purpose of this book is to present the historical development of the methodologies and theorems on optimization of frame structures.

The book is organized as follows:

In Chapter 1, the basic concepts and methodologies of optimization of trusses and frames are presented with illustrative examples. Traditional problems with constraints on limit loads, member stresses, compliance, and eigenvalues of vibration are described in detail. A brief introduction is also presented for multiobjective structural optimization, and the shape and topology optimization of trusses.

In Chapter 2, the method of design sensitivity analysis, which is a necessary tool for optimization using nonlinear programming, is presented for various response quantities, including static response, eigenvalue of vibration, transient response for dynamic load, and so on. All formulations are written in matrix-vector form without resort to variational formulation to support ready comprehension by researchers and engineers.

In Chapter 3, details of truss topology optimization are described, including historical developments and difficulties in problems with stress constraints and multiple eigenvalue constraints. Recently developed formulations by semidefinite programming and mixed integer programming are introduced. Applications to plane and spatial trusses are demonstrated.

Chapter 4 presents methods for configuration optimization for simultaneously optimizing the geometry and topology of trusses. Difficulties in optimization of regular trusses are extensively discussed, and an application to generating a link mechanism is presented.

Chapter 5 summarizes various results of optimization of building frames. Uniqueness of the optimal solution of a regular frame is first investigated, and applications of parametric programming are presented. Multiobjective optimization problems are also presented for application to seismic design, and a simple heuristic method based on local search is presented.

In Chapter 6, as a unique aspect of this book, optimization results are presented for spatial trusses and latticed domes. Simple applications of nonlinear programming and heuristic methods are first introduced, and the spatial variation of seismic excitation is addressed in the following sections. The trade-off designs between geometrical properties and stiffness under static loads are shown for arch-type frames and latticed domes described using parametric curve and surface.

Mathematical preliminaries and basic methodologies are summarized in the Appendix, so that readers can understand the details, if necessary, without the exposition of tedious mathematics presented in the main chapters. Various methodologies specifically utilized in some of the sections, e.g., the response spectrum approach for seismic response analysis, are also explained in the Appendix. Also, many small examples that can be solved by hand or using a simple program are presented in the main chapters. Therefore, this book is self-contained, and easily used as a textbook or sub-textbook in a graduate course.

The author would like to deliver his sincere appreciation to Prof. Tsuneyoshi Nakamura, Prof. Emeritus of Kyoto University, Japan, for supervising the author's study for master's degree and Ph.D. dissertation on structural optimization. Supervision by Prof. Jasbir S. Arora of The University of Iowa during the author's sabbatical leave is also acknowledged.

The numerical examples in this book are a compilation of the author's work on structural optimization at Kyoto University, Japan, during the period 1985–2010. The author would like to extend his appreciation to researchers for collaborations on the studies that appear as valuable contents in this book, namely, Prof. Naoki Katoh of the Dept. of Architecture and Architectural Engineering, Kyoto University; Prof. Shinji Nishiwaki of the Dept. of Mechanical Engineering and Science, Kyoto University; Prof. Hiroshi Tagawa of the Dept. of Environmental Engineering and Architecture, Nagoya University; Prof. Yoshihiro Kanno of the Dept. of Mathematical Informatics, University of Tokyo; Prof. Peng Pan of the Dept. of Civil Engineering, Tsinghua University, P. R. China; Dr. Takao Hagishita of Mitsubishi Heavy Industries; Mr. Yuji Kato of JSOL Corporation; Mr. Takuya Kinoshita, Mr. Shinnosuke Fujita, and Mr. Ryo Watada, graduate students in the Dept. of Architecture and Architectural Engineering, Kyoto University. The author would also like to thank again Prof. Yoshihiro Kanno of University of Tokyo for checking the details of the manuscript.

The assistance of Ms. Kari Budyk and Ms. Leong Li-Ming of CRC Press and Taylor & Francis in bringing the manuscript to its final form is heartily acknowledged.

January 2010   Makoto Ohsaki

Contents

Chapter 1

Various Formulations of Structural Optimization

Various formulations of optimization of finite dimensional structures are presented in this chapter. The concepts of structural optimization are first presented in Sec. 1.1 followed by historical review in Sec. 1.2. The basic formulations are presented in Sec. 1.3 with an illustrative example. The simple optimization approach to plastic design that is formulated as a linear programming problem is presented in Sec. 1.4. Optimization results under stress constraints are shown in Sec. 1.5. The approximate method called fully-stressed design (FSD) is presented in Sec. 1.6 with investigation of the relation between optimum design and FSD. The optimality criteria approach to a problem with displacement constraints is presented in Sec. 1.7. Problems concerning the compliance and frequency of free vibration as measures of static and dynamic stiffness are extensively studied in Secs. 1.8 and 1.9, respectively. An example of shape and topology optimization of a truss is presented in Sec. 1.10 as an introduction to Chaps. 3 and 4. The basic formulation of multiobjective structural optimization programming and various methodologies of heuristics are shown in Secs. 1.11 and 1.12, respectively, as an introduction to several sections in the following chapters. Finally, developments in simultaneous analysis and design are summarized in Sec. 1.13.

1.1 Overview of structural optimization

In the process of designing structures in various fields of engineering, the designers and engineers make their best decisions at every step in view of structural and non-structural aspects such as stiffness, strength, serviceability, constructability, and aesthetic property. In other words, they make their *optimal* decisions to realize their best designs; hence, the process of structural design may be regarded as an *optimum design* even though *optimality* is not explicitly pursued.

Structural optimization is regarded as an application of *optimization methods* to structural design. The typical structural optimization problem is formally formulated to minimize an objective function representing the structural

1

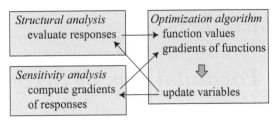

FIGURE 1.1:   Relations among structural analysis, optimization algorithm, and design sensitivity analysis for optimization using a nonlinear programming approach.

cost under constraints on mechanical properties of the structure. The total structural weight or volume is usually used for representing the structural cost. Even for the case in which the structural weight is not strongly related to the cost, it is very important that a feasible solution satisfying all the design requirements can be automatically found through the process of optimization. The mechanical properties include nodal displacements, member stresses, eigenvalues of vibration, and linear buckling loads. The structural optimization problem can be alternatively formulated to maximize a mechanical property under constraint on the structural cost.

Although there are many possible formulations for structural optimization, e.g., *minimum weight design* and *maximum stiffness design*, the term *structural optimization* or *optimum design* is usually used for representing all types of optimization problems corresponding to structural design.

In this book, we consider finite dimensional structures, such as frames and trusses, which are mainly used in civil and architectural engineering. In the typical process of structural optimization of finite dimensional structures, the cross-sectional properties, nodal locations, and member locations are chosen as design variables. There are many methods for structural optimization that are classified into

- Nonlinear programming based on the gradients (*sensitivity coefficients* or derivatives) of the objective and constraint functions, which is the most popular and straightforward approach.

- Heuristic approaches, including genetic algorithm and simulated annealing, that do not need gradient information.

In a nonlinear programming approach, the design variables are updated in the direction defined by the sensitivity coefficients of the objective function and constraints. The relations among structural analysis, optimization algorithm, and design sensitivity analysis for optimization using a nonlinear programming approach are illustrated in Fig. 1.1, where the arrows represent the direction of data flow; i.e., sensitivity analysis is carried out at each step of optimization to provide gradients of responses for the optimization algorithm, and structural analysis is needed for sensitivity analysis and function evaluation at an optimization step (see Chap. 2 and Appendix A.2.2 for details of sensitivity analysis and nonlinear programming, respectively).

There are several approaches to the classification of structural optimization problems. In the field of continuum structural optimization, shape optimization means the optimization of boundary shape, whereas the addition and/or removal of holes are allowed in topology optimization (Bendsøe and Sigmund 2003). In this book, we present various methodologies and results for optimization of finite dimensional structures, including rigidly jointed frames and pin-jointed trusses. Since optimization of trusses and frames was developed gradually in 1960s and 1970s by academic groups in different geographical locations, several different terminologies, e.g., *configuration, geometry*, and *layout*, were used for representing the similar processes of shape and topology optimization; see, e.g., Dobbs and Felton (1969), Svanberg (1981), Lin, Che, and Yu (1982), Imai and Schmit (1982), Zhou and Rozvany (1991), Twu and Choi (1992), Bendsøe, Ben-Tal, and Zowe (1994), Dems and Gatkowski (1995), Ohsaki (1997b), Bojczuk and Mróz (1999), Stadler (1999), Evgrafov (2006), and Achtziger (2007). On the other hand, optimization of cross-sectional areas of trusses was traditionally called *optimum design, design optimization*, or *structural optimization* (Hu and Shield 1961; Prager 1974a; Rozvany 1976). However, the term *sizing optimization* was often used recently to distinguish it from shape optimization (Grierson and Pak 1993; Lin, Che, and Yu 1982; Zou and Chan 2005; Schutte and Groenwold 2003), and structural optimization covers all areas related to optimization of structures.

In this chapter, we present a historical review and various formulations of optimization of finite dimensional structures.

1.2 History of structural optimization

The origin of structural optimization is sometimes credited to Galileo Galilei (1638), who investigated the optimal shape of a beam subjected to a static load. However, his approach was rather intuitive, and he did not establish any theoretical foundation of structural optimization.

The intrinsic properties of minimizing or maximizing functions or functionals in physical phenomena in nature were noticed from ancient times as various minimum/maximum principles. The theoretical basis of minimum principles as a foundation of modern optimization was investigated in the 18th century and established as the *calculus of variation*. The principle of minimum potential energy that leads to the shape of a hanging cable called *catenary* is extensively used nowadays for the design of flexible structures, e.g., cable nets and membrane structures (Krishna 1979). The surface of the minimum area for the specified boundary shape in three-dimensional space is called *minimal surface*, which is equivalent to the surface with vanishing mean curvature, and can be achieved by a membrane with a uniform tension field without external load or pressure. Therefore, the minimal surface is effectively used as the ideal self-equilibrium shape for designing a membrane structure that does not have bending stiffness (Otto 1967, 1969).

Papers by Michell (1904), Maxwell (1890), and Cilly (1900) are often cited as the first paper that mentioned the basic idea of topology optimization; see Sec. 3.1 for the history of topology optimization. However, the so-called *Michell truss* or *Michell structure* has an infinite number of members; hence, it did not lead to any practical development until the 1950s, when the properties of the optimal plastic design of frames were investigated (Foulkes 1954; Drucker and Shield 1961; Heyman 1959). We do not discuss the history of optimization of continuum structures such as plates and shells, because the scope of this book is limited to finite dimensional structures. A comprehensive literature review of early developments of structural optimization is found in Bradt (1986), which was originally published by the Polish Academy of Science, and includes about 300 entries up to the 1950s starting with the book by Galileo Galilei (1638), and more than 1800 entries for the period 1960–1980.

In the 1950s, optimality conditions were studied for the plastic design of frames (Foulkes 1954; Drucker and Shield 1961). In the 1960s, conditions or criteria of optimality were derived utilizing minimum principles for several performance measures of structures (Sewell 1987). Hu and Shield (1961) investigated the uniqueness of optimal plastic design. Taylor (1967) derived the optimality condition for a vibrating rod with specified natural frequency using Hamilton's principle or the *principle of least action*. Prager and Taylor (1968) developed optimality conditions for sandwich beams considering constraints on compliance, natural frequency, buckling load, and plastic limit load, using minimal total potential energy, Rayleigh's principle, and lower- and upperbound theorems of limit analysis, respectively. Prager (1972, 1974a) summarized the optimality conditions corresponding to various types of constraints, including the case of multiple constraints.

Plastic design of frames was extensively studied in the 1960s and 1970s, because analytical and/or computationally inexpensive methods can be used for this problem. Prager (1971) developed conditions for an optimal frame, subjected to alternative loads, exhibiting the so-called *Foulkes mechanism*. Adeli and Chyou (1987) presented a kinematic approach using automatic generation of independent mechanisms (see Hemp (1973) for various early developments in optimal plastic design).

In the 1970s, when the computer power was still not strong enough to use mathematical programming approaches to optimization of real-world structures, optimality criteria (OC) approaches were widely used for finite dimensional structures. The modern discrete OC approaches to trusses and frames were initiated by Venkayya, Khot, and Berke (1973). Dobbs and Nelson (1975) developed the OC approach to truss design. Reviews of OC approaches are found in Berke and Venkayya (1974) and Venkayya (1978).

Owing to the rapid development of computer hardware and software technologies, many numerical approaches were developed in the 1980s and 1990s to obtain optimization results for practical problems. Developments in this period can be found in many books, e.g., Arora (2007), Adeli (1994), Burns (2002), and Haftka, Gürdal, and Kamat (1990).

It should be noted that the preferred terminologies for structural optimization vary with age. As noted earlier, *structural optimization* of trusses covered only optimization of cross-sectional properties in the 1950s and 1960s. However, *sizing optimization* was recently used to distinguish it from shape and topology optimization. Optimality conditions were first called *Kuhn-Tucker conditions*; however, the name was corrected to *Karush-Kuhn-Tucker conditions* in the 1980s. Multiple load sets for formulation of constraints on static responses were called *alternative loads* until the 1970s; however, they are now usually called *multiple loading conditions* or multiple load sets. Furthermore, *framed structure* was used for representing finite dimensional structures, including pin-jointed trusses and rigidly jointed frames; however, they are classified into trusses and frames, respectively, in recent literature. In this book, we use up-to-date terminology, for consistency, even for describing the results of papers in the early stages of development.

1.3   Structural optimization problem

1.3.1   Continuous problem

If the design variables can vary continuously, i.e., can have real values, and the objective and constraint functions are continuous and differentiable with respect to the variables, the structural optimization problem can be formulated as a nonlinear programming (NLP) problem. Let $\mathbf{A} = (A_1, \ldots, A_m)^\top$ denote the vector of m design variables. For a sizing design optimization problem, $\mathbf{A}$ represents the cross-sectional areas of truss members, heights of the sections of frame members, etc. For a geometry optimization problem, $\mathbf{A}$ may represent the nodal coordinates of trusses and frames. All vectors are assumed to be column vectors throughout this book.

The number of design variables is often reduced using the approach called *design variable linking*, utilizing, e.g., the symmetry properties of the structure. The requirements to be considered in practical applications can also be used for reducing the number of variables; e.g., the beams in the same story of a building frame should have the same section. However, in the following, we assume that each variable can vary independently, and, for trusses and frames, A_i belongs to member i, for simplicity.

Consider an elastic finite dimensional structure subjected to static loads. The vector of state variables representing the nodal displacements is denoted by $\mathbf{U} = (U_1, \ldots, U_n)^\top$, where n is the number of degrees of freedom. In most of the design problems in various fields of engineering, the design requirements for responses such as stresses and displacements are given with inequality constraints specified by design codes:

$$H_j(\mathbf{U}(\mathbf{A}), \mathbf{A}) \leq 0, \quad (j = 1, \ldots, n^{\mathrm{I}}) \tag{1.1}$$

where n^{I} is the number of inequality constraints. Generally, there exist equality constraints on the response quantities; e.g., an eigenvalue of vibration should be exactly equal to the specified value. However, we consider inequality constraints only, for simple presentation of formulations.

The constraint function $H_j(\mathbf{U}(\mathbf{A}), \mathbf{A})$ depends on the design variables implicitly through the displacement (state variable) vector $\mathbf{U}(\mathbf{A})$ and also directly on the design variables. For example, the axial force N_i of the ith member of a truss is given using a constant n-vector $\mathbf{b}_i$, defining the stress-displacement relation as

$$N_i = A_i \mathbf{b}_i^\top \mathbf{U}(\mathbf{A}) \tag{1.2}$$

which depends explicitly on A_i and implicitly on $\mathbf{A}$ through $\mathbf{U}(\mathbf{A})$.

The upper and lower bounds, which are denoted by A_i^{U} and A_i^{L}, respectively, are usually given for the design variable A_i due to the restriction in fabrication and construction. The objective function, e.g., the total structural volume, is denoted by $F(\mathbf{A})$. Then the structural optimization problem is formulated as

$$\text{Minimize} \quad F(\mathbf{A}) \tag{1.3a}$$

$$\text{subject to} \quad H_j(\mathbf{U}(\mathbf{A}), \mathbf{A}) \leq 0, \quad (j = 1, \ldots, n^{\mathrm{I}}) \tag{1.3b}$$

$$A_i^{\mathrm{L}} \leq A_i \leq A_i^{\mathrm{U}}, \quad (i = 1, \ldots, m) \tag{1.3c}$$

Problem (1.3) is classified as an NLP problem, because $\mathbf{U}(\mathbf{A})$ is a nonlinear function of $\mathbf{A}$; see Appendix A.2.2 for details of NLP. The constraints (1.3c) are called *side constraints*, *bound constraints*, or *box constraints*, which are treated separately from the general inequality constraints (1.3b) in most of the optimization algorithms.

As is seen from the definition of constraints in (1.3b), the differential coefficients of $\mathbf{U}(\mathbf{A})$ with respect to $\mathbf{A}$, called design sensitivity coefficients, are needed when solving Problem (1.3) using a gradient-based NLP algorithm. For convenience in deriving the conditions to be satisfied at the optimal solution, the constraint function with respect to $\mathbf{A}$ only is defined as

$$\widetilde{H}_j(\mathbf{A}) = H_j(\mathbf{U}(\mathbf{A}), \mathbf{A}) \tag{1.4}$$

If the side constraints are treated separately from the general inequality constraints, the conditions for optimality are derived using the Lagrangian

$\psi(\mathbf{A}, \boldsymbol{\mu})$ defined as

$$\psi(\mathbf{A}, \boldsymbol{\mu}) = F(\mathbf{A}) + \sum_{j=1}^{n^{\mathrm{I}}} \mu_j \widetilde{H}_j(\mathbf{A}) \tag{1.5}$$

where $\boldsymbol{\mu} = (\mu_1, \ldots, \mu_{n^{\mathrm{I}}})^{\top} \ (\geq \mathbf{0})$ is the vector of Lagrange multipliers.

The necessary conditions for local optimality, which are called *Karush-Kuhn-Tucker conditions* or simply KKT conditions, are given as

$$\begin{cases} \dfrac{\partial \psi}{\partial A_i} \geq 0 & \text{for} \quad A_i = A_i^{\mathrm{L}} \\[2mm] \dfrac{\partial \psi}{\partial A_i} = 0 & \text{for} \quad A_i^{\mathrm{L}} < A_i < A_i^{\mathrm{U}} \\[2mm] \dfrac{\partial \psi}{\partial A_i} \leq 0 & \text{for} \quad A_i = A_i^{\mathrm{U}} \end{cases} \tag{1.6}$$

where

$$\frac{\partial \psi}{\partial A_i} = \frac{\partial F}{\partial A_i} + \sum_{j=1}^{n^{\mathrm{I}}} \mu_j \frac{\partial \widetilde{H}_j}{\partial A_i}, \quad (i = 1, \ldots, m) \tag{1.7}$$

$$\widetilde{H}_j \leq 0, \quad \mu_j \geq 0, \quad \mu_j \widetilde{H}_j = 0, \quad (i = 1, \ldots, n^{\mathrm{I}}) \tag{1.8}$$

The third equation in (1.8) is called *complementarity conditions* (see Appendix A.2.2.3 for details of the optimality conditions).

Conditions (1.6)–(1.8) are the necessary and sufficient conditions for local optimality, if all the objective and constraint functions are locally convex. Furthermore, (1.6)–(1.8) are sufficient conditions for global optimality, if all the objective and constraint functions are globally convex.

For problems with real variables and continuously differentiable functions, the optimal solutions are found using various approaches of mathematical programming. If the objective function and the constraints are linear functions of the design variables, the problem is formulated as a linear programming (LP) problem, and the optimal solutions are easily found using the standard approach called the simplex method (Luenberger 2003) or the relatively new approach called the interior-point method (Karmarkar 1984; Gondzio 1995).

If the objective and the constraint functions are nonlinear, various approaches of the NLP problem can be used (Fiacco and Cormic 1968; Mangasarian 1969; Pierre and Lowe 1975; Peressini, Sullivan, and Uhl 1988; Ben-Israel, Ben-Tal, and Zolbec 1981; Bersekas 1982). However, there is no approach that is applicable to any type of NLP problem; i.e., the most suitable method should be appropriately chosen for each problem at hand. Furthermore, the method should be selected with regard to the desired accuracy and computational cost for optimization. One of the most popular approaches is sequential quadratic programming (Gill, Murray, and Saunders 2002), which

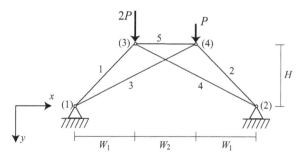

FIGURE 1.2: A five-bar plane truss.

is used for most of the examples of the application of NLP in this book. Readers may refer to Appendix A.2.2 for details of NLP.

Example 1.1

As a simple example of structural optimization, consider a five-bar plane truss, as shown in Fig. 1.2, subjected to vertical static loads. The intersecting members 3 and 4 are not connected with each other. The five bars are classified into Groups 1 and 2, consisting of members $\{1, 2, 5\}$ and $\{3, 4\}$, respectively. The members in Group i ($i = 1, 2$) have the same cross-sectional area A_i^g, and let $\mathbf{A}^g = (A_1^g, A_2^g)^\top$. The sum of the lengths of members in the ith group is denoted by L_i^g. Then the total structural volume $C(\mathbf{A}^g)$, which is taken as the objective function, is defined as

$$C(\mathbf{A}^g) = A_1^g L_1^g + A_2^g L_2^g \qquad (1.9)$$

For a simple illustration of the problem, the constraints are given, as follows, for the y-directional displacement U_3 of node 3, which is assumed to be positive, and the stress σ_4 of member 4, which is assumed to be negative:

$$U_3 \le U_3^U, \quad \sigma_4^L \le \sigma_4 \qquad (1.10)$$

where U_3^U and σ_4^L are the specified upper bound of U_3 and the lower bound of σ_4, respectively. The constraints are formulated using the function $\tilde{H}_1(\mathbf{A}^g)$ of $\mathbf{A}^g$ only:

$$\tilde{H}_1(\mathbf{A}^g) = U_3(\mathbf{A}^g) - U_3^U \le 0, \quad \tilde{H}_2(\mathbf{A}^g) = -\sigma_4^L - \sigma_4(\mathbf{A}^g) \le 0 \qquad (1.11)$$

Let $W_1 = W_2 = H = 1$ m in Fig. 1.2. The elastic modulus is 200 kN/mm², and $P = 10.0$ kN. The bounds for the displacement and stress are $U_3^U = 1.25$ mm and $\sigma_4^L = -0.06$ kN/mm².

The set of solutions satisfying $U_3 = U_3^U$ and $\sigma_4 = \sigma_4^L$ is shown in the solid lines in Fig. 1.3 that are drawn in the design variable space. The gray area

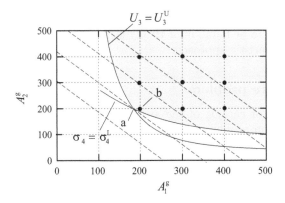

FIGURE 1.3: Feasible region and optimal solutions of the five-bar truss.

is the feasible region satisfying the two constraints with equality. From (1.9), we obtain

$$A_2^g = \frac{1}{L_2^g} C(\mathbf{A}^g) - \frac{L_1^g}{L_2^g} A_1^g \qquad (1.12)$$

The solution on each dotted line in Fig. 1.3 has the same values of C. Therefore, if A_1^g and A_2^g can take real values, the point 'a' with the coordinates $(A_1^g, A_2^g) = (184.33, 198.90)$ in the design variable space corresponds to the optimal solution.

In order to verify the optimality of the solution, the sensitivity coefficients are obtained at the optimal solution as

$$\frac{\partial U_3(\mathbf{A}^g)}{\partial A_1^g} = -0.013112, \quad \frac{\partial U_3(\mathbf{A}^g)}{\partial A_2^g} = -0.018015,$$
$$\frac{\partial \sigma_4(\mathbf{A}^g)}{\partial A_1^g} = 0.45650, \quad \frac{\partial \sigma_4(\mathbf{A}^g)}{\partial A_2^g} = 0.20541 \qquad (1.13)$$

The sensitivity coefficients of the objective function are easily computed from the member lengths as

$$\frac{\partial C}{\partial A_1^g} = 3828.4, \quad \frac{\partial C}{\partial A_2^g} = 4472.1 \qquad (1.14)$$

Then, from the second equation in (1.6) with $i = 1$ and 2, while A_i is replaced by A_i^g, the positive Lagrange multipliers are found as $\lambda_1 = 1.8680 \times 10^5$ and $\lambda_2 = 2.2694 \times 10^7$. Hence, the optimality conditions are satisfied at the solution $(A_1^g, A_2^g) = (184.33, 198.90)$.

As is seen in the above example, the optimal solution can be found for a simple truss graphically in the design variable space, if we have only two design variables. However, for larger structures with more design variables, the optimal solutions are to be found numerically using a mathematical programming approach or a heuristic approach.

1.3.2 Discrete problem

Suppose a list or catalog of the available standard sections is given for a sizing optimization problem of a frame, and the list $\mathcal{A}_i$ of the cross-sectional properties of the ith member is given as

$$\mathcal{A}_i = \{(A_i^1, I_i^1, Z_i^1), \ldots, (A_i^r, I_i^r, Z_i^r)\} \tag{1.15}$$

where A_i^j is the cross-sectional area, I_i^j is the second moment of inertia, Z_i^j is the section modulus of the jth candidate section for member i, and r is the number of available sections, which is the same for all members, for brevity. Note that other properties such as fully-plastic moment should be included if elastoplastic responses are to be considered; see Appendix A.8 for examples of section lists.

Suppose that $J_i = j\ (i = 1, \ldots, m)$ indicates that the jth section in the list is assigned to the ith member, where m is the number of members. This way, the mechanical properties of the frame are defined by the vector $\mathbf{J} = (J_1, \ldots, J_m)^\top$ of integer variables. Hence, the nodal displacement vector is a function of $\mathbf{J}$ that is denoted by $\mathbf{U}(\mathbf{J})$. The objective and the constraint functions are also functions of $\mathbf{J}$, which are written as $F(\mathbf{J})$ and $\widetilde{H}_j(\mathbf{J}) = H_j(\mathbf{J}, \mathbf{U}(\mathbf{J}))$, respectively. Then the optimization problem with inequality constraints only is formulated as

$$\text{Minimize} \quad F(\mathbf{J}) \tag{1.16a}$$

$$\text{subject to} \quad \widetilde{H}_j(\mathbf{J}) \le 0, \quad (j = 1, \ldots, n^\mathrm{I}) \tag{1.16b}$$

$$J_i \in \{1, \ldots, r\}, \quad (i = 1, \ldots, m) \tag{1.16c}$$

Since Problem (1.16) is an integer programming problem, which is equivalently called a combinatorial optimization problem, various methods, e.g., the branch-and-bound method and the branch-and-cut method, can be used (Horst and Tuy 1985; Horst, Pardalos, and Thoai 1995) (see Sec. 3.5 for application of the branch-and-bound method to topology optimization of trusses).

For the example of the five-bar truss in Fig. 1.2, suppose A_1^g and A_2^g can take only integer values 100, 200, ... Then, the feasible designs satisfying (1.10) are plotted in the filled circle in Fig. 1.3, and the optimal solution exists at point 'b'.

Since the state variables are continuous functions of the design variables, a structural optimization problem turns out to be a mixed integer nonlinear programming (MINLP) problem (Floudas 1995) if the formulation of simultaneous analysis and design, see Sec. 1.13, is used considering the nodal displacements as independent variables. Arora (2002) classified the structural optimization problems into the following six categories:

1. Continuous design variables; functions are twice continuously differentiable (standard NLP problem).

2. Mixed design variables; functions are twice continuously differentiable; discrete variables can have non-discrete values during the solution process (functions can be evaluated at non-discrete points). A configuration optimization problem of a truss with discrete cross-sectional areas and continuous nodal coordinates belongs in this category.

3. Mixed design variables; functions are non-differentiable; discrete variables can have non-discrete values during the solution process. A configuration optimization problem of a truss with discrete cross-sectional area, continuous nodal coordinates, and nodal cost defined as a non-differential function of cross-sectional areas belongs in this category (see Sec. 4.3).

4. Mixed design variables; functions may or may not be differentiable; some of the discrete variables must have only discrete value in the solution process. A configuration optimization problem with a list of candidate topologies and continuous nodal coordinates belongs in this category.

5. Mixed design variables; functions may or may not be differentiable; some of the discrete variables are linked to others; assignment of a value to one variable specifies values for others. A frame optimization problem with discrete cross-sectional properties such as second moment of inertia linked with cross-sectional area belongs to this category.

6. Combinatorial problems; purely discrete non-differentiable problems. Optimization problems for selection of materials, location of supports, etc. belong to this category.

Arora, Huang, and Hsieh (1994) summarized various methods of optimization with discrete variables.

1.4 Plastic design

Optimal plastic design is the simplest and classical problem of optimization of trusses and frames, which was extensively studied in the 1960s. Consider a truss consisting of a perfectly rigid-plastic material; i.e., the strain before yielding is negligibly small, and the stress after yielding is constant at the yield stress, which is assumed to be the same for all members. The truss is subjected to a vector of quasistatic proportional loads $\mathbf{P} = \Lambda \mathbf{P}^0$ defined by the load factor Λ and the constant load pattern vector $\mathbf{P}^0$.

The axial force vector is given as $\mathbf{N} = (N_1, \ldots, N_m)^\top$, where m is the number of members. Let n denote the number of degrees of freedom. The equilibrium equations are formulated in terms of the $n \times m$ equilibrium matrix $\mathbf{D}$ as

$$\mathbf{DN} = \Lambda\mathbf{P}^0 \tag{1.17}$$

Let $\mathbf{N}^\mathrm{P} = (N_1^\mathrm{P}, \ldots, N_m^\mathrm{P})^\top$ denote the vector of tensile yield axial forces of the members. The yield axial force in compression is given for the ith member, ignoring member buckling, as $-N_i^\mathrm{P}$. Then the yield condition is written as

$$-N_i^\mathrm{P} \leq N_i \leq N_i^\mathrm{P}, \quad (i = 1, \ldots, m) \tag{1.18}$$

Note that N_i^P is proportional to the cross-sectional area A_i as $N_i^\mathrm{P} = A_i \sigma^\mathrm{P}$, where σ^P is the tensile yield stress.

First the plastic limit analysis problem is formulated as a linear programming (LP) problem. Utilizing the lower-bound theorem of plastic limit analysis (Shames and Cozzarelli 1997), we can obtain the plastic limit load factor through maximization of the load factor under constraints on the equilibrium equations and the yield conditions:

$$\text{Maximize} \quad \Lambda \tag{1.19a}$$

$$\text{subject to} \quad -\mathbf{N}^\mathrm{P} \leq \mathbf{N} \leq \mathbf{N}^\mathrm{P} \tag{1.19b}$$

$$\mathbf{DN} = \Lambda\mathbf{P}_0 \tag{1.19c}$$

which is an LP problem with variables Λ and $\mathbf{N}$. Therefore, the plastic limit load can easily be obtained using a standard method of LP such as the simplex method.

The problem of minimizing the total structural volume under constraint on limit load factor can also be formulated as an LP problem. Since N_i^P is proportional to A_i, the optimal design that minimizes the total structural volume can be obtained by minimizing $\mathbf{N}^{\mathrm{P}\top}\mathbf{L}$, where $\mathbf{L} = (L_1, \ldots, L_m)^\top$ is the vector of member lengths. The upper and lower bounds for N_i^P are denoted by N_i^{pU} and N_i^{pL}, respectively, with the vectors $\mathbf{N}^{\mathrm{pU}} = (N_1^{\mathrm{pU}}, \ldots, N_m^{\mathrm{pU}})^\top$ and $\mathbf{N}^{\mathrm{pL}} = (N_1^{\mathrm{pL}}, \ldots, N_m^{\mathrm{pL}})^\top$. The specified limit load factor is denoted by Λ^P. Then, the optimization problem is formulated as

$$\text{Minimize} \quad \mathbf{N}^{\mathrm{P}\top}\mathbf{L} \tag{1.20a}$$

$$\text{subject to} \quad -\mathbf{N}^\mathrm{P} \leq \mathbf{N} \leq \mathbf{N}^\mathrm{P} \tag{1.20b}$$

$$\mathbf{DN} = \Lambda^\mathrm{P}\mathbf{P}^0 \tag{1.20c}$$

$$\mathbf{N}^{\mathrm{pL}} \leq \mathbf{N}^\mathrm{P} \leq \mathbf{N}^{\mathrm{pU}} \tag{1.20d}$$

where the variables are $\mathbf{N}$ and $\mathbf{N}^\mathrm{P}$. Because Problem (1.20) is also an LP problem, this problem was extensively studied in the 1960s and is still important for application to the plastic design of trusses. Note that the plastic collapse mechanisms can be found as the Lagrange multipliers at the optimal

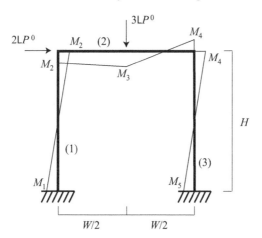

FIGURE 1.4: A simple plane frame.

solution of Problem (1.20), or by solving the dual of Problem (1.20) that is formulated on the basis of the upper-bound theorem of plastic limit analysis, which states that the smallest load factor corresponding to admissible strain and displacement rates defines the collapse load.

The plastic design problem of a frame with concentrated plastic hinges can also be formulated as an LP problem, as follows, if the interaction between the axial force and bending moment on the yield condition is ignored (Adeli and Chyou 1987):

Example 1.2
Consider a plane frame, as shown in Fig. 1.4, subjected to a proportional horizontal load $2\Lambda P^0$ and a vertical load $3\Lambda P^0$ simultaneously. The bending moments at the member ends and the center of the beam are denoted by $M_1, \ldots, M_5$, as shown in Fig. 1.4, which illustrates the state where $M_1, \ldots, M_5$ are all positive. The numbers in parentheses are member numbers.

The equilibrium equations are given as

$$
\begin{aligned}
\frac{M_2 + M_1}{H} + \frac{M_4 + M_5}{H} &= 2\Lambda P^0, \\
\frac{M_2 - M_3}{W/2} - \frac{M_3 + M_4}{W/2} &= 3\Lambda P^0
\end{aligned}
\tag{1.21}
$$

The fully-plastic moment of member i is denoted by M_i^{P}. The yield conditions are then given as

$$
\begin{aligned}
-M_1^{\mathrm{P}} &\leq M_1 \leq M_1^{\mathrm{P}}, \quad -M_1^{\mathrm{P}} \leq M_2 \leq M_1^{\mathrm{P}}, \\
-M_2^{\mathrm{P}} &\leq M_2 \leq M_2^{\mathrm{P}}, \quad -M_2^{\mathrm{P}} \leq M_3 \leq M_2^{\mathrm{P}}, \quad -M_2^{\mathrm{P}} \leq M_4 \leq M_2^{\mathrm{P}}, \\
-M_3^{\mathrm{P}} &\leq M_4 \leq M_3^{\mathrm{P}}, \quad -M_3^{\mathrm{P}} \leq M_5 \leq M_3^{\mathrm{P}}
\end{aligned}
\tag{1.22}
$$

We assume that the tensile yield stress σ^{P} and the compressive yield stress $-\sigma^{\mathrm{P}}$ are the same, respectively, for all members, and the cross-section of each member is modeled as a *sandwich section*; i.e., the half of the cross-sectional area A_i is concentrated at each flange, and M_i^{P} is proportional to A_i as $M_i^{\mathrm{P}} = A_i r_i \sigma^{\mathrm{P}}$, where r_i is the distance between the flanges.

Hence, the objective function that is proportional to the total structural volume of the frame is formulated as a function of $\mathbf{M}^{\mathrm{P}} = (M_1^{\mathrm{P}}, M_2^{\mathrm{P}}, M_3^{\mathrm{P}})^{\top}$:

$$F(\mathbf{M}^{\mathrm{P}}) = M_1^{\mathrm{P}} H + M_2^{\mathrm{P}} W + M_3^{\mathrm{P}} H \qquad (1.23)$$

Since both the objective function and the constraints are linear functions of the variables $M_1, \ldots, M_5$ and $\mathbf{M}^{\mathrm{P}}$, the optimal solution can easily be found by solving an LP problem. For a simple case with $P^0 = 1$, $\Lambda^{\mathrm{P}} = 1$, and $H = W = 1$, we obtain the optimal solution as $M_1^{\mathrm{P}} = 3/8$ and $M_2^{\mathrm{P}} = M_3^{\mathrm{P}} = 5/8$ with $M_1 = M_2 = 3/8$, $M_3 = M_4 = M_5 = 5/8$, and $F(\mathbf{M}^{\mathrm{P}}) = 13/8$.

It is well known that the Foulkes mechanism satisfying the following conditions exists at the optimal solution of a frame for the case in which M_i^{P} is proportional to A_i (Foulkes 1954):

$$\begin{cases} \theta_i = \lambda L_i & \text{for} \quad M_i^{\mathrm{P}} > M_i^{\mathrm{PL}} \\ \theta_i \leq \lambda L_i & \text{for} \quad M_i^{\mathrm{P}} = M_i^{\mathrm{PL}} \end{cases} \qquad (1.24)$$

where θ_i is the sum of absolute values of the rotation rate of the plastic hinges in the ith member, and λ is a positive constant. Note that the upper bound for M_i^{P} is not considered, for brevity. Condition (1.24) suggests that the plastic energy dissipation rate per unit volume is the same for members with $M_i^{\mathrm{P}} > M_i^{\mathrm{PL}}$.

There have been many studies on plastic design since the 1960s (Tam and Jennings 1989). Multiple (alternative) loads are considered in some papers, e.g., Prager (1967) and Chan (1969). Munro and Chuang (1986) presented a fuzzy LP approach for the case in which uncertainty exists in the loads. A probabilistic LP approach to limit design under uncertainty was developed by Gavarini and Veneziano (1972).

1.5   Stress constraints

In view of structural design procedure in civil engineering based on allowable stress design criteria, it is very important to obtain an optimal design that satisfies stress and displacement constraints against design loads. In this section, we consider stress constraints only, for simple presentation of the optimization procedure. An approach to optimization of a truss under displacement constraints is demonstrated in Sec. 1.8. Another important aspect in structural design is that several loads, including static loads (self-weight, service load, snow load, etc.) and dynamic loads (wind load, seismic load, etc.), should be considered, and, in the practical design process, the dynamic loads are represented by equivalent static loads. Furthermore, the self-weight and service load are classified as long-term loads, while others are short-term loads. Therefore, different bounds should be given for the stresses against each loading condition.

Consider n^{P} loading conditions (load patterns), and let the superscript k denote the variables and parameters corresponding to the kth loading condition. The upper and lower bounds for σ_i^k are denoted by σ_i^{kU} and σ_i^{kL}, respectively. Then the optimization problem for minimizing the total structural volume of a truss under stress constraints is formulated as

$$\text{Minimize} \quad \sum_{i=1}^{m} A_i L_i \tag{1.25a}$$

$$\text{subject to} \quad \sigma_i^{kL} \leq \sigma_i^k \leq \sigma_i^{kU}, \quad (i = 1, \ldots, m; \ k = 1, \ldots, n^{\mathrm{P}}) \tag{1.25b}$$

$$A_i^{L} \leq A_i \leq A_i^{U}, \quad (i = 1, \ldots, m) \tag{1.25c}$$

where A_i^{L} and A_i^{U} are the lower and upper bounds for A_i, respectively. Note again that the n^{P} load patterns are applied independently, and the stress constraints are assigned for each loading condition.

Example 1.3
Optimum designs are found for a 10-bar truss, as shown in Fig. 1.5, subjected to vertical loads P_1 and P_2, where the numbers with and without parentheses are node numbers and member numbers, respectively (Katoh, Ohsaki, and Tani 2002). Note that the intersecting members are not connected at their centers. A small lower bound $A_i^{L} = 0.1$ mm^2 is given for A_i to prevent instability of the truss, while the upper bound is not given for A_i. The bounds for the stresses are $\sigma_i^{kU} = 0.2$ N/mm^2 and $\sigma_i^{kL} = -0.2$ N/mm^2. Optimal solutions are obtained using the optimization software package SNOPT Ver. 7.2 (Gill, Murray, and Saunders 2002), which utilizes the sequential quadratic programming; see Appendix A.2.2.5.

First, consider a single loading condition $(P_1, P_2) = (0.0, 100.0 \text{ kN})$. The optimal cross-sectional areas and the optimal objective value are shown in the second column in Table 1.1. The optimal solution is also illustrated in Fig. 1.6, where the width of each member is proportional to its cross-sectional area. Note that A_i is equal to its lower bound in members 4, 5, 6, 8, and 10, which may be removed to obtain the statically determinate truss of optimal topology after fixing the unstable node 4. The stress is equal to its upper or lower bound in each member with $A_i > A_i^{L}$. This process of topology optimization is called the ground structure approach; it is extensively studied in Chap. 3.

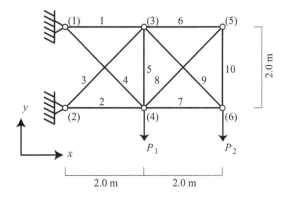

FIGURE 1.5:   A 10-bar truss.

TABLE 1.1:   Optimal cross-sectional areas and structural
volume of the 10-bar truss under stress constraints.

Member number	A_i (mm^2)	
	Single loading	Multiple loading
1	999.931	825.107
2	500.069	674.893
3	707.010	459.771
4	0.100	421.531
5	0.100	211.499
6	0.100	0.100
7	499.937	499.909
8	0.100	0.129
9	707.017	706.978
10	0.100	0.100
Total volume (mm^3)	8.00051×10^6	8.91591×10^6

Next, we obtain the optimal solution under stress constraints against multi-
ple loading conditions $(P_1, P_2) = (0, 100.0 \text{ kN})$ and $(100.0 \text{ kN}, 0)$. The optimal
cross-sectional areas, which are also illustrated in Fig. 1.7, and the objective
value are listed in the last column of Table 1.1. The optimal objective value
is 8.91591×10^6 mm^3, which is larger than that for the single loading condition. Only the members 6 and 10 connected to node 5 satisfy $A_i = A_i^L$,
and node 5 cannot be removed, because the cross-sectional area of member
8 is larger than its lower bound. Note that a very strict tolerance of 10^{-10}
is assigned for the constraints and optimality conditions in SNOPT. In fact,
we can confirm that the stresses for the first load $(P_1, P_2) = (0, 100.0 \text{ kN})$ are
$\sigma_6^1 = \sigma_{10}^1 = 0.18265$ and $\sigma_8^1 = -0.2$; the member 8 is fully stressed. Since the
equilibrium condition $N_8 = -\sqrt{2}(N_6 + N_{10})$ should be satisfied for member
forces N_i of members 6, 8, and 10, the cross-sectional area should be larger

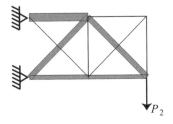

FIGURE 1.6: Optimal design of the 10-bar truss under single loading condition $(P_1, P_2) = (0.0, 100.0 \text{ kN})$.

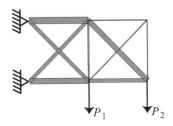

FIGURE 1.7: Optimal solution for multiple loading conditions $(P_1, P_2) = (0, 100.0 \text{ kN})$ and $(100.0 \text{ kN}, 0)$.

than its lower bound if the stresses of members 6 and 10 are close to their upper or lower bounds. Hence, it should be noted that the member with $A_i = A_i^L$ sometimes cannot be removed in the conventional ground structure approach of topology optimization, and the optimal topology may consist of many members with small cross-sectional areas.

The result in Fig. 1.7 demonstrates that the optimal truss under multiple loading conditions is statically indeterminate even when the removal of members 6, 8, and 10 is allowed. It has been confirmed that the stress is equal to its upper or lower bound against at least one loading condition for a member with $A_i > A_i^L$.

1.6 Fully-stressed design

1.6.1 Stress-ratio approach

Consider again a simple optimization problem of a truss with stress constraints, and suppose only the lower bound A_i^L is given for the cross-sectional area A_i of the ith member. In a practical design process, obtaining a feasible

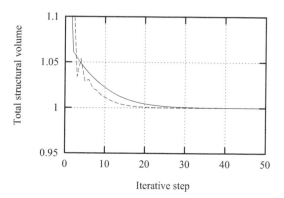

FIGURE 1.8: Convergence history of the total structural volume of FSD of the 10-bar truss; solid line: $r = 1$, dashed line: $r = 1.5$.

solution is sometimes more important than minimizing an objective function. Furthermore, the results of the example in the previous section suggest that an optimal design can be obtained by finding the cross-sectional areas so that the stress of a member with $A_i^{\mathrm{L}} < A_i$ is equal to its upper or lower bound for at least one of the n^{P} loading conditions. The design satisfying this condition is called a *fully-stressed design* (FSD). Note that the inequality constraints $\sigma_i^{k\mathrm{L}} \leq \sigma_i^k \leq \sigma_i^{k\mathrm{U}}$ $(k = 1, \dots, n^{\mathrm{P}})$ are to be satisfied by the member with $A_i = A_i^{\mathrm{L}}$. In the fully-stressed design approach, a design satisfying these conditions is obtained by iteratively modifying the design variables.

For a simple case of a single loading condition with $\sigma_i^{1\mathrm{L}} = -\sigma_i^{1\mathrm{U}}$ for all members, the FSD can be obtained by the following simple iterative algorithm for updating the cross-sectional areas:

$$A_i^{(k+1)} = A_i^{(k)} \left(\frac{|\sigma_i^1|}{\sigma_i^{1\mathrm{U}}} \right)^r, \quad (i = 1, \dots, m) \tag{1.26}$$

where $A_i^{(k)}$ is the value of A_i at the kth step of iteration, and r is the parameter for controlling the convergence property, which is usually between 1 and 2. Note that $A_i^{(k+1)}$ is replaced by A_i^{L} if $A_i < A_i^{\mathrm{L}}$ is satisfied as the result of application of (1.26).

The design update rule (1.26) called the *stress-ratio approach*, assumes that the modification of the cross-sectional area of a member does not have any strong effect on the axial forces of the members. For example, the axial forces of a statically determinate truss are determined only from the equilibrium equations and are independent of the cross-sectional areas. In this case, the stress of a member against the specified set of loads is inversely proportional to its cross-sectional area, and the FSD can be found within only one step of application of (1.26) with $r = 1$.

The stress-ratio approach can also be effectively used for building frames, for which constraints are given for the stress at each edge of the section at the member ends due to the bending moment and axial force. It is very convenient for investigating the nearly optimal load paths of a plane regular frame from the loaded nodes to the supports. If the variable A_i defines the size of the section with the dimension of length, e.g., height and width of the wide-flange section, the appropriate value of the parameter r in (1.26) ranges between 1/3 and 1/2 (Mueller, Liu, and Burns 2002).

Example 1.4
An FSD is found for the 10-bar truss in Fig. 1.5 under single loading condition $(P_1, P_2) = (0.0, 100.0 \text{ kN})$. A small lower bound, 0.1 mm^2, which is the same as the value in Example 1.3 in Sec. 1.5, is given for A_i to prevent instability of the truss and to compare the results. Fig. 1.8 shows the convergence history of the total structural volume divided by the value 8.91591×10^6 in Table 1.1 for the optimal solution under stress constraints subjected to the same single loading condition. The histories of A_1 and A_6 are also plotted in Figs. 1.9 (a) and (b), respectively, where only a small range is plotted for A_6, because A_6 is equal to the lower bound, 0.1 mm^2, at the converged FSD. As is seen from these figures, the total structural volume and the cross-sectional areas converge monotonically to the optimal values within 30 steps, if $r = 1$. The convergence property is improved if a larger value, 1.5, is assigned for r; i.e., an approximate optimal solution can be found within 20 steps; however, some oscillation is observed at the early stage of iteration. Note that the total structural volume converges to the optimal value under stress constraints, and the cross-sectional areas of the FSD are the same as the optimal values in Table 1.1. This way, the optimal truss under stress constraints for a single loading condition can easily be found by the simple stress-ratio approach (1.26) if the absolute values of the upper- and lower-bound stresses are the same.

The relation between the FSD and optimum design under stress constraints has been extensively studied since the 1960s (Razani 1965; Kicher 1966; Patnaik and Dayaratnam 1970; McNeil 1971; Chern and Prager 1972; Nagtegaal 1973; Gunnlaugsson and Martin 1973) and was revisited mainly in the community of applied mathematics in the 1990s (Bendsøe and Sigmund 2003). However, it seems that the FSDs are not clearly defined each case with $A_i^L > 0$ and $A_i^L = 0$. Here we do not assign an upper bound for the cross-sectional area, and define the FSD as follows (Nagtegaal 1973):

- If $A_i^L > 0$, then the stress σ_i^k of a member with $A_i > A_i^L$ should be equal to σ_i^{kU} or σ_i^{kL} for at least one loading condition; whereas $\sigma_i^{kL} \leq \sigma_i^k \leq \sigma_i^{kU}$ should be satisfied by a member with $A_i = A_i^L$.

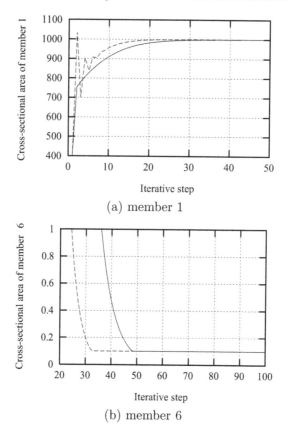

(a) member 1

(b) member 6

FIGURE 1.9: Convergence history of cross-sectional areas of FSD of the 10-bar truss; solid line: $r = 1$, dashed line: $r = 1.5$.

- If $A_i^L = 0$, then the stress σ_i^k of a member with $A_i > 0$ should be equal to σ_i^{kU} or σ_i^{kL} for at least one loading condition; whereas no constraint exists on the stress of a nonexistent member with $A_i = 0$.

Therefore, there is a discontinuity in the FSDs between the cases with $A_i^L = 0$ and $A_i^L = e$, where e is a small positive value. Note that the case with $A_i^L = 0$ corresponds to the topology optimization problem that is extensively investigated in Sec. 3.5.3.

1.6.2   Single loading condition

First we consider a truss subjected to a single loading condition, and let n denote the number of degrees of freedom. Suppose the truss consisting of m members is statically indeterminate; i.e., $n < m$ with n being the number of degrees of freedom. The vectors of nodal displacements $\mathbf{U}^1 = (U_1^1, \ldots, U_n^1)^\top$

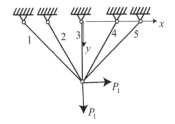

FIGURE 1.10: A statically indeterminate five-bar truss.

and member strains $\boldsymbol{\varepsilon}^1 = (\varepsilon_1^1, \ldots, \varepsilon_m^1)^\top$ should satisfy the compatibility conditions

$$\boldsymbol{\varepsilon}^1 = \mathbf{C}\mathbf{U}^1 \tag{1.27}$$

where $\mathbf{C}$ is an $m \times n$ matrix that is defined by the kinematic relations only. Suppose the truss is stable and $\mathbf{C}$ is full-rank; i.e., the rank of $\mathbf{C}$ is equal to n because $n < m$. Hence, we can eliminate $\mathbf{U}^1$ using (1.27) to express $m - n$ components of $\boldsymbol{\varepsilon}^1$ with respect to the remaining n components.

Therefore, using the constitutive relation $\sigma_i^1 = E\varepsilon_i^1$ with the elastic modulus E, the $m - n$ equations are obtained for the stresses $\sigma_1^1, \ldots, \sigma_m^1$. Hence, the stress can be independently assigned only for n members, and the stresses of all the members of a statically indeterminate truss cannot generally be equal to the upper or lower bound. Consequently, for a truss to be fully-stressed, the cross-sectional areas of at least $m - n$ members should be equal to their lower bounds, and the stresses of the remaining n members should be appropriately assigned so that $\sigma_i^1 = \sigma_i^{1\mathrm{L}}$ or $\sigma_i^1 = \sigma_i^{1\mathrm{U}}$ for those members, and $\sigma_i^{1\mathrm{L}} \leq \sigma_i^1 \leq \sigma_i^{1\mathrm{U}}$ for the remaining $m - n$ members with $A_i = A_i^{\mathrm{L}}$.

Example 1.5
Consider a symmetric five-bar truss, as shown in Fig. 1.10, where $n = 2$ and $m = 5$, i.e., the truss is statically indeterminate, and we can assign two member stresses independently. In the following examples, the units of load and length are omitted for brevity. The truss is symmetric with respect to the y-axis. The angles of members 1 and 2 from the x-axis are $\pi/4$ and $\pi/3$, respectively. If σ_1^1 and σ_3^1 are chosen as the independent variables, the stresses of the remaining members are given as

$$\begin{aligned}
\sigma_2^1 &= \frac{\sqrt{3}}{2}\sigma_1^1 + \frac{3 - \sqrt{3}}{4}\sigma_3^1, \\
\sigma_4^1 &= -\frac{\sqrt{3}}{2}\sigma_1^1 + \frac{3 + \sqrt{3}}{4}\sigma_3^1, \\
\sigma_5^1 &= \sigma_3^1 - \sigma_1^1
\end{aligned} \tag{1.28}$$

which show that the stresses cannot have the same absolute value for all members. Therefore, generally $A_i = A_i^{\mathrm{L}}$ should be satisfied for three members

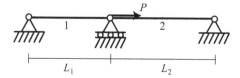

FIGURE 1.11:   A statically indeterminate two-bar truss.

so that the truss is fully stressed. Hence, the FSD is statically determinate if $A_i^L = 0$ for all members. Note that the stress of a nonexistent member can be computed from the strain because the two nodes (supports) connected to any member of this truss exist (see Sec. 3.5.3 for more details).

For example, the total structural volume is minimized under conditions $P_1 = 1$, $P_2 = 0$, and $A_i^L = 0$ for all members. If the bounds of stresses are given as $\sigma_i^U = -\sigma_i^L = 1/\sqrt{2}$ for all members, the optimal cross-sectional areas are obtained as $A_1 = A_5 = 1$ and $A_2 = A_3 = A_4 = 0$. Then the stresses are obtained as $\sigma_1 = 1/\sqrt{2}$. Accordingly, $\sigma_2 = \sqrt{6}/4$, $\sigma_3 = 0$, $\sigma_4 = -\sqrt{6}/4$, $\sigma_5 = -1/\sqrt{2}$, and the truss is statically determinate and fully stressed. For this example, the truss is fully stressed even if a very small positive value e is assigned for A_i^L of all members because $\sigma_i^L \leq \sigma_i \leq \sigma_i^U$ is satisfied by the nonexistent member 3.

Example 1.6

As another illustrative example, consider a statically indeterminate truss, as shown in Fig. 1.11, that has two colinearly located members, and assume $P > 0$. The bounds for the stress are given as $\sigma_i^{1L} = -\sigma_i^{1U}$, where σ_1^{1U} and σ_2^{1U} are not necessarily the same. The lower bounds for A_i are given as $A_1^L = A_2^L = e$, where e has a sufficiently small positive value.

If $L_2 = 2L_1$ and $\sigma_1^{1U} = \sigma_2^{1U} = \overline{\sigma}$ for a specified positive value $\overline{\sigma}$, then the optimization for minimizing the total structural volume V leads to $A_1 \simeq P/\overline{\sigma}$ and $A_2 = e$, because member 2 is longer than member 1, and $\sigma_1^1 = \overline{\sigma}$ and $\sigma_2^1 = -\overline{\sigma}/2$ are satisfied from the compatibility condition. Hence, the optimal solution is fully stressed. By contrast, if $\sigma_1^{1U} = \overline{\sigma}$ and $\sigma_2^{1U} = 4\overline{\sigma}$, then the optimization leads to $A_1 = e$ and $A_2 \simeq P/(4\overline{\sigma})$, because the larger length of member 2 is compensated by the larger absolute value of the allowable stress; consequently, V is approximately equal to $PL/(2\overline{\sigma})$. In fact, if we assume $A_2 = e$, then $A_1 \simeq P/\overline{\sigma}$ and, accordingly, V is approximately equal to $PL/\overline{\sigma}$, which is larger than $PL/(2\overline{\sigma})$.

At the optimal solution with $A_2 \simeq P/(4\overline{\sigma})$, $\sigma_1^1 = 8\overline{\sigma}$, and $\sigma_2^1 = -4\overline{\sigma}$ are satisfied. Hence, the optimal solution is not fully stressed; i.e., the stress constraint is violated by member 1. Therefore, the optimal solution may not be fully stressed if $A_i^L > 0$ and the stress bounds are not the same for all the members of a statically indeterminate truss. However, if $A_i^L = 0$, then the

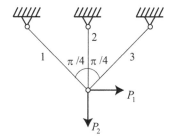

FIGURE 1.12: A three-bar truss (Type 1).

optimal solution is $(A_1, A_2) = (0, P/(4\bar{\sigma}))$, which is fully stressed, because the stress constraint need not be satisfied by the nonexistent member 1.

1.6.3 Multiple loading conditions

Next we consider the truss under multiple loading conditions. The optimization problem under stress constraints is formulated as (1.25) without upper-bound cross-sectional area. Let n^{A} denote the number of members for which the stress is equal to its lower or upper bound for at least one loading condition. If the truss is statically determinate, the axial force is independent of the cross-sectional areas, and $n^{\mathrm{A}} = m$ should be satisfied; i.e., the truss is fully stressed.

For a statically indeterminate truss, we can specify stresses for at most $n \times n^{\mathrm{P}}$ members, because the stresses of n members can be specified for each loading condition, as demonstrated in Example 1.5. Therefore, the stress can be equal to its lower or upper bound for all members if $n^{\mathrm{P}} \geq m/n$ (Patnaik and Dayaratnam 1970). However, it is well known that the optimal solution under multiple loading conditions is not generally fully stressed even for the case where $\sigma_i^{1\mathrm{L}} = -\sigma_i^{1\mathrm{U}}$ and $\sigma_i^{1\mathrm{U}}$ is the same for all members; i.e., $\sigma_i^{1\mathrm{L}} \leq \sigma_i^1 \leq \sigma_i^{1\mathrm{U}}$ may be satisfied by a member with $A_i > 0$ (Kicher 1966; Patnaik and Dayaratnam 1970; McNeil 1971; Gunnlaugsson and Martin 1973; Patnaik and Hopkins 1998).

For a topology optimization problem with $A_i^{\mathrm{L}} = 0$ for all members, the stress constraints may be violated by the nonexistent members, and the optimal truss may be statically determinate even for the multiple loading conditions; see Sec. 3.5 for details.

Example 1.7

Consider a three-bar truss (Type 1), as shown in Fig. 1.12, which is subjected to two independent loads, P_1 and P_2, respectively. From $n^{\mathrm{P}} = 2$, $m = 3$, and $n = 2$, we can see $n^{\mathrm{P}} \geq m/n$ is satisfied. Suppose $P_1 = P_2 = 10$, and the lengths of members 1, 2, and 3 are $\sqrt{2}$, 1, and $\sqrt{2}$, respectively. The

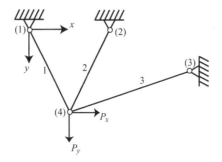

FIGURE 1.13: A three-bar truss (Type 2).

lower-bound cross-sectional areas are given as $A_i^L = 0.1$ for all members; i.e., removal of a member is not allowed.

The optimal solution for $\sigma_i^U = -\sigma_i^L = 10$ for all members is $(A_1, A_2, A_3) = (\sqrt{2}/2, 1/2, \sqrt{2}/2)$, where the total structural volume is 2.5. Note that the value of A_i^L does not have any effect on the optimal solution if it is positive and not more than $1/2$. The stresses are computed as $(\sigma_1^1, \sigma_2^1, \sigma_3^1) = (10, 0, -10)$ and $(\sigma_1^2, \sigma_2^2, \sigma_3^2) = (5, 10, 5)$. Therefore, the optimal truss is fully stressed. However, if $A_i^L = 0$, then a member may be removed to obtain a statically determinate optimal truss, and the stress constraint may be violated by a nonexistent member. In fact, if we remove member 2, then the optimal solution is $(A_1, A_3) = (\sqrt{2}/2, \sqrt{2}/2)$, where the total structural volume is 2.0, which is less than 2.5 for the three-bar truss. If we assume that the elastic modulus E is equal to 1, then the vertical displacement of node 2 is 20. Therefore, the strain of member 2, although it does not exist, is 20, which is double the upper-bound stress.

Example 1.8

Next, consider a three-bar truss (Type 2), as shown in Fig. 1.13, where the coordinates of nodes and supports 1–4 are (0,0), (1000,0), (2000,500), and (500,1000), respectively. Three independent load sets $(P_x, P_y) = (5, 10)$, $(-5, 10)$, and $(-20, 10)$ are applied. The upper-bound stress is 0.2 and $A_i^L = 0$ for all members. The optimal cross-sectional areas and the maximum absolute value of stress of each member under three loading conditions are listed in the second column of Table 1.2. As is seen, the optimal truss is statically indeterminate. Because the absolute value of the stress of member 2 is less than 0.2 for any loading condition, the optimal truss is not fully stressed. In fact, the optimal solutions of two-bar trusses after removal of members 1, 2, and 3, respectively, have larger objective values, as shown in the third, fourth, and fifth columns of Table 1.2, than that of the three-bar truss in the second column.

TABLE 1.2: Optimal cross-sectional areas, total structural volume, and maximum absolute values of stresses of the three-bar truss (Type 2) and corresponding two-bar trusses.

	three-bar	two-bar		
		2, 3	1, 3	1, 2
A_1	42.717	0	63.888	83.853
A_2	36.063	89.443	0	139.75
A_3	84.219	94.868	112.94	0
$\max |\sigma_1^k|$	0.20000	0.60000	0.20000	0.20000
$\max |\sigma_2^k|$	0.19709	0.20000	0.26429	0.20000
$\max |\sigma_3^k|$	0.20000	0.20000	0.20000	0.30000
Total volume	2.3539	2.5000	2.5000	2.5000

1.7 Optimality criteria approach

An optimization method that directly solves the optimality criteria (optimality conditions) is called the *optimality criteria approach* (OC approach) (Venkayya, Khot, and Berke 1973; Berke and Venkayya 1974; Dobbs and Nelson 1975; Khot, Berke, and Venkayya 1978). This approach is very effective for the case where the optimality conditions are written in a simple manner with explicit expressions of sensitivity coefficients with respect to the design variables and the state variables. Furthermore, this approach is more efficient in view of computational time and required memory than the gradient-based nonlinear programming (NLP) approaches; see Sec. 2.2 for sensitivity analysis of static responses, and Appendix A.2.2 for details of optimality conditions for general NLP problems.

Another advantage of the OC approach is that the computer program is very simple. Therefore, in the 1960s and 1970s, when computer power was not sufficient for computing sensitivity coefficients of the responses of moderately large structures many times for optimization, various studies on theoretical and computational aspects of the OC approaches were presented. For problems with general equality and inequality constraints, the OC approach is classified as a dual approach of NLP (Fleury 1979, 1980). Since the purpose here is to find a solution that satisfies the constraints and optimality conditions, it is possible to use a Newton-Raphson iteration for solving these nonlinear equations (Khot, Berke, and Venkayya 1978). However, the recursive formulas, as presented below, are generally used in an OC approach.

For the problem under stress constraints only, an approximate optimal solution can be easily found using the stress-ratio approach of fully-stressed design, as discussed in the previous section. Therefore, in this section, we consider an optimization problem of a truss under displacement constraints.

Suppose, for simplicity, a constraint is given only for the jth displacement component as

$$U_j \leq U_j^{\mathrm{U}} \tag{1.29}$$

where U_j is assumed to be positive. The objective function is the total structural volume, which is a monotonically increasing function of the cross-sectional areas $\mathbf{A} = (A_1, \ldots, A_m)^{\top}$. On the other hand, the nodal displacements are generally decreasing functions of $\mathbf{A}$, if the loads do not depend on $\mathbf{A}$. Therefore, the displacement constraint (1.29) is considered to be active, i.e., satisfied with equality, at the optimal solution assuming that the lower bounds A_i^{L} for the cross-sectional areas are sufficiently small. The upper bounds A_i^{U} are assumed to be sufficiently large to ensure the existence of a feasible solution. In this case, the following condition is obtained from the optimality condition (1.6) for members with $A_i^{\mathrm{L}} < A_i < A_i^{\mathrm{U}}$:

$$L_i + \mu_j \frac{\partial U_j}{\partial A_i} = 0 \tag{1.30}$$

where μ_j (≥ 0) is the Lagrange multiplier for the constraint (1.29).

The term $\partial U_j / \partial A_i$ in (1.30) is the sensitivity coefficient of U_j with respect to A_i that can be obtained efficiently using the adjoint variable method described in Sec. 2.2.2, because we have only one displacement component to be constrained. Let n denote the number of degrees of freedom. The axial force and the $n \times n$ stiffness matrix with respect to the global coordinates of the ith member are denoted by N_i and $\mathbf{K}_i$, respectively. The displacement vector against the specified loads is denoted by $\mathbf{U}$. The values corresponding to the virtual unit load at the jth displacement component are indicated by the superscript $(\cdot)^j$. Then the following relation is derived from the adjoint variable method of design sensitivity analysis of static response:

$$\begin{aligned} \frac{\partial U_j}{\partial A_i} &= -\mathbf{U}^{j\top} \frac{\partial \mathbf{K}_i}{\partial A_i} \mathbf{U} \\ &= -\frac{L_i}{A_i^2 E} N_i^j N_i \end{aligned} \tag{1.31}$$

where E is the elastic modulus.

Define Z_i as

$$Z_i = \mu_j \frac{N_i^j N_i}{A_i^2 E} \tag{1.32}$$

Then the optimality condition (1.30) is written as

$$Z_i = 1 \tag{1.33}$$

Let the superscript $(\cdot)^{(k)}$ denote a value at the kth step of iteration. For a statically determinate truss, N_i^j and N_i are independent of A_i. Therefore, assuming that μ_j is constant, the cross-sectional areas can be updated from (1.32) as follows:

$$A_i^{(k+1)} = (Z_i^{(k)})^{\frac{1}{2}} A_i^{(k)} \tag{1.34}$$

For a statically indeterminate truss, the cross-sectional area is updated by

$$A_i^{(k+1)} = (Z_i^{(k)})^r A_i^{(k)} \tag{1.35}$$

where r is a parameter between 0 and 1 for controlling the convergence property. Note that A_i is replaced with A_i^{L} and A_i^{U}, respectively, if $A_i < A_i^{\mathrm{L}}$ or $A_i > A_i^{\mathrm{U}}$ after application of (1.35).

Next, we derive the update rule of the Lagrange multiplier. From (1.32) and (1.33), we obtain

$$A_i = \sqrt{\mu_j \frac{N_i^j N_i}{E}} \tag{1.36}$$

By using the principle of virtual unit load and the active constraint $U_j = U_j^{\mathrm{U}}$, we have

$$U_j^{\mathrm{U}} = \sum_{i=1}^{m} \frac{L_i}{A_i E} N_i^j N_i \tag{1.37}$$

Through incorporation of A_i in (1.36) into (1.37), the Lagrange multiplier μ_j is updated as

$$\mu_j^{(k+1)} = \left(\sum_{i=1}^{m} \frac{L_i \sqrt{N_i^j N_i}}{U_j^{\mathrm{U}} \sqrt{E}} \right)^2 \tag{1.38}$$

Then we move to the next step of iteration. This way, the solution satisfying the constraint and optimality conditions is found by an iterative approach.

Alternatively, a linear approximation can be used for recursively updating A_i and μ_j (Khot, Berke, and Venkayya 1978). Multiplying $(1 - \alpha) A_i^{(k)}$ on both sides of (1.33) with a parameter $0 < \alpha < 1$, letting

$$(1 - \alpha) A_i^{(k)} = A_i^{(k+1)} - \alpha A_i^{(k)} \tag{1.39}$$

and rearranging the equation, we have the following update rule for A_i:

$$A_i^{(k+1)} = A_i^{(k)} [\alpha + (1 - \alpha) Z_i^{(k)}] \tag{1.40}$$

Linear approximation of U_j leads to the following requirement for the displacement constraint at the $(k + 1)$st step:

$$U_j^{(k)} + \sum_{i=1}^{m} \frac{\partial U_j}{\partial A_i} (A_i^{(k+1)} - A_i^{(k)}) = U_j^{\mathrm{U}} \tag{1.41}$$

Then, from (1.31), (1.37), (1.40), and (1.41), we obtain the following recursive formula for μ_j:

$$\mu_j \sum_{i=1}^{m} \frac{L_i(N_i^j N_i)^2}{E^3 A_i^3} = \frac{(2-\alpha)(U_j^{(k)} - U_j^U)}{1-\alpha} \tag{1.42}$$

The OC approach assumes that the member forces against the applied loads and the virtual unit load are insensitive to variation of the design variables, which is the same as the assumption for the stress-ratio approach (1.26) for fully-stressed design. This is better achieved if the responses are approximated with respect to the reciprocals of the cross-sectional areas (Schmit and Farshi 1974; Zhou and Haftka 1995). Let $a_i = 1/A_i$, and regard A_i as a function of a_i. Then we have

$$\frac{\partial A_i}{\partial a_i} = -\frac{1}{a_i^2} \tag{1.43}$$

and the sensitivity coefficient of U_j with respect to a_i is obtained from (1.31), (1.43), and $A_i = 1/a_i$ as

$$\begin{aligned} \frac{\partial U_j}{\partial a_i} &= \frac{\partial U_j}{\partial A_i} \frac{\partial A_i}{\partial a_i} \\ &= -\frac{L_i}{E} N_i^j N_i \end{aligned} \tag{1.44}$$

which does not explicitly depend on a_i. Although the sensitivity of the total structural volume turns out to be dependent on a_i, convergence of the recursive formulation is improved by using a_i as a design variable.

Because the number of analyses for computing the displacements against virtual unit load is proportional to the number of active or nearly active displacement constraints, this approach is effective for the case with a small number of active displacement constraints. For extension of the OC approach to problems with stress constraints, a pair of unit self-equilibrium forces is applied at the two ends of each member with an active stress constraint. However, the OC approach can be successfully combined with the fully-stressed design approach for problems with stress and displacement constraints.

Pereyra, Lawver, and Isenberg (2003) used an OC approach with a penalty function to optimize a building frame. For continuum structures such as beams, plates, and shells, a continuum-type optimality criteria (COC) approach was developed in the 1960s (Prager and Taylor 1968; Olhoff and Taylor 1979; Rozvany 1989). Furthermore, OC and COC were combined to a discretized form of COC approach termed DCOC (Zhou and Rozvany 1992, 1993; Rozvany and Zhou 1994).

Since the OC approach is simple and easy to implement, it is widely applied in many areas of heuristics, e.g., evolutionary structural optimization (ESO) (Yang, Xie, Steven, and Querin 1999a; Xie and Steven 1993) and cellular automaton (Canyurt and Hajela 2005).

1.8   Compliance constraint

1.8.1   Problem formulation and sensitivity analysis

The *compliance* is equivalent to the external work by static loads, which is also equivalent to twice the strain energy if the linear elastic response is considered. For the case without forced displacement, a smaller compliance leads to a stiffer structure against the specified loads.

Consider a truss with m members and n degrees of freedom. Let $\mathbf{K}(\mathbf{A})$ denote the $n \times n$ stiffness matrix, which is a function of the vector $\mathbf{A} = (A_1, \ldots, A_m)^\top$ of the cross-sectional areas. The displacement vector against the load vector $\mathbf{P}$, which is independent of $\mathbf{A}$, is denoted by $\mathbf{U}(\mathbf{A})$. The stiffness (equilibrium) equation for computing $\mathbf{U}(\mathbf{A})$ is written as

$$\mathbf{K}(\mathbf{A})\mathbf{U}(\mathbf{A}) = \mathbf{P} \tag{1.45}$$

Using (1.45), the compliance $W(\mathbf{U}(\mathbf{A}), \mathbf{A})$ is written as

$$
\begin{aligned}
W(\mathbf{U}(\mathbf{A}), \mathbf{A}) &= \mathbf{U}^\top \mathbf{P} \\
&= \mathbf{U}^\top \mathbf{K}\mathbf{U} \\
&= 2\left(\mathbf{U}^\top \mathbf{P} - \frac{1}{2}\mathbf{U}^\top \mathbf{K}\mathbf{U}\right)
\end{aligned}
\tag{1.46}
$$

i.e., the compliance is also equivalent to the total potential energy multiplied by -2. Note that the second argument $\mathbf{A}$ in $W(\mathbf{U}(\mathbf{A}), \mathbf{A})$ indicates the explicit dependence on $\mathbf{A}$ through $\mathbf{K}(\mathbf{A})$. By differentiating $W(\mathbf{U}(\mathbf{A}), \mathbf{A})$ in the last expression in (1.46), and using (1.45), we find that the partial derivative of the compliance with respect to $\mathbf{U}$ vanishes as

$$\frac{\partial W}{\partial \mathbf{U}} = 2(\mathbf{P} - \mathbf{K}\mathbf{U}) = \mathbf{0} \tag{1.47}$$

Because $\mathbf{U}$ is an implicit function of $\mathbf{A}$, the compliance is written as a function of $\mathbf{A}$ only as

$$\widetilde{W}(\mathbf{A}) = W(\mathbf{U}(\mathbf{A}), \mathbf{A}) \tag{1.48}$$

Let W^{U} denote the specified upper bound for $\widetilde{W}(\mathbf{A})$. Then the optimization problem for minimizing the total structural volume $V(\mathbf{A})$ under a compliance constraint is formulated as

$$\text{Minimize} \quad V(\mathbf{A}) = \sum_{i=1}^{m} A_i L_i \tag{1.49a}$$

$$\text{subject to} \quad \widetilde{W}(\mathbf{A}) \leq W^{\mathrm{U}} \tag{1.49b}$$

$$A_i \geq A_i^{\mathrm{L}}, \quad (i = 1, \ldots, m) \tag{1.49c}$$

where L_i is the length of member i, A_i^{L} is the lower bound for A_i, and the upper bound for A_i is not given for simplicity.

The $n \times n$ stiffness matrix of a truss can be written as a linear function of $\mathbf{A}$ using constant $n \times n$ matrices $\mathbf{K}_i$ $(i = 1, \dots, m)$ as

$$\mathbf{K} = \sum_{i=1}^{m} A_i \mathbf{K}_i \tag{1.50}$$

where

$$\mathbf{K}_i = \frac{\partial \mathbf{K}}{\partial A_i} \tag{1.51}$$

By using the last expression of (1.46) as well as (1.47) and (1.51), we obtain the sensitivity coefficient (differential coefficient) of compliance with respect to A_i as

$$
\begin{aligned}
\frac{\partial \widetilde{W}}{\partial A_i} &= \frac{\partial W}{\partial A_i} + \left(\frac{\partial W}{\partial \mathbf{U}}\right)^{\top} \frac{\partial \mathbf{U}}{\partial A_i} \\
&= \frac{\partial W}{\partial A_i} \\
&= -\mathbf{U}^{\top} \mathbf{K}_i \mathbf{U}
\end{aligned}
\tag{1.52}
$$

Although simpler derivations are possible for the sensitivity coefficient, (1.52) is important for understanding the characteristics of the compliance. Let ε_i denote the strain of the ith member. Since $\mathbf{U}^{\top} \mathbf{K}_i \mathbf{U}$ is twice the strain energy per unit cross-sectional area of the ith member, $\partial \widetilde{W}/\partial A_i$ is equivalently written as

$$\frac{\partial \widetilde{W}}{\partial A_i} = -E \varepsilon_i^2 L_i \tag{1.53}$$

where E is the elastic modulus (see Sec. 2.2 for details of static sensitivity analysis).

Alternatively, by differentiating (1.45) with respect to A_i and using (1.51), we obtain the sensitivity coefficient of $\mathbf{U}$ with respect to A_i as

$$\frac{\partial \mathbf{U}}{\partial A_i} = -\mathbf{K}^{-1} \mathbf{K}_i \mathbf{U} \tag{1.54}$$

Therefore, using the form $\widetilde{W}(\mathbf{A}) = \mathbf{U}^{\top}(\mathbf{A}) \mathbf{K}(\mathbf{A}) \mathbf{U}(\mathbf{A})$, the sensitivity coefficient of the compliance is given as

$$
\begin{aligned}
\frac{\partial \widetilde{W}}{\partial A_i} &= \mathbf{U}^{\top} \mathbf{K}_i \mathbf{U} + 2 \mathbf{U}^{\top} \mathbf{K} \frac{\partial \mathbf{U}}{\partial A_i} \\
&= -\mathbf{U}^{\top} \mathbf{K}_i \mathbf{U}
\end{aligned}
\tag{1.55}
$$

which is the same as (1.52). Furthermore, differentiation of (1.45) with respect to A_i leads to

$$\mathbf{K}_i \mathbf{U} + \mathbf{K} \frac{\partial \mathbf{U}}{\partial A_i} = \mathbf{0} \tag{1.56}$$

By premultiplying $\mathbf{U}^\top$ to (1.56), and using (1.45) and the first expression of the compliance in (1.46), we obtain the same result as (1.52).

1.8.2 Optimality conditions

The Lagrange multiplier for the constraint (1.49b) is denoted by μ (≥ 0). If we consider the lower-bound constraint (1.49c) separately, the Lagrangian is formulated as

$$\psi(\mathbf{A}, \mu) = \sum_{i=1}^{m} A_i L_i + \mu(\widetilde{W}(\mathbf{A}) - W^{\mathrm{U}}) \tag{1.57}$$

Since $\mathbf{P}$ does not depend on $\mathbf{A}$, the stiffness against $\mathbf{P}$ increases as A_i is increased, and accordingly, $\widetilde{W}(\mathbf{A})$ generally decreases as $A_1, \ldots, A_m$ are increased. Furthermore, $V(\mathbf{A})$ is an increasing function of $A_1, \ldots, A_m$. Therefore, the constraint $\widetilde{W}(\mathbf{A}) \leq W^{\mathrm{U}}$ is satisfied with equality at the optimal solution. Considering the side constraints for the variables, the necessary conditions for optimality are generally obtained using the derivatives of $\psi(\mathbf{A}, \mu)$ as in (1.6); hence, from (1.53), we have

$$\begin{cases} E\varepsilon_i^2 = \lambda & \text{for } A_i > A_i^{\mathrm{L}} \\ E\varepsilon_i^2 \leq \lambda & \text{for } A_i = A_i^{\mathrm{L}} \end{cases} \tag{1.58}$$

where $\lambda = 1/\mu > 0$. It is seen from (1.58) that the absolute value of the strain, or the strain energy per unit volume, is the same for the members with $A_i > A_i^{\mathrm{L}}$; i.e., the optimal solution is fully stressed. This way, the optimality conditions are expressed in a simple form without sensitivity coefficients of the displacements.

Note that the conditions derived above using the Lagrangian are *necessary conditions* for optimality. In the following, we show, on the basis of the *principle of minimum total potential energy*, that the conditions (1.58) are also *sufficient conditions* for global optimality. Although sufficient conditions can be derived from the convexity of $\widetilde{W}(\mathbf{A})$, as demonstrated in Sec. 1.8.4, we show below the conventional proof by Prager (1972).

Suppose the optimality conditions (1.58) are satisfied by the solution $\widehat{\mathbf{A}} = (\widehat{A}_1, \ldots, \widehat{A}_m)^\top$. Let $\mathbf{A}^* = (A_1^*, \ldots, A_m^*)^\top$ denote a feasible solution that is different from $\widehat{\mathbf{A}}$ and has the same compliance value W^{U}. The values corresponding to $\widehat{\mathbf{A}}$ and $\mathbf{A}^*$ are indicated by $\widehat{(\cdot)}$ and $(\cdot)^*$, respectively, as $\widehat{\mathbf{U}}$, $\widehat{\varepsilon}_i$, $\mathbf{U}^*$, ε_i^*, etc. Then we have

$$\mathbf{U}^{*\top}\mathbf{P} - \frac{1}{2}\sum_{i=1}^{m} E\varepsilon_i^{*2} A_i^* L_i = \widehat{\mathbf{U}}^\top\mathbf{P} - \frac{1}{2}\sum_{i=1}^{m} E\widehat{\varepsilon}_i^2 \widehat{A}_i L_i = \frac{W^{\mathrm{U}}}{2} \tag{1.59}$$

Since $\widehat{\varepsilon}_1, \ldots, \widehat{\varepsilon}_m$ are the strains corresponding to an admissible displacement vector $\widehat{\mathbf{U}}$ that is not necessarily equal to the correct displacement vector $\mathbf{U}^*$

for $\mathbf{A}^*$, the following relation is obtained from the principle of minimum total potential energy:

$$\widehat{\mathbf{U}}^{\top}\mathbf{P} - \frac{1}{2}\sum_{i=1}^{m}E\widehat{\varepsilon}_i^2 A_i^* L_i \leq \mathbf{U}^{*\top}\mathbf{P} - \frac{1}{2}\sum_{i=1}^{m}E\varepsilon_i^{*2} A_i^* L_i \qquad (1.60)$$

By using (1.59), we obtain the following relation from (1.60):

$$\widehat{\mathbf{U}}^{\top}\mathbf{P} - \frac{1}{2}\sum_{i=1}^{m}E\widehat{\varepsilon}_i^2 A_i^* L_i \leq \widehat{\mathbf{U}}^{\top}\mathbf{P} - \frac{1}{2}\sum_{i=1}^{m}E\widehat{\varepsilon}_i^2 \widehat{A}_i L_i \qquad (1.61)$$

which is rewritten as

$$\sum_{i=1}^{m}(A_i^* L_i - \widehat{A}_i L_i)E\widehat{\varepsilon}_i^2 \geq 0 \qquad (1.62)$$

From (1.62), we have

$$\sum_{i=1}^{m}\lambda(A_i^* - \widehat{A}_i)L_i - \sum_{i=1}^{m}(A_i^* - \widehat{A}_i)(\lambda - E\widehat{\varepsilon}_i^2)L_i \geq 0 \qquad (1.63)$$

Let $\mathcal{I}$ denote the set of indices of members with $\widehat{A}_i = A_i^{\mathrm{L}}$ in the solution satisfying the optimality conditions (1.58). Then, using (1.58), (1.63) is reduced to

$$\sum_{i=1}^{m}\lambda(A_i^* - \widehat{A}_i)L_i - \sum_{i\in\mathcal{I}}(A_i^* - \widehat{A}_i)(\lambda - E\widehat{\varepsilon}_i^2)L_i \geq 0 \qquad (1.64)$$

Furthermore, $A_i^* \geq \widehat{A}_i = A_i^{\mathrm{L}}$ is satisfied for $i \in \mathcal{I}$; therefore, from (1.58), we have

$$\sum_{i\in\mathcal{I}}(A_i^* - \widehat{A}_i)(\lambda - E\widehat{\varepsilon}_i^2)L_i \geq 0 \qquad (1.65)$$

Finally, from (1.64), (1.65), and $\lambda > 0$, the following relation is derived:

$$V(\mathbf{A}^*) - V(\widehat{\mathbf{A}}) = \sum_{i=1}^{m}(A_i^* - \widehat{A}_i)L_i \geq 0 \qquad (1.66)$$

Therefore, if (1.58) holds, no solution satisfying $W(\mathbf{A}^*) = W^{\mathrm{U}}$ has a smaller objective value than $V(\widehat{\mathbf{A}})$, which means that (1.58) is also a *sufficient condition* for global optimality.

The necessary and sufficient conditions can also be obtained for problems with constraints on linear buckling load and eigenvalue of vibration in a similar manner using Rayleigh's principle for the case where the associated eigenvalue problems have simple lowest eigenvalues (Prager 1974a).

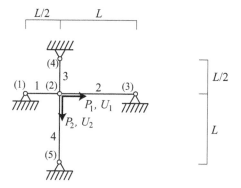

FIGURE 1.14: A four-bar truss.

Example 1.9

Consider, as a small example, a four-bar truss as shown in Fig. 1.14. The lower-bound cross-sectional areas are given as $A_i^L = \overline{A}$ for all members. Let $P_1 = 2P$ and $P_2 = P$, and define U_1 and U_2 as shown in Fig. 1.14. The total potential energy is given as

$$\Pi = \frac{A_1 E}{L}U_1^2 + \frac{A_2 E}{2L}U_1^2 + \frac{A_3 E}{L}U_2^2 + \frac{A_4 E}{2L}U_2^2 - 2PU_1 - PU_2 \qquad (1.67)$$

By differentiating Π with respect to U_1 and U_2, respectively, we obtain the equilibrium equations as

$$\frac{E}{L}(2A_1 + A_3)U_1 = 2P, \quad \frac{E}{L}(2A_2 + A_4)U_2 = P \qquad (1.68)$$

The sensitivity coefficients of compliance are obtained as

$$\frac{\partial \widetilde{W}}{\partial A_1} = -\frac{2EU_1^2}{L}, \quad \frac{\partial \widetilde{W}}{\partial A_2} = -\frac{EU_1^2}{L}, \quad \frac{\partial \widetilde{W}}{\partial A_3} = -\frac{2EU_2^2}{L}, \quad \frac{\partial \widetilde{W}}{\partial A_4} = -\frac{EU_2^2}{L} \qquad (1.69)$$

Therefore, with the reciprocal λ of the Lagrange multiplier

$$\lambda = \frac{4EU_1^2}{L^2} = \frac{4EU_2^2}{L^2} \qquad (1.70)$$

in (1.58), the following conditions are satisfied at the optimal solution:

$$-\frac{\partial \widetilde{W}}{\partial A_1} = -\frac{\partial \widetilde{W}}{\partial A_3} = \frac{\lambda L}{2}, \quad -\frac{\partial \widetilde{W}}{\partial A_2} \leq \lambda L, \quad -\frac{\partial \widetilde{W}}{\partial A_4} \leq \lambda L \qquad (1.71a)$$

$$A_1 \geq \overline{A}, \quad A_2 = \overline{A}, \quad A_3 \geq \overline{A}, \quad A_4 = \overline{A} \qquad (1.71b)$$

which leads to

$$U_1 = U_2 = \frac{L}{2}\sqrt{\frac{\lambda}{E}}, \quad \widetilde{W} = \frac{3PL}{2}\sqrt{\frac{\lambda}{E}} \qquad (1.72)$$

Suppose W^U and $\overline{A}$ are given with a coefficient c as

$$W^U = cPL, \quad \overline{A} = \frac{P}{2Ec} \tag{1.73}$$

Then we have

$$U_1 = U_2 = \frac{cL}{3}, \quad A_1 = \frac{11P}{4Ec}, \quad A_3 = \frac{5P}{4Ec}, \quad V = \frac{3PL}{Ec} \tag{1.74}$$

When the topology is optimized by letting $\overline{A} = 0$, the displacements of the optimal solution are the same as those in (1.74), and the cross-sectional areas and the total structural volume are obtained as

$$A_1 = \frac{3P}{Ec}, \quad A_3 = \frac{3P}{2Ec}, \quad A_2 = A_4 = 0, \quad V = \frac{9PL}{4Ec} \tag{1.75}$$

Example 1.10
As a numerical example, an optimal solution is found for the 10-bar truss in Fig. 1.5 under a compliance constraint (Katoh, Ohsaki, and Tani 2002). The loads are given as $(P_1, P_2) = (0.0, 100.0 \text{ kN})$, and the upper bound for the compliance is 0.3 kNm. Other parameters are the same as those in Example 1.3 for optimization under stress constraints against a single loading condition. Optimal solutions are obtained using the optimization software package SNOPT Ver. 7.2 (Gill, Murray, and Saunders 2002). The optimal cross-sectional areas are similar to those in Fig. 1.6, where the values of $A_1, \ldots, A_{10}$ (mm^2) are 1599.93, 800.06, 1131.27, 0.10, 0.10, 0.10, 799.94, 0.10, 1131.28, 0.10. The optimal objective value is 1.28005×10^7 mm^3. As is seen from Fig. 1.6, the members are located so as to efficiently transmit the load to the supports. Note that $A_i = A_i^L$ is satisfied by the thin members, and the optimal solution is fully stressed, as is verified by the optimality conditions (1.58).

1.8.3 Reformulation of the optimization problem

As we have observed in Sect. 1.8.2, the optimal solution under a compliance constraint is equivalent to the fully-stressed design with $\sigma_i^L = -\sigma_i^U$ for all members with appropriately chosen σ_i^U, because the absolute values of stresses are the same for the members with $A_i > A_i^L$ at the optimal solution. Furthermore, as we can intuitively recognize, minimization of compliance under volume constraint is equivalent, after an appropriate scaling of the objective function, to minimization of the volume under a compliance constraint, because the compliance is a decreasing function and the total structural volume is an increasing function of the cross-sectional areas. Therefore, utilizing these properties, many approaches have been presented for equivalent reformulation of the problem.

Consider the following problem of minimizing the compliance under constraint on the total structural volume:

$$\text{Minimize} \quad \widetilde{W}(\mathbf{A}) = \mathbf{P}^\top \mathbf{U}(\mathbf{A}) \tag{1.76a}$$

$$\text{subject to} \quad V(\mathbf{A}) = \sum_{i=1}^{m} A_i L_i \leq V^\mathrm{U} \tag{1.76b}$$

$$A_i \geq A_i^\mathrm{L}, \quad (i = 1, \dots, m) \tag{1.76c}$$

where V^U is the specified upper bound for the total structural volume.

Let λ (≥ 0) be the Lagrange multiplier for the constraint (1.76b). With the use of (1.53), the Lagrangian without the side constraints (1.76c) and its sensitivity coefficient with respect to A_i are written as

$$\psi(\mathbf{A}, \lambda) = \widetilde{W}(\mathbf{A}) + \lambda \left(\sum_{i=1}^{m} A_i L_i - V^\mathrm{U} \right) \tag{1.77a}$$

$$\frac{\partial \psi}{\partial A_i} = -E\varepsilon_i^2 L_i + \lambda L_i, \quad (i = 1, \dots, m) \tag{1.77b}$$

Therefore, from (1.77b) and the convexity of the compliance that is shown in Sec. 1.8.4, it is easily seen that the optimality conditions for Problem (1.76) are the same as (1.58), and Problem (1.76) is equivalent to Problem (1.49) if W^U and V^U are appropriately assigned (Achtziger 1997). For the four-bar truss in Example 1.9,

$$V^\mathrm{U} = \frac{3PL}{Ec} \tag{1.78}$$

for Problem (1.76) leads to the same optimal solution as that of Problem (1.49).

Consider the simple case where $A_i^\mathrm{L} = 0$ for all members, which corresponds to the topology optimization problem that is extensively studied in Chap. 3. We can also use the approach of *simultaneous analysis and design* (SAND), as follows, with the member cross-sectional areas and the nodal displacements as variables (Bendsøe and Sigmund 2003) by explicitly incorporating the stiffness equation (1.45) that is reformulated using (1.50) into the constraints:

$$\text{Minimize} \quad \mathbf{P}^\top \mathbf{U} \tag{1.79a}$$

$$\text{subject to} \quad \sum_{i=1}^{m} A_i \mathbf{K}_i \mathbf{U} = \mathbf{P} \tag{1.79b}$$

$$\sum_{i=1}^{m} A_i L_i \leq V^\mathrm{U} \tag{1.79c}$$

$$A_i \geq 0, \quad (i = 1, \dots, m) \tag{1.79d}$$

where the variables are $\mathbf{A}$ and $\mathbf{U}$ (see Sec. 1.13 for details of SAND).

A diagonal matrix $\mathbf{S}$ that consists of the extensional stiffness of each member is given as

$$\mathbf{S} = \mathrm{diag}(EA_1/L_1, \ldots, EA_m/L_m) \tag{1.80}$$

The vector of elongations of members is denoted by $\mathbf{d} = (d_1, \ldots, d_m)^\top$. Then the compliance is expressed as $\mathbf{d}^\top \mathbf{S} \mathbf{d}$. Let $\mathbf{D}$ and $\mathbf{N} = (N_1, \ldots, N_m)^\top$ denote the equilibrium matrix and the vector of axial forces. The equilibrium equation $\mathbf{D}\mathbf{N} = \mathbf{P}$ and the constitutive relation $\mathbf{N} = \mathbf{S}\mathbf{d}$ are given instead of the stiffness equation (1.79b). Hence, Problem (1.76) can be reformulated into the following form:

$$\text{Minimize} \quad \mathbf{d}^\top \mathbf{S} \mathbf{d} \tag{1.81a}$$

$$\text{subject to} \quad \mathbf{D}\mathbf{N} = \mathbf{P} \tag{1.81b}$$

$$\mathbf{N} = \mathbf{S}\mathbf{d} \tag{1.81c}$$

$$\sum_{i=1}^{m} A_i L_i \leq V^{\mathrm{U}} \tag{1.81d}$$

$$A_i \geq 0, \quad (i = 1, \ldots, m) \tag{1.81e}$$

where the variables are $\mathbf{d}$, $\mathbf{N}$, and $\mathbf{A}$. Note that the compatibility conditions between $\mathbf{d}$ and $\mathbf{U}$ are ignored; however, they are satisfied at the optimal solution, as shown below.

The Lagrangian for this problem is given as

$$\psi = \mathbf{d}^\top \mathbf{S} \mathbf{d} + \boldsymbol{\mu}^\top (\mathbf{D}\mathbf{N} - \mathbf{P}) + \boldsymbol{\eta}^\top (\mathbf{N} - \mathbf{S}\mathbf{d})$$
$$+ \lambda \left(\sum_{i=1}^{m} A_i L_i - V^{\mathrm{U}} \right) - \sum_{i=1}^{m} \kappa_i A_i \tag{1.82}$$

where $\boldsymbol{\mu}$, $\boldsymbol{\eta}$, λ (≥ 0), and κ_i (≥ 0) are the Lagrange multipliers. From the stationary conditions of the Lagrangian as well as the complementarity conditions $\kappa_i A_i = 0$, $A_i \geq 0$, and $\kappa_i \geq 0$, we have

$$2\mathbf{S}\mathbf{d} - \mathbf{S}\boldsymbol{\eta} = \mathbf{0}, \tag{1.83a}$$

$$\mathbf{D}^\top \boldsymbol{\mu} + \boldsymbol{\eta} = \mathbf{0} \tag{1.83b}$$

$$\mathbf{d}^\top \frac{\partial \mathbf{S}}{\partial A_i} \mathbf{d} + \lambda L_i = 0 \quad \text{for} \quad A_i > 0 \tag{1.83c}$$

$$\mathbf{d}^\top \frac{\partial \mathbf{S}}{\partial A_i} \mathbf{d} + \lambda L_i \geq 0 \quad \text{for} \quad A_i = 0 \tag{1.83d}$$

We can see from these equations that the same optimality conditions as (1.58) are obtained, and there always exists a set of vectors $\mathbf{U} = (1/2)\boldsymbol{\mu}$ and $\mathbf{d} = (1/2)\boldsymbol{\eta}$ satisfying the compatibility conditions (1.83b) at the optimal solution of Problem (1.81).

Since the optimal solution for this problem is fully stressed, it can be obtained numerically in the same manner as the stress-ratio approach for fully-stressed design. Let e_i be defined as

$$e_i = \frac{1}{L_i} \mathbf{d}^\top \frac{\partial \mathbf{S}}{\partial A_i} \mathbf{d} \tag{1.84}$$

which is twice the strain energy per unit volume of member i. The mean value of e_i among the members with $A_i > 0$ is denoted by $\bar{e}$. Then the cross-sectional areas are updated by

$$A_i^{(k+1)} = A_i^{(k)} \left(\frac{e_i^{(k)}}{\bar{e}^{(k)}} \right)^\alpha \frac{V^{\mathrm{U}}}{V(\mathbf{A}^{(k)})} \tag{1.85}$$

where the superscript $(\cdot)^{(k)}$ denotes the value at the kth iteration, and α is a parameter that is slightly less than 1, e.g., 0.8 (Pedersen and Pedersen 2009).

By expressing $\mathbf{d}$ with respect to $\mathbf{N}$, we can formulate Problem (1.81) in terms of the axial force vector $\mathbf{N}$ (Achtziger 1997):

$$\text{Minimize} \quad \sum_{A_i \notin \mathcal{I}} \frac{L_i N_i^2}{E_i A_i} \tag{1.86a}$$

$$\text{subject to} \quad \mathbf{DN} = \mathbf{P} \tag{1.86b}$$

$$N_i = 0 \quad \text{for all} \ i \in \mathcal{I} \tag{1.86c}$$

$$\sum_{i=1}^{m} A_i L_i \leq V^{\mathrm{U}} \tag{1.86d}$$

$$A_i \geq 0, \quad (i = 1, \ldots, m) \tag{1.86e}$$

where $\mathcal{I}$ is the set of indices of members satisfying $A_i = 0$.

Let $\mathcal{A}$ denote the feasible region of $\mathbf{A}$ satisfying the constraints (1.79c) and (1.79d). Then Problem (1.79) can be rewritten using the last expression of (1.46) and the principle of minimum total potential energy as

$$\max_{\mathbf{A} \in \mathcal{A}} \min_{\mathbf{U}} \left[\frac{1}{2} \mathbf{U}^\top \left(\sum_{i=1}^{m} A_i \mathbf{K}_i \right) \mathbf{U} - \mathbf{P}^\top \mathbf{U} \right] \tag{1.87}$$

Because Problem (1.87) is linear with respect to $\mathbf{A}$ and convex with respect to $\mathbf{U}$, global optimality is guaranteed for both problems of minimization with respect to $\mathbf{U}$ and maximization with respect to $\mathbf{A}$. Therefore, the inner minimization problem and the outer maximization problem can be exchanged as

$$\min_{\mathbf{U}} \max_{\mathbf{A} \in \mathcal{A}} \left[\frac{1}{2} \mathbf{U}^\top \left(\sum_{i=1}^{m} A_i \mathbf{K}_i \right) \mathbf{U} - \mathbf{P}^\top \mathbf{U} \right] \tag{1.88}$$

In the conventional ground structure approach for topology optimization, ideally all the pairs of nodes should be connected by members that can exist;

see Sec. 3.3. Hence, the number of components in $\mathbf{A}$ is far greater than that of $\mathbf{U}$. Therefore, it is desirable to reformulate the problem to include only $\mathbf{U}$ as the variables. It is seen from the optimality condition (1.58) that $(\mathbf{U}^\top \mathbf{K}_i \mathbf{U})/L_i$, which is twice the strain energy per unit volume of member i, has the same value for the members with $A_i > 0$ at the optimal solution. We also have the relation

$$\sum_{i=1}^{m} A_i L_i = V^{\mathrm{U}} \tag{1.89}$$

because the volume constraint is satisfied with equality at the optimal solution. Therefore, the total strain energy of the optimal solution is written as $V^{\mathrm{U}}(\mathbf{U}^\top \mathbf{K}_i \mathbf{U})/(2L_i)$, where i represents an arbitrary member satisfying $A_i > 0$. Then, Problem (1.88) is reformulated with respect to $\mathbf{U}$ only as a variable vector (Ben-Tal and Bendsøe 1993):

$$\min_{\mathbf{U}} \max_{i=1,\ldots,m} \left(\frac{V^{\mathrm{U}}}{2L_i} \mathbf{U}^\top \mathbf{K}_i \mathbf{U} - \mathbf{P}^\top \mathbf{U} \right) \tag{1.90}$$

Since the function in the parentheses in (1.90) is convex with respect to $\mathbf{U}$, and the pointwise maximum of convex functions is also convex, as shown in Appendix A.1.1, Problem (1.90) is an unconstrained minimization problem of a convex function with respect to $\mathbf{U}$. However, the objective function is nonsmooth, because we have to find the maximum of m different convex functions.

The nonsmoothness of Problem (1.88) can be alleviated by using the standard reformulation with an arbitrary variable ν as

$$\text{Minimize} \quad \nu \tag{1.91a}$$

$$\text{subject to} \quad \frac{V^{\mathrm{U}}}{2L_i} \mathbf{U}^\top \mathbf{K}_i \mathbf{U} - \mathbf{P}^\top \mathbf{U} \leq \nu, \quad (i = 1, \ldots, m) \tag{1.91b}$$

which is alternatively written with an auxiliary variable $\tau \; (> 0)$ (Ben-Tal, Kočvara, and Zowe 1993):

$$\text{Minimize} \quad \tau - \mathbf{P}^\top \mathbf{U} \tag{1.92a}$$

$$\text{subject to} \quad \frac{V^{\mathrm{U}}}{2L_i} \mathbf{U}^\top \mathbf{K}_i \mathbf{U} \leq \tau, \quad (i = 1, \ldots, m) \tag{1.92b}$$

Problem (1.92) is equivalently written as follows after scaling $\mathbf{P}$ and $\mathbf{U}$ by $\sqrt{\tau}$:

$$\text{Minimize} \quad -\mathbf{P}^\top \mathbf{U} \tag{1.93a}$$

$$\text{subject to} \quad \frac{V^{\mathrm{U}}}{2L_i} \mathbf{U}^\top \mathbf{K}_i \mathbf{U} \leq 1, \quad (i = 1, \ldots, m) \tag{1.93b}$$

Using the relation $\mathbf{U}^\top \mathbf{K}_i \mathbf{U} = E \varepsilon_i^2 L_i$ in (1.53), Problem (1.93) can be reformulated in a form of fully-stressed design (Achtziger, Bendsøe, Ben-Tal, and Zowe 1992; Beckers and Fleury 1997; Achtziger 1997):

$$\text{Minimize} \quad -\mathbf{P}^\top \mathbf{U} \tag{1.94a}$$

$$\text{subject to} \quad -1 \le \sqrt{\frac{V^U E}{2}} \frac{\mathbf{b}_i^\top \mathbf{U}}{L_i} \le 1, \quad (i = 1, \dots, m) \tag{1.94b}$$

where $\mathbf{b}_i^\top$ defines the strain-displacement relation as

$$\varepsilon_i = \frac{\mathbf{b}_i^\top \mathbf{U}}{L_i} \tag{1.95}$$

Because Problem (1.94) is an LP problem with n variables, n inequality constraints in (1.94b) are satisfied with equality; i.e., the member stresses are obtained such that they take upper- or lower-bound values in n members. Then, from the optimality conditions, the cross-sectional areas of the remaining $m - n$ members are 0, and those of m members are computed to satisfy the equilibrium equations. Hence, the optimal truss is statically determinate and fully stressed for the topology optimization problem with $A_i^L = 0$ for all members.

Example 1.11
Consider again the simple four-bar truss, as shown in Fig. 1.9 in Example 1.14, with $V^U = 9PL/(4Ec)$. Let the term in the parentheses in (1.90) be denoted by f_i, i.e., for this example,

$$f_i = \left(\frac{9PL}{4Ec} \right) \left(\frac{1}{2L_i} \right) \left(\frac{E}{L_i} d_i^2 \right) - 2PU_1 - PU_2 \tag{1.96}$$

From the relations $L_1 = L_3 = L/2$, $L_2 = L_4 = L$, $d_1 = -d_2 = U_1$, and $d_2 = -d_4 = U_2$, we can see that $f_1 > f_2$ and $f_3 > f_4$ are always satisfied. Therefore, minimization of the maximum of f_1 and f_3 leads to $f_1 = f_3$, from which $U_1 = U_2$ is obtained. Incorporating $U_2 = U_1$ into (1.96), we have the stationary condition of f_3 with respect to U_1 as

$$\left(\frac{9PL}{4Ec} \right) \left(\frac{2}{L} \right) \left(\frac{2E}{L} \right) (2U_1) - 3P = 0 \tag{1.97}$$

from which we obtain $U_1 = U_2 = cL/3$, which agrees with the results in (1.74). It is easily seen that the same solution can be found by solving Problems (1.92), (1.93), and (1.94), respectively.

1.8.4   Convexity of compliance

The convexity of the compliance $\widetilde{W}(\mathbf{A})$ of a truss can be proved in several manners. The sufficiency of the conditions (1.58) in Sec. 1.8.2 can be derived from the convexity of $\widetilde{W}(\mathbf{A})$.

The conventional approach for the proof of convexity utilizes the minimum principle of total potential energy:

Proof 1.1

Let $\Pi(\mathbf{A}, \mathbf{U})$ denote the total potential energy that is conceived as a function of $\mathbf{A}$ and $\mathbf{U}$ as

$$\Pi(\mathbf{A}, \mathbf{U}) = \frac{1}{2}\mathbf{U}^\top \mathbf{K}(\mathbf{A})\mathbf{U} - \mathbf{P}^\top \mathbf{U} \tag{1.98}$$

From the minimum principle of total potential energy, the correct $\mathbf{U}$ for the specified $\mathbf{A}$ is found by minimizing $\Pi(\mathbf{A}, \mathbf{U})$ with respect to $\mathbf{U}$:

$$\mathbf{U}(\mathbf{A}) = \underset{\mathbf{U}}{\operatorname{argmin}} \, \Pi(\mathbf{A}, \mathbf{U}) \tag{1.99}$$

which leads to $\mathbf{KU} = \mathbf{P}$. Therefore, the compliance $\widetilde{W}(\mathbf{A}) = W(\mathbf{U}(\mathbf{A}), \mathbf{A})$ is written as

$$\widetilde{W}(\mathbf{A}) = -2 \min_{\mathbf{U}} \Pi(\mathbf{A}, \mathbf{U}) \tag{1.100}$$

Let $\mathbf{A}^{\mathrm{I}}$ and $\mathbf{A}^{\mathrm{II}}$ denote two different solutions. The displacement vectors for $\mathbf{A}^{\mathrm{I}}$ and $\mathbf{A}^{\mathrm{II}}$ are denoted by $\mathbf{U}^{\mathrm{I}}$ and $\mathbf{U}^{\mathrm{II}}$, respectively. A linear combination of $\mathbf{A}^{\mathrm{I}}$ and $\mathbf{A}^{\mathrm{II}}$ is given using a parameter $0 \leq \alpha \leq 1$ as $\widehat{\mathbf{A}} = \alpha \mathbf{A}^{\mathrm{I}} + (1 - \alpha)\mathbf{A}^{\mathrm{II}}$. From the linearity of $\Pi(\mathbf{A}, \mathbf{U})$ with respect to $\mathbf{A}$, which is easily seen from (1.50), we have

$$\begin{aligned} \widetilde{W}(\widehat{\mathbf{A}}) &= \widetilde{W}(\alpha \mathbf{A}^{\mathrm{I}} + (1 - \alpha)\mathbf{A}^{\mathrm{II}}) \\ &= -2 \min_{\mathbf{U}} \left[\alpha \Pi(\mathbf{A}^{\mathrm{I}}, \mathbf{U}) + (1 - \alpha)\Pi(\mathbf{A}^{\mathrm{II}}, \mathbf{U}) \right] \end{aligned} \tag{1.101}$$

Since $\mathbf{U}$ cannot minimize $\Pi(\mathbf{A}^{\mathrm{I}}, \mathbf{U})$ and $\Pi(\mathbf{A}^{\mathrm{II}}, \mathbf{U})$ simultaneously in general, we have

$$\begin{aligned} &\min_{\mathbf{U}} \left[\alpha \Pi(\mathbf{A}^{\mathrm{I}}, \mathbf{U}) + (1 - \alpha)\Pi(\mathbf{A}^{\mathrm{II}}, \mathbf{U}) \right] \\ &\geq \min_{\mathbf{U}^{\mathrm{I}}, \mathbf{U}^{\mathrm{II}}} \left[\alpha \Pi(\mathbf{A}^{\mathrm{I}}, \mathbf{U}^{\mathrm{I}}) + (1 - \alpha)\Pi(\mathbf{A}^{\mathrm{II}}, \mathbf{U}^{\mathrm{II}}) \right] \end{aligned} \tag{1.102}$$

Therefore, from (1.101) and (1.102), we obtain

$$\begin{aligned} \widetilde{W}(\widehat{\mathbf{A}}) &\leq -2 \min_{\mathbf{U}^{\mathrm{I}}, \mathbf{U}^{\mathrm{II}}} \left[\alpha \Pi(\mathbf{A}^{\mathrm{I}}, \mathbf{U}^{\mathrm{I}}) + (1 - \alpha)\Pi(\mathbf{A}^{\mathrm{II}}, \mathbf{U}^{\mathrm{II}}) \right] \\ &= \alpha \widetilde{W}(\mathbf{A}^{\mathrm{I}}) + (1 - \alpha)\widetilde{W}(\mathbf{A}^{\mathrm{II}}) \end{aligned} \tag{1.103}$$

which shows the convexity of $\widetilde{W}(\mathbf{A})$. $\qquad\square$

Let $\mathbf{I}_n$ denote the $n \times n$ identity matrix. Differentiating the obvious relation $\mathbf{KK}^{-1} = \mathbf{I}_n$ with respect to A_i, we obtain

$$\frac{\partial \mathbf{K}}{\partial A_i} \mathbf{K}^{-1} + \mathbf{K}\frac{\partial \mathbf{K}^{-1}}{\partial A_i} = \mathbf{O} \tag{1.104}$$

From (1.50) and (1.104), the sensitivity coefficient of $\mathbf{K}^{-1}$ with respect to A_i is expressed as

$$\frac{\partial \mathbf{K}^{-1}}{\partial A_i} = -\mathbf{K}^{-1} \frac{\partial \mathbf{K}}{\partial A_i} \mathbf{K}^{-1}$$

$$= -\mathbf{K}^{-1} \mathbf{K}_i \mathbf{K}^{-1} \qquad (1.105)$$

Svanberg (1984, 1994) presented a proof based on the positive semidefiniteness of the Hessian of $\widetilde{W}(\mathbf{A})$, assuming that $\widetilde{W}(\mathbf{A})$ is twice continuously differentiable with respect to $\mathbf{A}$ and $\mathbf{K}$ is nonsingular:

Proof 1.2
From $\widetilde{W}(\mathbf{A}) = \mathbf{P}^{\top}\mathbf{U}$ and $\mathbf{K}(\mathbf{A})\mathbf{U} = \mathbf{P}$, the compliance can be rewritten without using $\mathbf{U}$ as

$$\widetilde{W}(\mathbf{A}) = \mathbf{P}^{\top} \mathbf{K}^{-1}(\mathbf{A})\mathbf{P} \qquad (1.106)$$

where $\mathbf{K}^{-1}$ is a flexibility matrix which is symmetric and positive definite. Differentiating $\widetilde{W}(\mathbf{A})$ with respect to A_i and using (1.105), we obtain

$$\frac{\partial \widetilde{W}}{\partial A_i} = \mathbf{P}^{\top} \frac{\partial \mathbf{K}^{-1}}{\partial A_i} \mathbf{P}$$

$$= -\mathbf{P}^{\top} \mathbf{K}^{-1} \mathbf{K}_i \mathbf{K}^{-1} \mathbf{P} \qquad (1.107)$$

Further differentiating with respect to A_j, and using $\mathbf{U} = \mathbf{K}^{-1}\mathbf{P}$, the Hessian of $\widetilde{W}(\mathbf{A})$ is obtained as

$$\frac{\partial^2 \widetilde{W}(\mathbf{A})}{\partial A_i \partial A_j} = -\mathbf{P}^{\top} \frac{\partial \mathbf{K}^{-1}}{\partial A_j} \mathbf{K}_i \mathbf{K}^{-1} \mathbf{P} + \mathbf{P}^{\top} \mathbf{K}^{-1} \mathbf{K}_i \frac{\partial \mathbf{K}^{-1}}{\partial A_j} \mathbf{P}$$

$$= \mathbf{P}^{\top} \mathbf{K}^{-1} \mathbf{K}_j \mathbf{K}^{-1} \mathbf{K}_i \mathbf{K}^{-1} \mathbf{P} + \mathbf{P}^{\top} \mathbf{K}^{-1} \mathbf{K}_i \mathbf{K}^{-1} \mathbf{K}_j \mathbf{K}^{-1} \mathbf{P} \qquad (1.108)$$

$$= \mathbf{U}^{\top} \mathbf{K}_j \mathbf{K}^{-1} \mathbf{K}_i \mathbf{U} + \mathbf{U}^{\top} \mathbf{K}_i \mathbf{K}^{-1} \mathbf{K}_j \mathbf{U}$$

For an arbitrary vector $\mathbf{y} = (y_1, \ldots, y_m)^{\top}$, define the vector $\mathbf{z}$ as

$$\mathbf{z} = \sum_{i=1}^{m} y_i \mathbf{K}_i \mathbf{U} \qquad (1.109)$$

Note that $\mathbf{z}$ is a kind of equivalent nodal load vector corresponding to $\mathbf{U}$ if y_i is regarded as the cross-sectional area of the ith member. Then, we can show the following relation from (1.108) and symmetry of $\mathbf{K}^{-1}$:

$$\frac{1}{2} \sum_{i=1}^{m} \sum_{j=1}^{m} \left(\frac{\partial^2 \widetilde{W}}{\partial A_i \partial A_j} \right) y_i y_j$$

$$= \sum_{i=1}^{m} \sum_{j=1}^{m} (y_j \mathbf{U}^{\top} \mathbf{K}_j) \mathbf{K}^{-1} (y_i \mathbf{U}^{\top} \mathbf{K}_i) \qquad (1.110)$$

$$= \mathbf{z}^{\top} \mathbf{K}^{-1} \mathbf{z}$$

Since $\mathbf{K}$ of a stable truss is positive definite, and accordingly, $\mathbf{K}^{-1}$ is positive definite, (1.110) implies that the Hessian of $\widetilde{W}(\mathbf{A})$ is positive semidefinite. Note that the Hessian in (1.108) is not always positive definite, because there may exist a member that does not deform under displacement $\mathbf{U}$; i.e., $\mathbf{z}$ may be a null vector for a nonzero vector $\mathbf{y}$. This concludes the proof of convexity of $\widetilde{W}(\mathbf{A})$.   □

Svanberg (1994) also showed that the compliance is a concave function of the reciprocals of the cross-sectional areas, which ensures global convergence of the convex linearization method (CONLIN) (Nguyen, Strodiot, and Fleury 1987) applied to the compliance optimization problem.

Finally, Stolpe and Svanberg (2001a) presented a proof based on the convexity of a function defined as the pointwise maximum of a collection of convex functions:

Proof 1.3

Let $G(\mathbf{A}, \mathbf{U})$ be defined as

$$G(\mathbf{A}, \mathbf{U}) = 2 \left(\mathbf{P}^\top \mathbf{U} - \frac{1}{2} \mathbf{U}^\top \mathbf{K}(\mathbf{A}) \mathbf{U} \right) \tag{1.111}$$

Then, for every $\mathbf{A}$,

$$\widetilde{W}(\mathbf{A}) = \max_{\mathbf{U}} G(\mathbf{A}, \mathbf{U}) \tag{1.112}$$

Suppose the extensional stiffness of member i is a concave function (including linear function) of A_i, which is valid for a truss. Then, for each fixed $\mathbf{U}$, $G(\mathbf{A}, \mathbf{U})$ is convex with respect to $\mathbf{A}$. Because a function defined as a pointwise maximum of a collection of convex functions is convex, $\widetilde{W}(\mathbf{A})$ is a convex function (see Appendix A.1.1 for the definition of pointwise maximum).   □

For a frame that has a solid section, the compliance is not generally a convex function of the cross-sectional areas; e.g., the bending stiffness of a section with constant height/width ratio is proportional to the square of the cross-sectional area with a positive coefficient. Therefore, the stiffness matrix is a convex function of the cross-sectional parameters, and the requirement for Proof 1.3 is not satisfied.

1.8.5   Other topics on compliance optimization

As we have seen, the stiffest structure under specified static loads can be obtained through minimization of the compliance. By contrast, if forced displacements are applied, the work done by the reactions and the displacements should be maximized so as to find the stiffest structure. For geometrically nonlinear problems, Buhl, Pedersen, and Sigmund (2000) defined *end compliance*, which is the external work at the final state.

Contrary to the attempt to design the stiffest structure, optimization approaches have been developed to generate a *compliant mechanism* that has flexibility against the applied loads to produce a specified large deformation. Nishiwaki, Min, Yoo, and Kikuchi (2001) used mutual energy for retaining stiffness at the output node against the input load, which is utilized by Saxena and Ananthasuresh (2000) for frame mechanisms. Ohsaki and Nishiwaki (2005) introduced stiffness constraints for the initial undeformed and the final deformed state for generating a compliant bar-joint system that has multiple stable equilibrium states utilizing snapthrough behavior.

1.9 Frequency constraints

The fundamental natural frequency (or period) is an important performance measure of structures in the various fields of engineering, because it defines the stiffness against dynamic loads, e.g., earthquake loads, wind loads, in civil engineering. Furthermore, it is important that the higher-order frequencies are also appropriately assigned so as to prevent resonance to the possible dynamic loads. In this section, basic formulations are presented for optimization of trusses under frequency constraints. Detailed results of topology optimization under multiple eigenvalue constraints are shown in Sec. 3.9.

Let $\mathbf{M}^{\mathrm{s}}(\mathbf{A})$ and $\mathbf{M}^0$ denote the $n \times n$ mass matrices corresponding to the structural mass and nonstructural mass, respectively, with n being the number of degrees of freedom. Note that $\mathbf{M}^{\mathrm{s}}(\mathbf{A})$ represents the mass of the structural members, which is a function of the vector $\mathbf{A}$ of the cross-sectional areas that are the design variables. In the following, the argument $\mathbf{A}$ is omitted for brevity. The eigenvalue problem of free vibration is formulated as

$$\mathbf{K}\mathbf{\Phi}_r = \Omega_r(\mathbf{M}^0 + \mathbf{M}^{\mathrm{s}})\mathbf{\Phi}_r, \quad (r = 1, \ldots, n) \tag{1.113}$$

where Ω_r and $\mathbf{\Phi}_r$ are the rth eigenvalue and eigenmode, respectively, which are functions of $\mathbf{A}$. The vector $\mathbf{\Phi}_r$ is ortho-normalized by

$$\mathbf{\Phi}_r^\top \mathbf{M}\mathbf{\Phi}_s = \delta_{rs}, \quad (r = 1, \ldots, n) \tag{1.114}$$

where δ_{rs} is the Kronecker delta.

The objective here is to find an optimal solution of minimizing the total structural volume under constraint such that the fundamental eigenvalue Ω_1 is not less than the specified lower bound Ω^{L}. Hence, the optimization problem is formulated as

$$\text{Minimize} \quad V = \sum_{i=1}^{m} A_i L_i \tag{1.115a}$$

$$\text{subject to} \quad \Omega_1 \geq \Omega^{\mathrm{L}}, \quad (r = 1, \ldots, n) \tag{1.115b}$$

$$A_i \geq A_i^{\mathrm{L}}, \quad (i = 1, \ldots, m) \tag{1.115c}$$

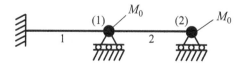

FIGURE 1.15: A two-bar lumped mass structure.

where m is the number of members, A_i^L is the lower bound for A_i, and the upper bound for A_i is not given for simplicity.

Define $n \times n$ matrices $\mathbf{K}_i$ and $\mathbf{M}_i$ as

$$\mathbf{K}_i = \frac{\partial \mathbf{K}}{\partial A_i}, \quad \mathbf{M}_i = \frac{\partial \mathbf{M}^{\mathrm{s}}}{\partial A_i} \tag{1.116}$$

For a truss, we have

$$\mathbf{K} = \sum_{i=1}^{m} A_i \mathbf{K}_i, \quad \mathbf{M} = \sum_{i=1}^{m} A_i \mathbf{M}_i + \mathbf{M}^0 \tag{1.117}$$

If the fundamental eigenvalue is simple, the optimality conditions are derived using the Lagrange multiplier approach. The Lagrangian ignoring the side constraints (1.115c) is given as

$$\psi(\mathbf{A}, \mu) = \sum_{i=1}^{m} A_i L_i + \mu(\Omega^L - \Omega_1) \tag{1.118}$$

where μ (≥ 0) is the Lagrange multiplier.

Since the total structural volume is an increasing function of $\mathbf{A}$, and Ω_1 is generally an increasing function of $\mathbf{A}$, the constraint (1.115b) is satisfied with equality assuming that sufficiently small values are assigned for A_i^L. Therefore, incorporating the sensitivity coefficients of Ω_1 with respect to A_i presented in Sec. 2.3 to the KKT conditions (1.6), we obtain the following optimality conditions:

$$\begin{cases} \mathbf{\Phi}_1^\top \mathbf{K}_i \mathbf{\Phi}_1 - \Omega_1 \mathbf{\Phi}_1^\top \mathbf{M}_i \mathbf{\Phi}_1 = L_i/\mu & \text{for } A_i > A_i^L \\ \mathbf{\Phi}_1^\top \mathbf{K}_i \mathbf{\Phi}_1 - \Omega_1 \mathbf{\Phi}_1^\top \mathbf{M}_i \mathbf{\Phi}_1 \leq L_i/\mu & \text{for } A_i = A_i^L \end{cases} \tag{1.119}$$

where $\mu > 0$ and the normalization condition (1.114) has been used. Note that $\mathbf{\Phi}_1^\top \mathbf{K}_i \mathbf{\Phi}_1$ and $\Omega_1 \mathbf{\Phi}_1^\top \mathbf{M}_i \mathbf{\Phi}_1$ are conceived as twice the strain energy and kinetic energy, respectively, per unit area corresponding to $\mathbf{\Phi}_1$.

Example 1.12
Consider a two-bar structure, as shown in Fig. 1.15. The two bars have the same elastic modulus E and length L, and the structural mass is ignored; i.e.,

$\mathbf{M}^s = \mathbf{O}$. The stiffness matrix and mass matrix corresponding to nonstructural mass are given as

$$\mathbf{K} = \frac{E}{L}\begin{pmatrix} A_1 + A_2 & -A_2 \\ -A_2 & A_2 \end{pmatrix}, \quad \mathbf{M}^0 = \begin{pmatrix} M_0 & 0 \\ 0 & M_0 \end{pmatrix} \qquad (1.120)$$

The components of $\boldsymbol{\Phi}_1$ are written as $\boldsymbol{\Phi}_1 = (\Phi_{1,1}, \Phi_{1,2})^\top$. Then, from the optimality conditions, we have

$$(\Phi_{1,1})^2 = (\Phi_{1,1} - \Phi_{1,2})^2 = \lambda \qquad (1.121)$$

with a positive constant λ. Then, using the normalization condition (1.114), we obtain

$$\boldsymbol{\Phi}_1 = \frac{1}{\sqrt{5M_0}}\begin{pmatrix} 1 \\ 2 \end{pmatrix} \qquad (1.122)$$

Hence, from (1.113) and (1.115b) satisfied with equality, the optimal cross-sectional areas are obtained as

$$A_1 = \frac{3M_0 L\Omega^{\mathrm{L}}}{E}, \quad A_2 = \frac{2M_0 L\Omega^{\mathrm{L}}}{E} \qquad (1.123)$$

From Rayleigh's principle, the following inequality is satisfied by an arbitrary admissible n-vector $\boldsymbol{\Psi}$ (see Appendix A.1.2):

$$\frac{\boldsymbol{\Psi}^\top \mathbf{K}\boldsymbol{\Psi}}{\boldsymbol{\Psi}^\top (\mathbf{M}^s + \mathbf{M}^0)\boldsymbol{\Psi}} \geq \Omega_1 \qquad (1.124)$$

where the equality holds if $\boldsymbol{\Phi}$ is the eigenmode corresponding to Ω_1. If Ω_1 is a concave function of $\mathbf{A}$, then (1.119) is the necessary and sufficient condition for global optimality. A proof of concavity is given below for the special case in which the structural mass is sufficiently smaller than the nonstructural mass; i.e., $\mathbf{M}^s = \mathbf{O}$ can be assumed.

Proof 1.4
Let Ω_1^{I} and Ω_1^{II} denote the lowest eigenvalues corresponding to the solutions $\mathbf{A}^{\mathrm{I}}$ and $\mathbf{A}^{\mathrm{II}}$, respectively, for which other variables are also denoted by the superscripts I and II. The solutions are interpolated between $\mathbf{A}^{\mathrm{I}}$ and $\mathbf{A}^{\mathrm{II}}$ using a parameter $0 \leq \alpha \leq 1$ as $\widehat{\mathbf{A}} = \alpha \mathbf{A}^{\mathrm{I}} + (1-\alpha)\mathbf{A}^{\mathrm{II}}$. The lowest eigenmode associated with the solution $\widehat{\mathbf{A}}$ is denoted by $\widehat{\boldsymbol{\Phi}}_1$. The following relations hold from Rayleigh's principle:

$$\frac{\widehat{\boldsymbol{\Phi}}_1^\top \mathbf{K}^{\mathrm{I}}\widehat{\boldsymbol{\Phi}}_1}{\widehat{\boldsymbol{\Phi}}_1^\top \mathbf{M}^0\widehat{\boldsymbol{\Phi}}_1} \geq \Omega_1^{\mathrm{I}}, \quad \frac{\widehat{\boldsymbol{\Phi}}_1^\top \mathbf{K}^{\mathrm{II}}\widehat{\boldsymbol{\Phi}}_1}{\widehat{\boldsymbol{\Phi}}_1^\top \mathbf{M}^0\widehat{\boldsymbol{\Phi}}_1} \geq \Omega_1^{\mathrm{II}} \qquad (1.125)$$

Therefore, the following inequality is derived for the lowest eigenvalue $\widehat{\Omega}_1$ of $\widehat{\mathbf{A}}$:

$$
\begin{aligned}
\widehat{\Omega}_1 &= \frac{\widehat{\boldsymbol{\Phi}}_1^\top [\alpha \mathbf{K}^\mathrm{I} + (1-\alpha)\mathbf{K}^\mathrm{II}]\widehat{\boldsymbol{\Phi}}_1}{\widehat{\boldsymbol{\Phi}}_1^\top \mathbf{M}^0 \widehat{\boldsymbol{\Phi}}_1} \\
&= \alpha \frac{\widehat{\boldsymbol{\Phi}}_1^\top \mathbf{K}^\mathrm{I} \widehat{\boldsymbol{\Phi}}_1}{\widehat{\boldsymbol{\Phi}}_1^\top \mathbf{M}^0 \widehat{\boldsymbol{\Phi}}_1} + (1-\alpha)\frac{\widehat{\boldsymbol{\Phi}}_1^\top \mathbf{K}^\mathrm{II} \widehat{\boldsymbol{\Phi}}_1}{\widehat{\boldsymbol{\Phi}}_1^\top \mathbf{M}^0 \widehat{\boldsymbol{\Phi}}_1} \quad (1.126) \\
&\geq \alpha \Omega_1^\mathrm{I} + (1-\alpha)\Omega_1^\mathrm{II}
\end{aligned}
$$

which proves that the lowest eigenvalue Ω_1 is a concave function of $\mathbf{A}$.

$\square$

Since Problem (1.115) is a standard nonlinear programming (NLP) problem, the optimal solution may be found by a gradient-based NLP algorithm in conjunction with the sensitivity analysis described in Sec. 2.3. However, it is well known that the lowest eigenvalue often becomes multiple as the result of optimization (Olhoff and Rasmussen 1977; Olhoff 1980; Haug, Choi, and Komkov 1986; Nakamura and Ohsaki 1988); see Sec. 3.9 for details. Therefore, the lower bounds should be given for all eigenvalues, and the optimization problem for a truss is formulated as

$$
\text{Minimize} \quad V = \sum_{i=1}^m A_i L_i \tag{1.127a}
$$

$$
\text{subject to} \quad \Omega_r \geq \Omega^\mathrm{L}, \quad (r = 1, \ldots, n) \tag{1.127b}
$$

$$
A_i \geq A_i^\mathrm{L}, \quad (i = 1, \ldots, m) \tag{1.127c}
$$

Although Problem (1.127) is also an NLP problem, the convergence of the algorithm is deteriorated due to the discontinuity of the sensitivity coefficients; see Sec. 2.3.2 for details. The optimal solution with multiple eigenvalues can be found without any difficulty by converting Problem (1.127) to a *semidefinite programming* (SDP) problem (Wolkowicz, Saigal, and Vandenberghe 2000; Ohsaki, Fujisawa, Katoh, and Kanno 1999) (see Appendix A.2.4 for details of SDP). Using Rayleigh's principle (1.124), the eigenvalue constraint is converted as

$$
\boldsymbol{\Psi}^\top [\mathbf{K} - \Omega^\mathrm{L}(\mathbf{M}^\mathrm{s} + \mathbf{M}^0)]\boldsymbol{\Psi} \geq 0 \tag{1.128}
$$

for an arbitrary n-vector $\boldsymbol{\Psi}$. Define an $n \times n$ matrix $\mathbf{X}$ as

$$
\mathbf{X} = \mathbf{K} - \Omega^\mathrm{L}(\mathbf{M}^\mathrm{s} + \mathbf{M}^0) \tag{1.129}
$$

We can see from (1.128) and (1.129) that $\mathbf{X}$ should be positive semidefinite. Hence, the optimization problem (1.127) is reformulated as an SDP problem:

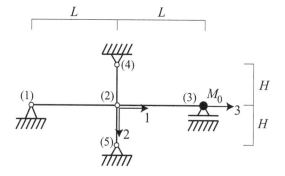

FIGURE 1.16: A four-bar truss with a concentrated mass.

$$\text{Minimize} \quad V = \sum_{i=1}^{m} A_i L_i \qquad (1.130\text{a})$$

$$\text{subject to} \quad \mathbf{X} = \sum_{i=1}^{m} (\mathbf{K}_i - \Omega^{\mathrm{L}} \mathbf{M}_i) A_i - \Omega^{\mathrm{L}} \mathbf{M}^0 \qquad (1.130\text{b})$$

$$A_i \geq A_i^{\mathrm{L}} \quad (i = 1, \dots, m) \qquad (1.130\text{c})$$

$$\mathbf{X} \succeq \mathbf{O} \qquad (1.130\text{d})$$

where $\mathbf{X}$ and $\mathbf{A} = (A_1, \dots, A_m)^{\top}$ are the variable matrix and vector, respectively, and $\mathbf{X} \succeq \mathbf{O}$ means $\mathbf{X}$ is positive semidefinite (see Sec. 3.9 for details of the SDP formulation of the truss topology optimization problem under eigenvalue constraints).

Example 1.13
As a simple example of the formulation of Problem (1.130), consider a four-bar truss, as shown in Fig. 1.16, for which a detailed investigation on multiple eigenvalues is presented in Sec. 3.9. The mass density of members, the lumped mass at node 3, and the elastic modulus are denoted by ρ, M_0, and E, respectively. The horizontal and vertical members have the same cross-sectional areas, respectively, denoted by A_1 and A_2. Suppose the lumped mass matrix is used for representing the structural mass of the members. Then the matrix $\mathbf{X}$ in Problem (1.130) is defined as

$$\mathbf{X} = \begin{pmatrix} 2E/L - \Omega^{\mathrm{L}} \rho L & 0 & -E/L \\ 0 & \rho L & 0 \\ -E/L & 0 & E/L - \Omega^{\mathrm{L}} \rho L/2 \end{pmatrix} A_1$$

$$+ \begin{pmatrix} -\Omega^{\mathrm{L}} \rho H & 0 & 0 \\ 0 & 2E/H - \Omega^{\mathrm{L}} \rho H & 0 \\ 0 & 0 & 0 \end{pmatrix} A_2 - \Omega^{\mathrm{L}} \begin{pmatrix} 0 & 0 & 0 \\ 0 & 0 & 0 \\ 0 & 0 & M_0 \end{pmatrix} \qquad (1.131)$$

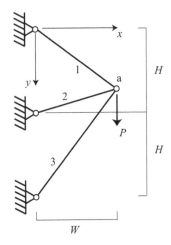

FIGURE 1.17:   A three-bar truss.

1.10   Configuration optimization of trusses

So far we mainly considered the optimization problems of trusses with fixed nodal locations, where the cross-sectional areas are the design variables. The topology of the truss can be optimized if zero lower bounds are assigned for the cross-sectional areas. By contrast, the nodal coordinates are modified in shape or geometry optimization. The effectiveness of configuration optimization, which combines shape and topology optimization, is illustrated below (Ohsaki 2003c); see Secs. 4.3, 4.4, and 6.6 for the numerical approaches.

Example 1.14
As a simple example, consider a three-bar truss, as shown in Fig. 1.17, subjected to a vertical load P at node 'a'. The distance of the node from the *wall*, on which three supports are located, has the fixed value at W. The objective function to be minimized is the vertical displacement U of node 'a'. The total structural volume is fixed at V, and the three members have the same cross-sectional area A, for simplicity. Therefore, A can be computed from V as

$$A = \frac{V}{L_1 + L_2 + L_3} \qquad (1.132)$$

where L_i ($i = 1, 2, 3$) is the length of the ith member.

Consider three cases in which H is equal to $2W$, W, and $W/2$, respectively, and suppose the y-coordinate Y of node 'a' is chosen as the design variable for each case. Because L_1, L_2, and L_3 depend on Y, the cross-sectional

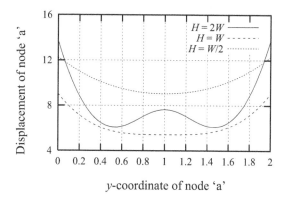

FIGURE 1.18: Relation between non-dimensional displacement U^* and non-dimensional nodal coordinate Y^* of the three-bar truss.

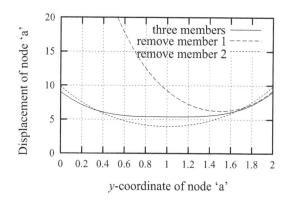

FIGURE 1.19: Relations between non-dimensional displacement U^* and non-dimensional nodal coordinate Y^* of the three-bar and two-bar trusses after removing a member.

area A computed from (1.132) is a function of Y, which is the only design variable. Fig. 1.18 shows the relations between Y and U for three cases with $H = 2W$, W, and $W/2$. Note that Y and U are converted to non-dimensional values $Y^* = Y/H$ and $U^* = EVU/(PW^2)$, respectively, with E being the elastic modulus. As is seen, U^* has two local minima with respect to Y^* for $H = 2W$, and U^* has the smallest value 5.4142 at $Y^* = 1$ for $H = W$. Hence, the nodal displacement under constraint on the total structural volume can be effectively reduced through geometry optimization considering the location of a node as a design variable.

Since this three-bar truss is statically indeterminate, the truss after removing one member is still stable. We consider the case with $H = W$, which has

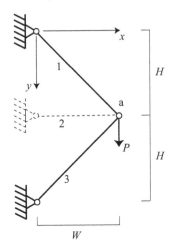

FIGURE 1.20:   Global optimal shape and topology of the three-bar truss.

the smallest objective value among the three cases. The relations between Y^* and U^* after removing members 1 and 2, respectively, are plotted in dashed and dotted lines in Fig. 1.19. As is seen, removal of member 2 leads to the smallest objective value $U^* = 4.0000$, and the corresponding global optimal topology is as shown in Fig. 1.20. This way, the objective value is drastically reduced by optimizing topology as well as the geometry.

1.11   Multiobjective structural optimization

1.11.1   Basic concepts

In the design process of structures in civil engineering, several performance measures against dynamic loads (seismic loads, wind loads) and static loads (service loads, self-weight) should be considered. Furthermore, cost should be defined in view of various aspects, including material, construction, manufacturing, and maintenance. Therefore, the optimization problem for practical application should be formulated as a multiobjective programming (MOP) problem (Cohon 1978) that has several performance measures as objective functions to be minimized or maximized.

Multiobjective programming arises in the following situations in the design process:

1. The structural weight or volume is to be minimized while the struc-

tural responses are reduced. In this case, the problem may be simply converted to a standard single-objective problem for minimizing the structural weight/volume under constraints on structural responses.

2. The initial cost as well as the maintenance cost should be minimized in the context of life-cycle design optimization; see Sec. 5.1.5.

3. The responses under various levels of seismic excitations are considered in the framework of performance-based design or performance-based engineering; see Sec. 5.5.

4. The mechanical performances as well as the aesthetic aspects and constructability should be considered in designing spatial frames and long-span structures; see Secs. 6.4 and 6.5.

There are numerous papers published on MOP approaches to structural design, e.g., Koski and Silvennoisen (1987), Stadler (1988), Marler and Arora (2004). Ohsaki, Nakamura, and Isshiki (1998) presented a trade-off design approach for an arch-type truss considering seismic responses and the shape preferred by the designer. Ohsaki, Ogawa, and Tateishi (2003) developed a design method of shells considering roundness of the surfaces and stiffness against the static loads. Li, Zhou, Duan, and Chen (1999) presented an approach to multiobjective design of steel frames based on *load and resistance factor design* (LRFD) criteria. Ghasemi and Farshchin (2009) presented an ant colony optimization method for multiobjective optimization of steel frames considering structural weight and frequency of free vibration.

One of the important aspects of MOP is that the competing performance measures can be formulated as either of the objective and constraint functions. The *soft* constraints that may preferably be satisfied can be naturally converted to an objective function (Ohsaki, Zhang, and Kimura 2005). Among the various methods of MOP, the constraint approach (Cohon 1978) and aspiration level approach (Nakayama 1995) for interactive optimization have no strict distinction between objective functions and constraints. The approach based on a fuzzy decision also treats objective and constraint functions equivalently (Rao 1987; Nakayama 1995; Dubois and Fortemps 1999; Massa, Lallemanda, and Tison 2009; Tonon 1999).

1.11.2 Problem formulation

For a simple case of minimizing the two objective functions $F_1(\mathbf{A})$ and $F_2(\mathbf{A})$, which are functions of the design variable vector $\mathbf{A}$, the multiobjective optimization problem is formulated as

$$\text{Minimize} \quad F_1(\mathbf{A}) \text{ and } F_2(\mathbf{A}) \tag{1.133a}$$
$$\text{subject to} \quad \mathbf{A} \in \mathcal{A} \tag{1.133b}$$

where $\mathcal{A}$ is the feasible region of $\mathbf{A}$. If there exists no feasible solution that improves both $F_1(\mathbf{A})$ and $F_2(\mathbf{A})$ from a feasible solution $\widehat{\mathbf{A}}$, then $\widehat{\mathbf{A}}$ is called the *Pareto optimal solution* or simply the Pareto solution. The basic definitions and formulations of MOP are summarized in Appendix A.4.

Example 1.15
As an illustrative example, consider a multiobjective structural optimization problem with discrete variables. Pareto optimal solutions are found for the five-bar truss shown in Fig. 1.2 in Sec. 1.3. Cross-sectional areas A_1^g and A_2^g of groups 1 and 2 consisting of members $\{1, 2, 5\}$, and $\{3, 4\}$, respectively, are selected from the list $\{100, 200, 300, 400, 500\}$; i.e., there are $5 \times 5 = 25$ possible combinations of design variables. Other parameters are the same as those in Example 1.1 in Sec. 1.3. Note that the units of length and force are mm and kN, which are omitted for brevity.

The total structural volume and compliance are used as the performance measures to define the objective functions. The objective values of the 25 solutions are plotted in Fig. 1.21 in objective function space, where the filled circles indicate a Pareto optimal solution. The values of the objective functions and the design variables of the Pareto optimal solutions are also listed in Table 1.3. As is seen, the solutions with $A_1^g = 500$ or $A_2^g = 100$ are selected as Pareto solutions. This way, the Pareto solutions can be easily obtained by enumeration if the number of design variables is small, and the computational cost for function evaluation is also small. However, for more complex structures with many design variables, a nonlinear programming algorithm for a continuous problem or a heuristic approach for a discrete problem should be used, as presented in Secs. 5.4, 6.4, and 6.5.

The solid line in Fig. 1.21 shows the set of Pareto solutions for continuous cross-sectional areas A_1^g and A_2^g with upper bound 500 and lower bound 100. In this example, the Pareto solutions of discrete variables are included in the set of continuous variables.

1.12 Heuristic approach

If the design variables are restricted to have only integer values, the design problem turns out to be a combinatorial optimization problem, for which integer programming approaches, such as the branch-and-bound method, can be effectively applied. However, the computational cost grows as an exponential function of problem size. It should be noted, however, in the practical application, that obtaining the global optimal solution is not very important. Heuristic approaches have been developed to obtain approximate optimal so-

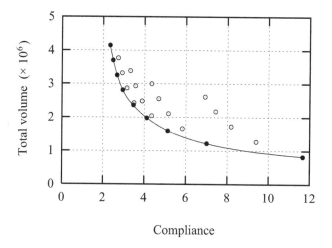

FIGURE 1.21: Pareto optimal solutions of the five-bar truss.

TABLE 1.3: Compliance and total structural volume of the Pareto optimal solutions of the five-bar truss with discrete cross-sectional areas.

A_1^g	A_2^g	Compliance	Total volume $(\times 10^6)$
100.0	100.0	0.58363	0.83006
200.0	100.0	0.35023	1.2129
300.0	100.0	0.25678	1.5957
400.0	100.0	0.20589	1.9786
500.0	100.0	0.17379	2.3614
500.0	200.0	0.14770	2.8086
500.0	300.0	0.13394	3.2559
500.0	400.0	0.12425	3.7031
500.0	500.0	0.11673	4.1503

lutions within reasonable computation times, although there is no theoretical proof of convergence to the optimal solution.

The most popular heuristic approach may be the genetic algorithm (GA), which is classified as a multipoint method or population-based method that has many solutions at each iterative step. There are several different terminologies for the classification of the heuristic approaches. For example, a GA can be categorized as an evolutionary approach, a statistical search, a probabilistic approach, a soft computing method, or a method of artificial life. However, in this book, we use *heuristic approach* for all the approximate optimization methods without rigorous proof for reaching the global optimal solution.

There are many papers on the application of GA to structural optimization

(Ohsaki 1995; Hajela and Lee 1995). Schutte and Groenwold (2003) used a particle swarm optimization method for sizing optimization of trusses. Since the computational cost for function evaluation by structural analysis is not very small, a multipoint approach may not be appropriate for optimization of large-scale structures. Therefore, in this book, we mainly use heuristic approaches that are based on local search and retain only one solution at each iterative process. Such methods are called single-point-search heuristics.

Balling (1991) applied simulated annealing (SA) to the optimization of building frames and showed that SA has better performance than a linearized branch-and-bound method. Lamberti (2008) proposed a multipoint SA and applied it to a truss optimization problem. Tagawa and Ohsaki (1999) used SA for configuration optimization of trusses. Pan, Ohsaki, and Tagawa (2007) used SA for optimizing the flange shape of a beam. The application of tabu search to structural optimization is found in Bennage and Dhingra (1995), Dhingra and Bennage (1995), Kargahi, Andersen, and Dessouky (2007), etc.; see Appendix A.3 for basic methodologies of heuristics.

Example 1.16
As a small example, an optimal solution is found using the greedy method for the five-bar truss in Fig. 1.2 in Sec. 1.3. The problem formulation and parameters are the same as those in Sec. 1.3; i.e., the total structural volume V is minimized under constraints on stress and displacement, where the design variables A_1^g and A_2^g have discrete values $\{100, 200, 300, 400, 500\}$. In the following, the units for length and force are mm and kN, respectively.

We start with the solution that has the smallest objective value and violates the constraints. Hence, we set the initial solution as $(A_1^g, A_2^g) = (100, 100)$ and carry out a local search by increasing either A_1^g and A_2^g, as illustrated with thick arrows in Fig. 1.22. The displacement and stress of the solution $(A_1^g, A_2^g) = (100, 100)$ are $U_3 = 2.362$ and $\sigma_4 = -0.116$, which do not satisfy the constraints $|U_3| \leq \overline{U}_3 = 1.25$ and $|\sigma_4| \leq \overline{\sigma}_4 = 0.06$.

Define the parameter α as the maximum ratio of the constraint functions to their upper bounds; i.e., α is equal to the larger value between $|U_3|/\overline{U}_3$ and $|\sigma_4|/\overline{\sigma}_4$. Therefore, the solutions are searched until $\alpha \leq 1$ is satisfied. At the initial solution, $\alpha = 1.93$ and $V = 8.30 \times 10^5$.

If A_1^g is increased by 100 to 200, then $\alpha = 1.44$ and $V = 1.21 \times 10^6$. On the other hand, if A_2^g is increased to 200, then $\alpha = 1.50$ and $V = 1.27 \times 10^6$. Let β denote the ratio of the decrease of α to the increase of V. Then it may be conceived that a larger value of β leads to more efficiency of the design modification. The values of β for increasing A_1^g and A_2^g, respectively, in this case are 1.29×10^{-7} and 0.977×10^{-7}. Therefore, we increase A_1^g to obtain $(A_1^g, A_2^g) = (200, 100)$. However, α is still greater than 1, and A_1^g or A_2^g should be increased again. By comparing the values of β for updating the solution to $(A_1^g, A_2^g) = (300, 100)$ and $(200, 200)$, we accept the latter solution, which

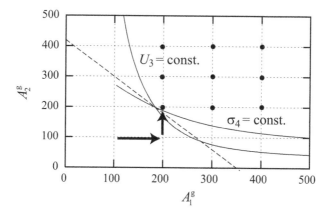

FIGURE 1.22: Local search of the five-bar truss.

satisfies the constraints. This way, the global optimal solution at point 'b' in Fig. 1.3 has been successfully found by the greedy method.

1.13 Simultaneous analysis and design

The structural optimization problem is usually formulated as Problem (1.3), where the nodal displacement vector $\mathbf{U}$, which is also called state variable vector, is regarded as implicit function of the design variables $\mathbf{A} = (A_1, \ldots, A_m)^\top$. This type of problem formulation is sometimes called *nested analysis and design*, abbreviated NAND (Arora and Wang 2005), or *design variable space approach* (Kirsch and Rozvany 1994). In this case, the sensitivity of the constraint functions is evaluated using the sensitivity of $\mathbf{U}$ with respect to $\mathbf{A}$ by carrying out the standard procedure of design sensitivity analysis described in Chap. 2. However, for a complex structure, the sensitivity analysis is very tedious and demands much effort in programming and computation.

As an alternative approach, the nodal displacements are also considered as independent variables, while the stiffness equations are explicitly assigned as equality constraints. This combined approach of analysis and design is called *simultaneous analysis and design* (SAND) (Haftka 1985; Sankaranarayanan, Haftka, and Kapania 1994; Wu and Arora 1987), and the problem of minimizing the objective function $F(\mathbf{A})$ under inequality constraints $H_j(\mathbf{U}, \mathbf{A}) \le 0$

$(j = 1, \ldots, n^{\mathrm{I}})$ is formulated as

$$\text{Minimize} \quad F(\mathbf{A}) \tag{1.134a}$$

$$\text{subject to} \quad H_j(\mathbf{U}, \mathbf{A}) \leq 0, \quad (j = 1, \ldots, n^{\mathrm{I}}) \tag{1.134b}$$

$$\mathbf{K}(\mathbf{A})\mathbf{U} = \mathbf{P} \tag{1.134c}$$

$$A_i^{\mathrm{L}} \leq A_i \leq A_i^{\mathrm{U}}, \quad (i = 1, \ldots, m) \tag{1.134d}$$

where A_i^{U} and A_i^{L} are the upper and lower bounds for A_i, $\mathbf{K}$ is the stiffness matrix, and $\mathbf{P}$ is the specified vector of external loads.

Using the SAND approach, optimal solutions can be found without carrying out structural analysis in a standard manner, and no procedure of design sensitivity analysis is needed. Although the number of variables increases, the nonlinearity of the governing equations is reduced, and the convergence properties may be improved. Taleb-Agha and Nelson (1975) optimized a truss with stress and displacement constraints under multiple loading conditions using SAND and convex linearization approaches. Wu and Arora (1987) presented a SAND formulation for a geometrically nonlinear problem. Shin, Haftka, and Plaut (1987) developed a method for eigenvalue optimization including bimodal cases. Orozco and Ghattas (1997) developed a reduced SAND approach utilizing sequential (recursive) quadratic programming for geometrically nonlinear structures.

The attempt to solve the analysis problem by optimization was initiated in the 1960s (Schmit and Fox 1965). Note that we can carry out plastic limit analysis utilizing a linear programming approach, as demonstrated in Sec. 1.4. However, for a linear analysis problem, the standard approach of solving equilibrium equations is more efficient than the optimization approach for minimizing the total potential energy. By contrast, for a geometrically nonlinear analysis and design problem, the computational cost can be reduced by utilizing the SAND formulation (Haftka and Kamat 1989).

The internal forces, such as axial forces and stresses, can also be treated as independent variables (Wang and Arora 2006a, 2006b). For example, the equilibrium equations and compatibility conditions of a truss are incorporated as equality constraints to formulate the problem in the following form:

$$\text{Minimize} \quad F(\mathbf{A}) \tag{1.135a}$$

$$\text{subject to} \quad H_j(\mathbf{U}, \mathbf{A}) \leq 0, \quad (j = 1, \ldots, n^{\mathrm{I}}) \tag{1.135b}$$

$$\mathbf{D}\mathbf{N} = \mathbf{P} \tag{1.135c}$$

$$\mathbf{S}(\mathbf{A})\mathbf{D}^{\top}\mathbf{U} = \mathbf{N} \tag{1.135d}$$

$$A_i^{\mathrm{L}} \leq A_i \leq A_i^{\mathrm{U}}, \quad (i = 1, \ldots, m) \tag{1.135e}$$

where $\mathbf{N}$ is the vector of member forces, $\mathbf{D}$ is the equilibrium matrix, and $\mathbf{S}$ is the diagonal matrix of which the ith diagonal term is EA_i/L_i with E and L_i being the elastic modulus and the length of the ith member, respectively.

Recently, the formulation of SAND has been extensively used for reformulation of topology optimization problems of plates and trusses (Bendsøe and Sigmund 2003), as we have seen in Sec. 1.8.3. On the basis of the SAND formulation, optimal topologies of large-scale trusses can be efficiently found utilizing an interior-point method (Jarre, Kočvara, and Zowe 1998). Because the number of variables is very large and large-scale linear equations should be solved at each step of optimization, the use of an iterative solver with multigrid smoothing may be very effective (Maar and Schulz 2000). The optimization problem with equilibrium constraints (1.134c) is generally categorized as a *mathematical program with equilibrium constraints* (MPEC) (Luo, Pang, and Ralph 1996).

Chapter 2

Design Sensitivity Analysis

As we have seen in various formulations of structural optimization in Chap. 1, the sensitivity coefficients of responses with respect to the design variables are needed for obtaining optimal designs using a gradient-based nonlinear programming approach. In this chapter, we first present an overview of sensitivity analysis in Sec. 2.1. The static and eigenvalue sensitivity analyses are presented in Secs. 2.2 and 2.3, respectively. Methods for linear buckling load, transient response, and geometrically nonlinear response are briefly presented in Secs. 2.4, 2.5, and 2.6, respectively. Finally, shape sensitivity analysis of trusses is presented in Sec. 2.7.

2.1 Overview of design sensitivity analysis

The term *sensitivity analysis* is mathematically defined as the procedure for computing the rate of variation of the solution with respect to a parameter defining the problem. Consider a problem of finding a solution to a set of n nonlinear equations defined as follows with a scalar parameter p:

$$\mathbf{G}(\mathbf{X}, p) = \mathbf{0} \tag{2.1}$$

where $\mathbf{X} = (X_1, \ldots, X_n)^\top$ is the variable vector, and $\mathbf{G}(\mathbf{X}, p)$ is supposed to be continuously differentiable with respect to $\mathbf{X}$ and p. Since the solution is defined by (2.1) for each specified value of p, it is regarded as a function of p that is denoted with tilde as $\widetilde{\mathbf{X}}(p)$. Suppose we have a solution $\widetilde{\mathbf{X}}(p^0)$ for $p = p^0$. Then the solution for $p = p^0 + \Delta p$ can be linearly approximated as

$$\mathbf{B}\left(\widetilde{\mathbf{X}}(p^0 + \Delta p) - \widetilde{\mathbf{X}}(p^0)\right) + \frac{\partial \mathbf{G}}{\partial p}\Delta p = \mathbf{0} \tag{2.2}$$

where the $n \times n$ matrix $\mathbf{B}$ is the Jacobian of $\mathbf{G}$, for which the (i, j)-component B_{ij} is defined as

$$B_{ij} = \frac{\partial G_i}{\partial X_j} \tag{2.3}$$

According to the *implicit function theorem*, (2.2) has a solution if $\mathbf{B}$ is nonsingular. Fig. 2.1 illustrates the process of linear estimation of the solution

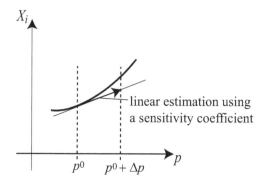

FIGURE 2.1:   Linear estimation of solution using a sensitivity coefficient.

with respect to the parameter variation by Δp. This approach is generally used for path-tracing analysis of geometrically nonlinear responses, where the load factor, a specific displacement component, or arc-length of the path is taken as the path parameter.

For the structural design problem, the displacements and stresses under static and/or dynamic loads depend on the design variables, including cross-sectional areas and nodal coordinates of the structure. The ratio of variation of the response to that of the design variable is called the *design sensitivity coefficient*, and the process of computing the design sensitivity coefficient is called *design sensitivity analysis* (Haug, Choi, and Komkov 1986; Adelman and Haftka 1986; Haftka and Adelman 1989; Choi and Kim 2004; van Keulen, Haftka, and Kim 2005). For the shape design, the variables are nodal coordinates, and the corresponding sensitivity coefficients and the process of computing them are called *shape sensitivity coefficients* and *shape sensitivity analysis*, respectively.

Various formulations of design sensitivity analysis have been developed mainly for continua utilizing variational principles, for which the sensitivity coefficients are to be computed for a functional, e.g., compliance, that can be formulated in an integral of a bilinear form of the responses. However, for finite-dimensional structures, including frames and trusses, the variational formulations are not needed, and the sensitivity coefficients can be found simply by differentiating the governing equations in matrix-vector form. Furthermore, even for the shape sensitivity analysis of continua such as plates and shells, it has been shown that differentiation of the matrix-vector form after discretization to finite elements leads to the same results obtained by differentiation in variational form followed by discretization to finite elements (Choi and Twu 1988).

The process of structural design may be classified into the steps of determining support locations, support conditions, nodal locations, member locations, and cross-sectional properties of members. The steps up to the determina-

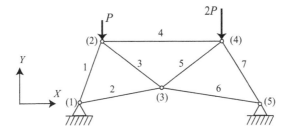

FIGURE 2.2: A seven-bar plane truss.

tion of member locations are classified as *shape design*, and the last step is called *stiffness design*. As an introduction to this chapter, the role of design sensitivity analysis in structural analysis and design is discussed using the simple seven-bar truss, as shown in Fig. 2.2. Consider a static analysis problem for computing the responses against the nodal loads. Let u_i and v_i denote the displacements in the X- and Y-directions, respectively, of node i. Then the deformation of the truss is defined by the nodal displacement vector $\mathbf{U} = (u_2, v_2, u_3, v_3, u_4, v_4)^\top$, which is also called the state variable vector. Because $\mathbf{U}$ varies with respect to the nodal coordinates and the member cross-sectional areas, it is conceived as a function of these variables.

Suppose the Y-directional displacement v_3 of node 3 exceeds the allowable limit defined by the design code. The displacement may be reduced by increasing the cross-sectional areas of the members. The most effective design modification may be obtained accurately by carrying out structural analysis for all possible modifications. However, this process demands much computational cost. Therefore, it is practically desirable to estimate the displacement linearly utilizing the sensitivity coefficients.

As we have seen in Chap. 1.1, sensitivity analysis plays a key role in structural optimization. The sensitivity coefficients are also effectively used to reduce the computational costs for a homology design (Yoshikawa, Elishakoff, and Nakagiri 1998) or inverse design for controlling the displacements to a specified values or ratios; e.g., equal vertical displacements can be attained at nodes 2 and 4 of the truss in Fig. 2.2 by adjusting the cross-sectional areas of members. It is well known for frame structures that the use of reciprocals of variables leads to more accurate linear estimation of responses than the case in which the cross-sectional areas are directly used as design variables (Noor and Lowder 1975; Storaasli and Sobieszczanski-Sobieski 1983; Haftka, Gürdal, and Kamat 1990). Higher-order sensitivity coefficients may be incorporated for more accurate estimation (Fleury 1989b; Fuchs 1993). Polynomial fitting algorithms may also be used for obtaining approximate responses (Haftka, Nachlas, Watson, Rizzo, and Desai 1987).

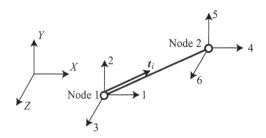

FIGURE 2.3:   Definition of coordinates, displacements, node numbers, and direction of a three-dimensional truss member.

2.2   Static responses

2.2.1   Direct differentiation method

In order to present the basic idea of design sensitivity analysis, the most standard approach of the *direct differentiation method* for static responses of a three-dimensional truss is presented in this section. Let A_i and L_i denote the cross-sectional area and length of the ith member, respectively. The global coordinates (X, Y, Z), member direction, local node numbers, and local displacement numbers are defined in Fig. 2.3. Let $\mathbf{t}_i$ denote the unit vector directed from node 1 to 2 of the ith members; i.e., the three components of $\mathbf{t}_i$ correspond to the directional cosines with respect to the X-, Y-, and Z-axes, respectively. A vector $\mathbf{d}$ is defined as

$$\mathbf{d}_i = (-\mathbf{t}_i^\top, \mathbf{t}_i^\top)^\top \tag{2.4}$$

The nodal displacement vector and nodal force vector of member i with respect to the global coordinates are denoted by $\mathbf{u}_i$ and $\mathbf{f}_i$, respectively, which have six components. Then $\mathbf{u}_i$ and $\mathbf{f}_i$ are related by the 6×6 member stiffness matrix $\mathbf{k}_i$ with respect to the global coordinates as

$$\mathbf{f}_i = \mathbf{k}_i \mathbf{u}_i \tag{2.5}$$

where

$$\mathbf{k}_i = \frac{A_i E}{L_i} \mathbf{d}_i \mathbf{d}_i^\top \tag{2.6}$$

with E being the elastic modulus. Then $\mathbf{k}_i$ of all members are assembled to construct the $n \times n$ global stiffness matrix $\mathbf{K}$, where n is the total number of degrees of freedom.

For design sensitivity analysis, the cross-sectional areas $\mathbf{A} = (A_1, \ldots, A_m)^\top$ are considered as design variables, where m is the number of members. The stiffness matrix is a function of $\mathbf{A}$, which is written as $\mathbf{K}(\mathbf{A})$. If the self-weight is considered, the nodal load vector $\mathbf{P} = (P_1, \ldots, P_n)^\top$ is a function of $\mathbf{A}$, which is written as $\mathbf{P}(\mathbf{A})$. The displacement vector $\mathbf{U}(\mathbf{A}) = (U_1, \ldots, U_n)^\top$ is obtained by solving the stiffness (equilibrium) equation

$$\mathbf{K}(\mathbf{A})\mathbf{U}(\mathbf{A}) = \mathbf{P}(\mathbf{A}) \qquad (2.7)$$

Differentiation of (2.7) with respect to A_i leads to

$$\frac{\partial \mathbf{K}}{\partial A_i}\mathbf{U} + \mathbf{K}\frac{\partial \mathbf{U}}{\partial A_i} = \frac{\partial \mathbf{P}}{\partial A_i} \qquad (2.8)$$

In the following, the argument $\mathbf{A}$ is omitted for brevity. By differentiating (2.6) with respect to A_i, we obtain

$$\frac{\partial \mathbf{k}_i}{\partial A_i} = \frac{E}{L_i}\mathbf{d}_i\mathbf{d}_i^\top \qquad (2.9)$$

which is constant and corresponds to the value of $\mathbf{k}_i$ for unit cross-sectional area. Hence, the sensitivity coefficients of $\mathbf{K}(\mathbf{A})$ with respect to A_i are easily obtained through an arithmetic operation, because $\mathbf{K}$ is an assemblage of $\mathbf{k}_i$. The sensitivity of $\mathbf{P}(\mathbf{A})$ is also easily computed, if the dependence of $\mathbf{P}(\mathbf{A})$ on $\mathbf{A}$ is defined through the self-weight that is an explicit linear function of $\mathbf{A}$.

Define $\mathbf{R}$ as

$$\mathbf{R} = \frac{\partial \mathbf{P}}{\partial A_i} - \frac{\partial \mathbf{K}}{\partial A_i}\mathbf{U} \qquad (2.10)$$

Then, using (2.8) and (2.10), the sensitivity coefficient of $\mathbf{U}$ with respect to A_i is obtained from

$$\mathbf{K}\frac{\partial \mathbf{U}}{\partial A_i} = \mathbf{R} \qquad (2.11)$$

As is seen from (2.7) and (2.11), the sensitivity vector $\partial \mathbf{U}/\partial A_i$ is computed as the displacement vector against the nodal load vector $\mathbf{R}$. The stiffness matrix $\mathbf{K}$ is decomposed to triangular and diagonal matrices by a Cholesky decomposition in the process of computing $\mathbf{U}$ from (2.7), if a *direct solver* of linear equations is used. Therefore, the computational cost of solving (2.11) is very small except for the case in which a recursive method, e.g., the conjugate gradient method, is used for solving the linear equations.

Example 2.1
Consider a two-bar truss, as shown in Fig. 2.4, subjected to horizontal and vertical loads P_1 and P_2, respectively. The nodal displacements are denoted by U_1 and U_2. The two bars have the cross-sectional areas A_1 and A_2. For the case where $P_1 = 0$ and $P_2 = P$, (2.7) is written as

$$\frac{E}{2\sqrt{2}L}\begin{pmatrix} A_1 + 2\sqrt{2}A_2 & A_1 \\ A_1 & A_1 \end{pmatrix}\begin{pmatrix} U_1 \\ U_2 \end{pmatrix} = \begin{pmatrix} 0 \\ P \end{pmatrix} \qquad (2.12)$$

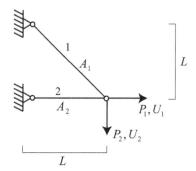

FIGURE 2.4: A two-bar truss.

from which we obtain

$$U_1 = -\frac{PL}{A_2 E}, \quad U_2 = \frac{(A_1 + 2\sqrt{2}A_2)PL}{A_1 A_2 E} \tag{2.13}$$

Let A_2 be a design variable, and suppose that P is independent of A_2. Eq. (2.8) is then written as

$$\frac{E}{L}\begin{pmatrix} 1 & 0 \\ 0 & 0 \end{pmatrix}\begin{pmatrix} U_1 \\ U_2 \end{pmatrix} + \frac{E}{2\sqrt{2}L}\begin{pmatrix} A_1 + 2\sqrt{2}A_2 & A_1 \\ A_1 & A_1 \end{pmatrix}\begin{pmatrix} \dfrac{\partial U_1}{\partial A_2} \\ \dfrac{\partial U_2}{\partial A_2} \end{pmatrix} = \begin{pmatrix} 0 \\ 0 \end{pmatrix} \tag{2.14}$$

By incorporating (2.13) into (2.14), we obtain

$$\frac{\partial U_1}{\partial A_2} = -\frac{PL}{A_2^2 E}, \quad \frac{\partial U_2}{\partial A_2} = \frac{PL}{A_2^2 E} \tag{2.15}$$

which agrees with the results of explicit differentiation of U_1 and U_2 in (2.13) with respect to A_2.

———————

The sensitivity coefficients can also be computed using the finite-difference approach. Let ΔA_i denote the small positive variation of A_i, and denote by $\Delta \mathbf{A}_i$ the m-vector of which the ith component is ΔA_i and other components are 0. An approximate sensitivity coefficient of $\mathbf{U}(\mathbf{A})$ is obtained by the central finite-difference approach as

$$\frac{\partial \mathbf{U}}{\partial A_i} \simeq \frac{\mathbf{U}(\mathbf{A} + \Delta \mathbf{A}_i) - \mathbf{U}(\mathbf{A} - \Delta \mathbf{A}_i)}{2\Delta A_i} \tag{2.16}$$

An approximate sensitivity coefficient can also be found by the forward finite-difference

$$\frac{\partial \mathbf{U}}{\partial A_i} \simeq \frac{\mathbf{U}(\mathbf{A} + \Delta \mathbf{A}_i) - \mathbf{U}(\mathbf{A})}{\Delta A_i} \tag{2.17}$$

or the backward finite-difference

$$\frac{\partial \mathbf{U}}{\partial A_i} \simeq \frac{\mathbf{U}(\mathbf{A}) - \mathbf{U}(\mathbf{A} - \Delta \mathbf{A}_i)}{\Delta A_i} \tag{2.18}$$

Although the finite-difference approaches are very simple, the number of response analyses (solution processes of (2.7)) for sensitivity evaluation with respect to m design variables is m for the forward or backward finite-difference approach, and $2m$ for the central finite-difference approach. Therefore, the computational cost for sensitivity analysis is drastically reduced if direct differentiation of the governing equations is utilized.

Furthermore, there are some inherent sources of errors in finite-difference approaches. Suppose a function is approximated by a polynomial of a variable. Then, the finite difference approximation assumes that the increment of the function consists of the first-order term only, and the second- and higher-order terms are ignored. This type of error is called *truncation error* and can be reduced using a small increment of a variable. However, a too small value of increment leads to underflow of the differences, and accuracy deteriorate due to *rounding error* (see for details, e.g., Forsythe and Wasow 1960; Richtmyer and Morton 1967). Therefore, it is not easy to decide on an appropriate value of the increment of the variable.

For plates and shells discretized to finite elements, it is not straightforward to differentiate the element stiffness matrix, especially with respect to the nodal coordinates. The computational cost can be reduced, while the differentiation of the stiffness matrix is avoided, if a finite-difference approach is used only for the element-level differentiation, and (2.11) for the direct differentiation method is solved in the system level to obtain the sensitivity coefficients of the displacements. This approach is called the *semi-analytical approach* (de Boer and van Keulen 2000) and is very effective for shape sensitivity analysis of complex structures with a large number of degrees of freedom.

Suppose that the sensitivity coefficients are to be found for the local responses of members. Let ε_j and N_j denote the strain and axial force of the jth member that are defined as

$$\varepsilon_j = \frac{1}{L_j} \mathbf{d}_i^\top \mathbf{u}_j \tag{2.19a}$$

$$N_j = \frac{A_j E}{L_j} \mathbf{d}_i^\top \mathbf{u}_j \tag{2.19b}$$

Since $\mathbf{u}_j$ consists of the components of $\mathbf{U}(\mathbf{A})$, the axial force is defined by the displacements $\mathbf{U}(\mathbf{A})$ and the design variable A_j. Hence, a local response is generally defined as $D(\mathbf{U}(\mathbf{A}), \mathbf{A})$, which is alternatively written as a function of $\mathbf{A}$ only:

$$\widetilde{D}(\mathbf{A}) = D(\mathbf{U}(\mathbf{A}), \mathbf{A}) \tag{2.20}$$

By differentiating (2.20) with respect to A_i, we obtain

$$\frac{\partial \tilde{D}}{\partial A_i} = \frac{\partial D}{\partial A_i} + \sum_{j=1}^{n} \frac{\partial D}{\partial U_j} \frac{\partial U_j}{\partial A_i}, \quad (i = 1, \cdots, m) \tag{2.21}$$

Because D is an explicit function of $\mathbf{U}$ and $\mathbf{A}$, as is the case of N_j in (2.19b), the partial differential coefficients $\partial D / \partial A_i$ and $\partial D / \partial U_j$ can be easily computed, and the sensitivity coefficient of $\tilde{D}$ can be obtained by incorporating those of displacements into (2.21). For ε_j and N_j, the following equations are derived by differentiating (2.19a) and (2.19b) with respect to A_i:

$$\frac{\partial \varepsilon_j}{\partial A_i} = \frac{1}{L_j} \mathbf{d}_i^\top \frac{\partial \mathbf{u}_j}{\partial A_i} \tag{2.22a}$$

$$\frac{\partial N_j}{\partial A_i} = \begin{cases} \dfrac{A_j E}{L_j} \mathbf{d}_i^\top \dfrac{\partial \mathbf{u}_j}{\partial A_i} & \text{for } i \neq j \\ \dfrac{E}{L_i} \mathbf{d}_i^\top \mathbf{u}_i + \dfrac{A_i E}{L_i} \mathbf{d}_i^\top \dfrac{\partial \mathbf{u}_i}{\partial A_i} & \text{for } i = j \end{cases} \tag{2.22b}$$

Note in (2.22b) that there exist two terms for $i = j$ due to explicit and implicit dependence on A_i, respectively, and only implicit dependence exists for $i \neq j$.

Since the computational cost in (2.21) is negligibly small, the total computational cost is governed by the cost for solving (2.11), which is proportional to the number of design variables. Therefore, the direct differentiation method is very effective when the sensitivity coefficients of many response quantities are needed with respect to a relatively small number of design variables. However, it is not efficient if the number of design variables is far larger than that of the responses, for which the sensitivity coefficients are to be computed. In such a case, the adjoint variable method presented in the following section is very effective.

2.2.2   Adjoint variable method

In the adjoint variable method, the sensitivity coefficients of the responses are found without computing those of all the nodal displacements. Our purpose here is to compute the sensitivity coefficients of the response $\tilde{D}(\mathbf{A}) = D(\mathbf{U}(\mathbf{A}), \mathbf{A})$, which represents the strain, stress, axial force of a member, etc. For this purpose, the *adjoint variable vector* $\overline{\mathbf{U}}$ is first introduced as

$$\mathbf{K}\overline{\mathbf{U}} = \frac{\partial D}{\partial \mathbf{U}} \tag{2.23}$$

where

$$\frac{\partial D}{\partial \mathbf{U}} = \left(\frac{\partial D}{\partial U_1}, \cdots, \frac{\partial D}{\partial U_n} \right)^\top \tag{2.24}$$

By premultiplying $\overline{\mathbf{U}}^\top$ on both sides of (2.11), we obtain

$$\overline{\mathbf{U}}^\top \mathbf{K} \frac{\partial \mathbf{U}}{\partial A_i} = \overline{\mathbf{U}}^\top \mathbf{R} \tag{2.25}$$

The following equation is also derived by premultiplying $(\partial \mathbf{U} / \partial A_i)^\top$ to (2.23):

$$\left(\frac{\partial \mathbf{U}}{\partial A_i} \right)^\top \mathbf{K} \overline{\mathbf{U}} = \left(\frac{\partial \mathbf{U}}{\partial A_i} \right)^\top \frac{\partial D}{\partial \mathbf{U}} \tag{2.26}$$

Furthermore, from (2.25), (2.26), and symmetry of $\mathbf{K}$,

$$\left(\frac{\partial \mathbf{U}}{\partial A_i} \right)^\top \frac{\partial D}{\partial \mathbf{U}} = \overline{\mathbf{U}}^\top \mathbf{R} \tag{2.27}$$

is derived, and finally from (2.21), (2.20), and (2.27), we obtain

$$\frac{\partial \widetilde{D}}{\partial A_i} = \frac{\partial D}{\partial A_i} + \overline{\mathbf{U}}^\top \mathbf{R} \tag{2.28}$$

Therefore, after solving (2.23) for $\overline{\mathbf{U}}$, the sensitivity coefficients of the response are obtained from (2.28) by a simple arithmetic operation. Note that $\partial D / \partial A_i$ corresponds to differentiation of the explicit term of A_i in D. This term vanishes for the responses such as displacements and strains, which are expressed directly using $\mathbf{U}$ only.

Consider a simple case of $D = U_k$; i.e., the sensitivity coefficients of the kth displacement component are to be found. In this case, $\partial D / \partial \mathbf{U}$ is a vector consisting of 0 in all components except 1 in the kth component, as is seen from (2.24). Therefore, from (2.23), $\overline{\mathbf{U}}$ is the displacement vector against the unit load in the kth component, which is denoted by $\overline{\mathbf{U}}^k$. Because $\mathbf{A}$ does not exist explicitly in the definition $D = U_k$, the following relation is obtained from (2.10), (2.28), and $\widetilde{D} = D = U_k$:

$$\frac{\partial U_k}{\partial A_i} = \overline{\mathbf{U}}^{k\top} \left(\frac{\partial \mathbf{P}}{\partial A_i} - \frac{\partial \mathbf{K}}{\partial A_i} \mathbf{U} \right) \tag{2.29}$$

Therefore, the sensitivity coefficient of the displacement component can be obtained easily from (2.29) through a simple arithmetic operation.

Since the computational cost is dominated by the computation of the adjoint variable vector in (2.23), the total computational cost is proportional to the number of response quantities for which the sensitivity coefficients are to be computed. Therefore, this approach is useful if the number of design variables is far larger than that of the response quantities.

We present below an explicit formulation of the adjoint variable method to compute the sensitivity coefficients of the kth displacement component of a plane truss. The relation between the local displacement vector $\mathbf{u}_i$ and the strain ε_i of the ith member is defined in (2.19a). From (2.6), the sensitivity coefficient of the member stiffness matrix $\mathbf{k}_i$ is obtained as (2.9). Assume, for simplicity, that $\mathbf{P}$ is independent of A_i. Then (2.29) is reduced to

$$
\frac{\partial U_k}{\partial A_i} = -\overline{\mathbf{U}}^{kT} \frac{\partial \mathbf{K}}{\partial A_i} \mathbf{U}
$$

$$
= -\overline{\mathbf{u}}_i^{kT} \frac{\partial \mathbf{k}_i}{\partial A_i} \mathbf{u}_i \tag{2.30}
$$

$$
= -\frac{E}{L_i} \overline{\mathbf{u}}_i^{kT} \mathbf{d}_i \mathbf{d}_i^T \mathbf{u}_i
$$

where $\overline{\mathbf{u}}_i^k$ is the local displacement vector of the ith member corresponding to $\overline{\mathbf{U}}^k$. The strain and axial force of member i corresponding to $\overline{\mathbf{U}}_i^k$ are denoted by $\overline{\varepsilon}_i^k$ and $\overline{N}_i^k$, respectively. Then (2.30) is rewritten as

$$
\frac{\partial U_k}{\partial A_i} = -E L_i \overline{\varepsilon}_i^k \varepsilon_i
$$

$$
= -\frac{L_i}{A_i^2 E} \overline{N}_i^k N_i \tag{2.31}
$$

Therefore, the sensitivity coefficients of the kth displacement component can be found easily from the strain $\overline{\varepsilon}_i^k$ or the axial force $\overline{N}_i^k$ against the unit virtual load at the kth displacement component.

Example 2.2
The sensitivity coefficient of U_2 is computed for the two-bar truss in Fig. 2.4. Eq. (2.23) is written as

$$
\frac{E}{2\sqrt{2}L} \begin{pmatrix} A_1 + 2\sqrt{2}A_2 & A_1 \\ A_1 & A_1 \end{pmatrix} \overline{\mathbf{U}}^2 = \begin{pmatrix} 0 \\ 1 \end{pmatrix} \tag{2.32}
$$

from which we obtain

$$
\overline{\mathbf{U}}^2 = \begin{pmatrix} -\dfrac{L}{A_2 E} \\[2mm] \dfrac{(A_1 + 2\sqrt{2}A_2)L}{A_1 A_2 E} \end{pmatrix} \tag{2.33}
$$

Furthermore, from (2.10) and (2.13),

$$
\mathbf{R} = \begin{pmatrix} -\dfrac{EU_1}{L} \\[2mm] 0 \end{pmatrix} \tag{2.34}
$$

is derived. Then, from (2.13), (2.28), and (2.34), we obtain

$$
\frac{\partial U_2}{\partial A_2} = -\frac{U_1}{A_2} = -\frac{PL}{A_2^2 E} \tag{2.35}
$$

which is the same as (2.15).

2.3 Eigenvalues of free vibration

2.3.1 Simple eigenvalue

We next present the methods of sensitivity analysis of eigenvalues and eigenmodes of free vibration. Let $\mathbf{M}(\mathbf{A})$ denote the $n \times n$ mass matrix of the structure, which is a function of the design variable vector $\mathbf{A}$. The eigenvalue problem of free vibration is formulated as

$$\mathbf{K}\boldsymbol{\Phi}_r = \Omega_r \mathbf{M}\boldsymbol{\Phi}_r, \quad (r = 1, \ldots, n) \tag{2.36}$$

where Ω_r and $\boldsymbol{\Phi}_r$ are the rth eigenvalue and eigenmode, respectively, which are functions of $\mathbf{A}$. The eigenmode $\boldsymbol{\Phi}_r$ is ortho-normalized with respect to $\mathbf{M}$ as

$$\boldsymbol{\Phi}_r^\top \mathbf{M}\boldsymbol{\Phi}_s = \delta_{rs}, \quad (r, s = 1, \ldots, n) \tag{2.37}$$

where δ_{rs} is the Kronecker delta.

Consider first the case where Ω_r is simple; i.e., $\Omega_{r-1} < \Omega_r < \Omega_{r+1}$ is satisfied. Differentiation of both sides of (2.36) with respect to A_i reads

$$\frac{\partial \mathbf{K}}{\partial A_i}\boldsymbol{\Phi}_r + \mathbf{K}\frac{\partial \boldsymbol{\Phi}_r}{\partial A_i} = \frac{\partial \Omega_r}{\partial A_i}\mathbf{M}\boldsymbol{\Phi}_r + \Omega_r\frac{\partial \mathbf{M}}{\partial A_i}\boldsymbol{\Phi}_r + \Omega_r\mathbf{M}\frac{\partial \boldsymbol{\Phi}_r}{\partial A_i} \tag{2.38}$$

Premultiplying $\boldsymbol{\Phi}_r^\top$ on both sides of (2.38), using (2.36), (2.37), and the symmetry of $\mathbf{M}$ and $\mathbf{K}$, we obtain

$$\frac{\partial \Omega_r}{\partial A_i} = \boldsymbol{\Phi}_r^\top \left(\frac{\partial \mathbf{K}}{\partial A_i} - \Omega_r \frac{\partial \mathbf{M}}{\partial A_i} \right) \boldsymbol{\Phi}_r \tag{2.39}$$

Note that $\mathbf{K}$ and $\mathbf{M}$ are explicit functions of A_i. Therefore, when the eigenvalue Ω_r and the eigenmode $\boldsymbol{\Phi}_r$ are known, the sensitivity coefficients of eigenvalues are computed from (2.39) through a simple arithmetic operation without resort to the sensitivity coefficients of the eigenmodes.

In civil and architectural engineering, the natural frequency and natural period are usually used rather than the eigenvalue. After obtaining the sensitivity coefficients of Ω_r, those of the rth frequency $f_r = \sqrt{\Omega_r}/(2\pi)$ and period $T_r = 2\pi/\sqrt{\Omega_r}$ are computed as

$$\frac{\partial f_r}{\partial A_i} = \frac{1}{8\pi^2 f_r} \frac{\partial \Omega_r}{\partial A_i}, \quad \frac{\partial T_r}{\partial A_i} = -\frac{\pi}{\sqrt{\Omega_r^3}} \frac{\partial \Omega_r}{\partial A_i} \tag{2.40}$$

Next, we carry out sensitivity analysis of eigenmodes (Yu, Liu, and Wang 1996). Differentiation of (2.37) with respect to A_i leads to

$$\boldsymbol{\Phi}_r^\top \frac{\partial \mathbf{M}}{\partial A_i} \boldsymbol{\Phi}_r + 2\boldsymbol{\Phi}_r^\top \mathbf{M}\frac{\partial \boldsymbol{\Phi}_r}{\partial A_i} = 0 \tag{2.41}$$

From (2.38) and (2.41),

$$
\begin{pmatrix} \mathbf{K} - \Omega_r \mathbf{M} & -\mathbf{M}\boldsymbol{\Phi}_r \\ -\boldsymbol{\Phi}_r^\top \mathbf{M} & 0 \end{pmatrix}
\begin{pmatrix} \dfrac{\partial \boldsymbol{\Phi}_r}{\partial A_i} \\[2mm] \dfrac{\partial \Omega_r}{\partial A_i} \end{pmatrix}
= \begin{pmatrix} \left(\Omega_r \dfrac{\partial \mathbf{M}}{\partial A_i} - \dfrac{\partial \mathbf{K}}{\partial A_i} \right) \boldsymbol{\Phi}_r \\[2mm] \dfrac{1}{2} \boldsymbol{\Phi}_r^\top \dfrac{\partial \mathbf{M}}{\partial A_i} \boldsymbol{\Phi}_r \end{pmatrix}
\tag{2.42}
$$

is derived. Therefore, the sensitivity coefficients of an eigenmode can be found by solving the set of $n + 1$ simultaneous linear equations (2.42).

However, it is not straightforward to see that (2.42) has a unique solution. For the case of a simple eigenvalue, the rank of $\mathbf{K} - \Omega_r \mathbf{M}$ is $n - 1$ with the kernel (null-space) spanned by $\boldsymbol{\Phi}_r$. Let $\mathbf{C}$ denote the matrix on the left-hand side of (2.42). Because the vector $\mathbf{M}\boldsymbol{\Phi}_r$ generally has the component of $\boldsymbol{\Phi}_r$, the $n \times (n+1)$ matrix $\mathbf{B}$, defined as follows, has full row rank that is equal to n:

$$
\mathbf{B} = \begin{pmatrix} \mathbf{K} - \Omega_r \mathbf{M} & \mathbf{M}\boldsymbol{\Phi}_r \end{pmatrix}
\tag{2.43}
$$

Furthermore, the vector $(-\boldsymbol{\Phi}_r^\top \mathbf{M} \quad 0)$ in the $(n+1)$st row of $\mathbf{C}$ cannot be expressed as a linear combination of the rows of the matrix $\mathbf{B}$, because $-\boldsymbol{\Phi}_r^\top \mathbf{M}$ has a component of the kernel of $\mathbf{K} - \Omega_r \mathbf{M}$. Hence, the matrix $\mathbf{C}$ has full rank $n + 1$, and (2.42) has a unique solution. Therefore, the sensitivity coefficients of eigenvalues and eigenmodes can be simultaneously computed from (2.42).

Suppose the sensitivity coefficient of an eigenvalue has already been computed using (2.39), and define a vector $\mathbf{b}$ as

$$
\mathbf{b} = \frac{\partial \Omega_r}{\partial A_i} \mathbf{M}\boldsymbol{\Phi}_r + \Omega_r \frac{\partial \mathbf{M}}{\partial A_i} \boldsymbol{\Phi}_r - \frac{\partial \mathbf{K}}{\partial A_i} \boldsymbol{\Phi}_r
\tag{2.44}
$$

If $\mathbf{b}$ is orthogonal to $\boldsymbol{\Phi}_r$, i.e., if $\mathbf{b}$ does not belong to the kernel of $\mathbf{K} - \Omega_r \mathbf{M}$, then the first equation in (2.42) has a solution. Furthermore, it can be easily confirmed that the orthogonality of $\mathbf{b}$ to $\boldsymbol{\Phi}_r$ leads to the relation (2.39); hence, $\mathbf{b}$ is actually orthogonal to $\boldsymbol{\Phi}_r$, and it is also possible to obtain the sensitivity coefficients of the eigenmodes after computing those of the eigenvalues.

In this approach, however, the coefficient matrix on the left-hand side of (2.42) depends on the order r of the eigenmode. Therefore, the computational cost is proportional to the number of eigenmodes for which the sensitivity coefficients are to be computed. Hence, this approach is not computationally efficient for the case in which the eigenmodes corresponding to many eigenvalues are to be computed.

Alternatively, the sensitivity coefficients of the eigenmode can be expanded as follows using the eigenmodes based on the *truncated modal method* (Fox and Kapoor 1968):

$$
\frac{\partial \boldsymbol{\Phi}_r}{\partial A_i} = \sum_{j=1}^{q} c_j \boldsymbol{\Phi}_j
\tag{2.45}
$$

where q modes are assumed to be used, and $c_1, \ldots, c_q$ are the coefficients.

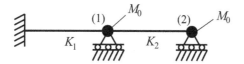

FIGURE 2.5: A two-bar lumped mass structure.

From (2.36) and (2.37), we have

$$\Phi_r^\top \mathbf{K} \Phi_s = \delta_{rs} \Omega_r, \quad (r, s = 1, \dots, q) \tag{2.46}$$

By premultiplying $\Phi_j^\top$ on the both sides of (2.38), and using (2.37), (2.45), and (2.46), we obtain

$$c_j(\Omega_j - \Omega_r) = -\Phi_j^\top \left(\frac{\partial \mathbf{K}}{\partial A_i} - \frac{\partial \Omega_r}{\partial A_i} \mathbf{M} - \Omega_r \frac{\partial \mathbf{M}}{\partial A_i} \right) \Phi_r \tag{2.47}$$

The coefficients c_j are found from (2.47) for $j \neq r$. The coefficient c_r is obtained from (2.37), (2.41), and (2.45) as

$$c_r = -\frac{1}{2} \Phi_r^\top \frac{\partial \mathbf{M}}{\partial A_i} \Phi_r \tag{2.48}$$

The computational cost is reduced by using only a few modes for expansion. However, as is seen from (2.48), if $\mathbf{M}$ does not depend on A_i, e.g., a lumped mass structure, then $c_r = 0$ and the vector of sensitivity coefficients of Φ_r is orthogonal to Φ_r. Therefore, higher modes are needed to accurately express the sensitivity coefficients of the lower eigenmodes. Note also that this method cannot be used to compute the sensitivity coefficients of an eigenmode associated with multiple eigenvalues, e.g., $\Omega_{r+1} = \Omega_r$, because $\Omega_j - \Omega_r$ in (2.47) vanishes for $j = r + 1$.

Example 2.3
Consider a two-bar structure, as shown in Fig. 2.5, which has the lumped mass M_0 at nodes 1 and 2. The structural mass of the bars is neglected for brevity. The extensional stiffnesses of bars 1 and 2 are denoted by K_1 and K_2, respectively. Eq. (2.36) is written as

$$\begin{pmatrix} K_1 + K_2 & -K_2 \\ -K_2 & K_2 \end{pmatrix} \Phi_r = \Omega_r \begin{pmatrix} M_0 & 0 \\ 0 & M_0 \end{pmatrix} \Phi_r, \quad (r = 1, 2) \tag{2.49}$$

The two eigenvalues are obtained as

$$\begin{aligned} \Omega_1 &= \frac{1}{2M_0} \left(K_1 + 2K_2 - \sqrt{K_1^2 + 4K_2^2} \right), \\ \Omega_2 &= \frac{1}{2M_0} \left(K_1 + 2K_2 + \sqrt{K_1^2 + 4K_2^2} \right) \end{aligned} \tag{2.50}$$

The sensitivity coefficients are to be evaluated at the design $(K_1, K_2) = (3K_0, 2K_0)$, where K_0 is the specified value. Note that this relation defines only the values of the current design; i.e., it does not mean that K_1 and K_2 are linked by K_0. The eigenvalues and eigenmodes for this design are given as

$$\Omega_1 = \frac{K_0}{M_0}, \quad \Omega_2 = \frac{6K_0}{M_0} \tag{2.51}$$

$$\boldsymbol{\Phi}_1 = \frac{1}{\sqrt{5M_0}} \begin{pmatrix} 1 \\ 2 \end{pmatrix}, \quad \boldsymbol{\Phi}_2 = \frac{1}{\sqrt{5M_0}} \begin{pmatrix} 2 \\ -1 \end{pmatrix} \tag{2.52}$$

Then the sensitivity coefficient of Ω_1 with respect to K_1 is obtained from (2.39) as

$$\frac{\partial \Omega_1}{\partial K_1} = \frac{1}{5M_0}(1,2)\begin{pmatrix} 1 & 0 \\ 0 & 0 \end{pmatrix}\begin{pmatrix} 1 \\ 2 \end{pmatrix} = \frac{1}{5M_0} \tag{2.53}$$

which agrees with the result obtained by differentiation of Ω_1 in (2.50).

Next, we compute the sensitivity coefficients of the lowest eigenmode $\boldsymbol{\Phi}_1 = (\Phi_{1,1}, \Phi_{1,2})^{\top}$ using (2.42). For the design $(K_1, K_2) = (3K_0, 2K_0)$, the matrices $\mathbf{K}$ and $\mathbf{K} - \Omega_1 \mathbf{M}$ are obtained as

$$\mathbf{K} = \begin{pmatrix} 5K_0 & -2K_0 \\ -2K_0 & 2K_0 \end{pmatrix}, \quad \mathbf{K} - \Omega_1 \mathbf{M} = \begin{pmatrix} 4K_0 & -2K_0 \\ -2K_0 & K_0 \end{pmatrix} \tag{2.54}$$

As is seen from (2.54), the matrix $\mathbf{K} - \Omega_1 \mathbf{M}$ is singular. From (2.42), we have

$$\begin{pmatrix} 4K_0 & -2K_0 & -\sqrt{\dfrac{M_0}{5}} \\[2ex] -2K_0 & K_0 & -2\sqrt{\dfrac{M_0}{5}} \\[2ex] -\sqrt{\dfrac{M_0}{5}} & -2\sqrt{\dfrac{M_0}{5}} & 0 \end{pmatrix} \begin{pmatrix} \dfrac{\partial \Phi_{1,1}}{\partial K_1} \\[2ex] \dfrac{\partial \Phi_{1,2}}{\partial K_1} \\[2ex] \dfrac{\partial \Omega_1}{\partial K_1} \end{pmatrix} = \begin{pmatrix} -\dfrac{1}{\sqrt{5M_0}} \\[2ex] 0 \\[2ex] 0 \end{pmatrix} \tag{2.55}$$

from which we obtain

$$\frac{\partial \Phi_{1,1}}{\partial K_1} = -\frac{4}{25K_0\sqrt{5M_0}}, \quad \frac{\partial \Phi_{1,2}}{\partial K_1} = \frac{2}{25K_0\sqrt{5M_0}}, \quad \frac{\partial \Omega_1}{\partial K_1} = \frac{1}{5M_0} \tag{2.56}$$

Note that the sensitivity of the eigenvalue agrees with the result in (2.53). The vector $\mathbf{b}$ in (2.44) is given as

$$\mathbf{b} = \frac{\partial \Omega_1}{\partial K_1}\begin{pmatrix} \sqrt{\dfrac{M_0}{5}} \\[2ex] 2\sqrt{\dfrac{M_0}{5}} \end{pmatrix} - \begin{pmatrix} -\dfrac{1}{\sqrt{5M_0}} \\[2ex] 0 \end{pmatrix} \tag{2.57}$$

It is easily confirmed that the sensitivity coefficient derived from the orthogonality between $\mathbf{b}$ and $\boldsymbol{\Phi}_1$ agrees with the result in (2.53).

The matrix $\mathbf{B}$ is found as

$$\mathbf{B} = \begin{pmatrix} 4K_0 & -2K_0 & -\sqrt{\dfrac{M_0}{5}} \\ -2K_0 & K_0 & -2\sqrt{\dfrac{M_0}{5}} \end{pmatrix} \tag{2.58}$$

which has full row rank 2.

Finally, we approximate the vector of the sensitivity coefficient of $\mathbf{\Phi}_1$ with respect to K_1 at $(K_1, K_2) = (3K_0, 2K_0)$ using $\mathbf{\Phi}_2$ only as

$$\frac{\partial \mathbf{\Phi}_1}{\partial K_1} = c_2 \mathbf{\Phi}_2 \tag{2.59}$$

From (2.47), (2.51), and (2.52), we obtain

$$\frac{1}{5M_0} \begin{pmatrix} -2 & 1 \end{pmatrix} \begin{pmatrix} 1 & 0 \\ 0 & 0 \end{pmatrix} \begin{pmatrix} 1 \\ 2 \end{pmatrix} + c_2 \frac{6K_0}{M_0} = c_2 \frac{K_0}{M_0} \tag{2.60}$$

which leads to

$$c_2 = \frac{2}{25K_0} \tag{2.61}$$

By incorporating c_2 in (2.61) into (2.59), we obtain the same results as (2.56), because, in this example, the mass matrix is independent of K_1, and the sensitivity of $\mathbf{\Phi}_1$ of the 2-degrees-of-freedom system can be exactly expressed by $\mathbf{\Phi}_2$.

2.3.2 Multiple eigenvalues

As discussed in Sec. 1.9, it is well known that the eigenvalues often become multiple as a result of optimization. In this case, the equations for sensitivity analysis of eigenvalues and eigenmodes presented in Sec. 2.3.1 cannot be used. Suppose the two lowest eigenvalues Ω_1 and Ω_2 have the same value at $A_i = A_i^*$, where A_i is a representative design variable. Fig. 2.6(a) illustrates the variation of Ω_1 with respect to A_i. Since Ω_1 is defined as the lowest eigenvalue, its variation is as plotted in the solid line in Fig. 2.6(a), and the two lowest eigenvalues coincide at $A_i = A_i^*$, at which the derivatives of the eigenvalues are not continuous functions of the design variable A_i. Therefore, only the directional derivative or the subdifferential can be defined, as shown in Fig. 2.6(b), at $A_i = A_i^*$; see Appendix A.1.4 for definitions of the directional derivative and the subdifferential.

For a design satisfying $\Omega_1 = \Omega_2$, the linear combination $\mathbf{\Psi}$ of the eigenmodes

$$\mathbf{\Psi} = a_1 \mathbf{\Phi}_1 + a_2 \mathbf{\Phi}_2 \tag{2.62}$$

is also an eigenmode corresponding to the eigenvalue $\Omega_1 (= \Omega_2)$, where a_1 and a_2 are the coefficients satisfying $a_1^2 + a_2^2 \neq 0$. Note that the orthogonality between $\mathbf{\Phi}_1$ and $\mathbf{\Phi}_2$ with respect to the mass matrix is satisfied.

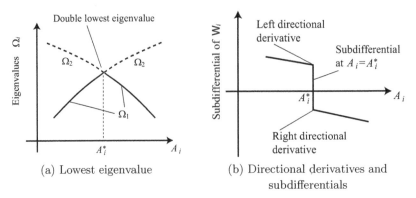

(a) Lowest eigenvalue   (b) Directional derivatives and
                                   subdifferentials

FIGURE 2.6:   Definition of directional derivatives and subdifferentials of
the lowest eigenvalue.

Define C_{rs} as

$$C_{rs} = \Phi_r^\top \left(\frac{\partial \mathbf{K}}{\partial A_i} - \Omega_r \frac{\partial \mathbf{M}}{\partial A_i} \right) \Phi_s, \quad (r,s = 1,2) \tag{2.63}$$

By premultiplying Ψ on both sides of (2.38) corresponding to $r = 1, 2$, and
using (2.36), (2.62), and (2.63), we obtain

$$\begin{pmatrix} C_{11} & C_{12} \\ C_{21} & C_{22} \end{pmatrix} \begin{pmatrix} a_1 \\ a_2 \end{pmatrix} = \frac{\partial \Omega_r}{\partial A_i} \begin{pmatrix} a_1 \\ a_2 \end{pmatrix} \tag{2.64}$$

Therefore, the two directional derivatives are found as the solution of the
eigenvalue problem (2.64) so that the derivatives and the coefficient vector
$(a_1, a_2)^\top$ have nonzero values (Haug and Choi 1982; Seyranian, Lund, and
Olhoff 1994).

Example 2.4
Consider the four-bar truss, as shown in Fig. 1.16, that was used in Example 1.13 in Sec. 1.9. The extensional stiffnesses of the horizontal and vertical
members are denoted by K_1 and K_2, respectively. The truss has a non-structural mass M_0, and the member mass is neglected. The eigenvalues and
eigenvectors of this 2-degrees-of-freedom truss are defined by

$$\begin{pmatrix} 2K_1 & 0 \\ 0 & 2K_2 \end{pmatrix} \Phi_r = \Omega_r \begin{pmatrix} M_0 & 0 \\ 0 & M_0 \end{pmatrix} \Phi_r, \quad (r = 1, 2) \tag{2.65}$$

The stiffness matrix $\mathbf{K}$ is differentiated with respect to K_1 as

$$\frac{\partial \mathbf{K}}{\partial K_1} = \begin{pmatrix} 2 & 0 \\ 0 & 0 \end{pmatrix} \tag{2.66}$$

If $K_1 = K_2 = K_0$, then the eigenvalues are duplicate as

$$\Omega_1 = \Omega_2 = \frac{2K_0}{M_0} \tag{2.67}$$

and the eigenmodes are found as

$$\Phi_1 = \begin{pmatrix} 1 \\ 0 \end{pmatrix}, \quad \Phi_2 = \begin{pmatrix} 0 \\ 1 \end{pmatrix} \tag{2.68}$$

Then, from (2.63), we have

$$C_{11} = 2, \quad C_{12} = C_{21} = 0, \quad C_{22} = 0 \tag{2.69}$$

which leads to

$$\frac{\partial \Omega_1}{\partial K_1} = 2 \text{ or } 0 \tag{2.70}$$

In fact, the increment of Ω_1 is 0 for $\Delta K_1 > 0$, and $2\Delta K_1$ for $\Delta K_1 < 0$.

Since the eigenmodes cannot be defined uniquely for duplicate eigenvalues, suppose the two eigenmodes that are mutually orthogonal with respect to the mass matrix are defined as

$$\Phi_1 = \frac{1}{2} \begin{pmatrix} \sqrt{2} \\ \sqrt{2} \end{pmatrix}, \quad \Phi_2 = \frac{1}{2} \begin{pmatrix} \sqrt{2} \\ -\sqrt{2} \end{pmatrix} \tag{2.71}$$

In this case, the coefficients in (2.63) are obtained as

$$C_{11} = 1, \quad C_{12} = C_{21} = 1, \quad C_{22} = 1 \tag{2.72}$$

and (2.64) is written as

$$\begin{pmatrix} 1 & 1 \\ 1 & 1 \end{pmatrix} \begin{pmatrix} a_1 \\ a_2 \end{pmatrix} = \frac{\partial \Omega_1}{\partial K_1} \begin{pmatrix} a_1 \\ a_2 \end{pmatrix} \tag{2.73}$$

which leads to the same result as (2.70).

———————

Generally, for a symmetric structure, the matrix $\mathbf{C} = (C_{ij})$ on the left-hand side of (2.64) is diagonal if the design variables are linked so that symmetry is preserved. In this case, the directional derivatives can be easily found from (2.39) for the simple eigenvalue using the symmetric (antisymmetric) eigenmodes corresponding to the multiple eigenvalues. For example, in Example 2.4, $\Phi_1 = (1,0)^\top$ and $\Phi_2 = (0,1)^\top$ in (2.39) lead to the sensitivity coefficients 2 and 0, respectively, which are the same as those in (2.70).

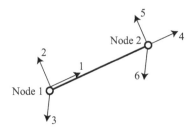

FIGURE 2.7: Definition of local displacement numbers.

2.4 Linear buckling load

For structures such as frames with slender members and thin-walled shells, buckling is a very important performance measure in the design process. Although buckling is related to geometrically nonlinear behavior, the buckling load can be computed with good accuracy as a solution of a linear eigenvalue problem, if the deformation at buckling called *prebuckling deformation* is sufficiently small (see, e.g., Bažant and Cedolin (1991) and Kollár (1999) for details of buckling analysis of frames).

Consider a truss subjected to the proportional loads $\Lambda \mathbf{P}_0$, where Λ is the load factor, and $\mathbf{P}_0$ is the constant vector of load pattern. The load factor Λ^c at which the structure becomes unstable as Λ is increased from 0 is called the *linear buckling load factor* if the prebuckling deformation is negligibly small, and, hence, the stresses are proportional to the load factor.

The displacement vector $\mathbf{U}_0$ against $\mathbf{P}_0$ is computed from

$$\mathbf{K}\mathbf{U}_0 = \mathbf{P}_0 \tag{2.74}$$

The axial force N_{0i} of member i of a truss corresponding to $\mathbf{U}_0$ is given by using the member displacement vector $\mathbf{u}_{0i}$ with respect to the global coordinates as

$$N_{0i} = \frac{EA_i}{L_i}\mathbf{d}_i^\top \mathbf{u}_{0i} \tag{2.75}$$

The reduction of stiffness due to compressive axial force is expressed by using the *geometrical stiffness matrix*. Suppose the local displacement numbers are defined as shown in Fig. 2.7. Then the geometrical stiffness matrix $\mathbf{k}_{Gi}$ of member i with respect to the local coordinates is given for the load $\mathbf{P}_0$ as

$$\mathbf{k}_{Gi} = \frac{N_{0i}}{L_i} \begin{pmatrix} 0 & 0 & 0 & 0 & 0 & 0 \\ 0 & 1 & 0 & 0 & -1 & 0 \\ 0 & 0 & 1 & 0 & 0 & -1 \\ 0 & 0 & 0 & 0 & 0 & 0 \\ 0 & -1 & 0 & 0 & 1 & 0 \\ 0 & 0 & -1 & 0 & 0 & 1 \end{pmatrix} \tag{2.76}$$

Note that N_{0i} has a negative value in the compressive state. Because the elastic axial stiffness is sufficiently larger than the geometrical stiffness, the effect of axial force on the axial stiffness is usually neglected. However, we can incorporate its effect by formulating $\mathbf{k}_{Gi}$ as

$$
\mathbf{k}_{Gi} = \frac{N_{0i}}{L_i}
\begin{pmatrix}
1 & 0 & 0 & -1 & 0 & 0 \\
0 & 1 & 0 & 0 & -1 & 0 \\
0 & 0 & 1 & 0 & 0 & -1 \\
-1 & 0 & 0 & 1 & 0 & 0 \\
0 & -1 & 0 & 0 & 1 & 0 \\
0 & 0 & -1 & 0 & 0 & 1
\end{pmatrix}
\tag{2.77}
$$

which is compatible with the definition of Green's strain and does not have any stiffness against rigid-body rotations.

By assembling the member geometrical stiffness matrix to the total structure to formulate the $n \times n$ geometrical stiffness matrix $\mathbf{K}_G(\mathbf{N}_0)$ as a function of $\mathbf{N}_0$, the *tangent stiffness matrix* is given as $\mathbf{K} + \Lambda \mathbf{K}_G(\mathbf{N}_0)$, and the buckling load is obtained from the linear buckling analysis as

$$
(\mathbf{K} + \Lambda \mathbf{K}_G)\mathbf{\Phi} = \mathbf{0} \tag{2.78}
$$

In the following, the argument $\mathbf{N}_0$ is omitted for brevity.

Note that $\mathbf{K}_G$ is an indefinite matrix, although the linear stiffness matrix $\mathbf{K}$ is positive definite, because the structure may buckle against the proportional loads in the opposite direction. However, we are interested in a positive buckling load factor for the load pattern $\mathbf{P}_0$ in the specified direction. Therefore, using the positive definiteness of $\mathbf{K}$ in (2.78), the buckling mode $\mathbf{\Phi}$ is normalized as

$$
\mathbf{\Phi}^\top \mathbf{K}_G \mathbf{\Phi} = -1 \tag{2.79}
$$

Note that $\mathbf{\Phi}$ can alternatively be normalized with respect to $\mathbf{K}$ as $\mathbf{\Phi}^\top \mathbf{K} \mathbf{\Phi} = 1$; however, the normalization in (2.79) leads to a simple form of sensitivity coefficient shown below. By differentiating (2.78) with respect to the design variable A_i, keeping in mind that $\mathbf{K}_G$ depends on $\mathbf{N}_0$, we have

$$
\frac{\partial \mathbf{K}}{\partial A_i}\mathbf{\Phi} + \mathbf{K}\frac{\partial \mathbf{\Phi}}{\partial A_i} + \frac{\partial \Lambda}{\partial A_i}\mathbf{K}_G\mathbf{\Phi} + \Lambda \sum_{j=1}^{m}\left(\frac{\partial \mathbf{K}_G}{\partial N_{0j}}\frac{\partial N_{0j}}{\partial A_i}\right)\mathbf{\Phi} + \Lambda \mathbf{K}_G\frac{\partial \mathbf{\Phi}}{\partial A_i} = \mathbf{0} \tag{2.80}
$$

By premultiplying $\mathbf{\Phi}$ on (2.80) and using (2.78), (2.79), and the symmetry of $\mathbf{K}$ and $\mathbf{K}_G$, we obtain

$$
\frac{\partial \Lambda}{\partial A_i} = \mathbf{\Phi}^\top \frac{\partial \mathbf{K}}{\partial A_i}\mathbf{\Phi} + \Lambda \sum_{j=1}^{m}\mathbf{\Phi}^\top\left(\frac{\partial \mathbf{K}_G}{\partial N_{0j}}\frac{\partial N_{0j}}{\partial A_i}\right)\mathbf{\Phi} \tag{2.81}
$$

It is seen from (2.81) that the sensitivity coefficients of the buckling mode are not needed for computing those of the linear buckling load factor, which is similar to the process of sensitivity analysis of eigenvalues of vibration in Sec. 2.3. However, the sensitivity coefficients of the axial forces should be computed using (2.75), as shown in Sec. 2.2, and are needed for a statically indeterminate truss. For a statically determinate truss, by contrast, the sensitivity coefficients of axial forces vanish, because the axial force is independent of the cross-sectional areas.

2.5   Transient responses

2.5.1   Direct differentiation method

Consider a linearly damped finite-dimensional structure subjected to time-varying loads $\mathbf{P}(t)$. The displacement vector $\mathbf{U}(t)$ is also a function of time t. Let $\mathbf{C}$ denote the damping matrix. Differentiation with respect to t is indicated by a dot. The equation of motion is written as

$$\mathbf{M}\ddot{\mathbf{U}}(t) + \mathbf{C}\dot{\mathbf{U}}(t) + \mathbf{K}\mathbf{U}(t) = \mathbf{P}(t) \tag{2.82}$$

with the initial conditions

$$\mathbf{U}(0) = \mathbf{U}_0, \quad \dot{\mathbf{U}}(0) = \dot{\mathbf{U}}_0 \tag{2.83}$$

where $\mathbf{U}_0$ and $\dot{\mathbf{U}}_0$ are the specified initial displacement and velocity vectors, respectively.

Suppose the dynamic transient response analysis has been completed to find $\mathbf{U}(t)$, $\dot{\mathbf{U}}(t)$, and $\ddot{\mathbf{U}}(t)$. By differentiating both sides of (2.82) with respect to A_i, we obtain

$$\mathbf{M}\frac{\partial \ddot{\mathbf{U}}}{\partial A_i} + \mathbf{C}\frac{\partial \dot{\mathbf{U}}}{\partial A_i} + \mathbf{K}\frac{\partial \mathbf{U}}{\partial A_i} = \mathbf{Q} \tag{2.84}$$

where

$$\mathbf{Q} = \frac{\partial \mathbf{P}}{\partial A_i} - \frac{\partial \mathbf{M}}{\partial A_i}\ddot{\mathbf{U}} - \frac{\partial \mathbf{C}}{\partial A_i}\dot{\mathbf{U}} - \frac{\partial \mathbf{K}}{\partial A_i}\mathbf{U} \tag{2.85}$$

It is assumed that the initial conditions are independent of the design variables as

$$\frac{\partial \mathbf{U}}{\partial A_i} = \mathbf{0}, \quad \frac{\partial \dot{\mathbf{U}}}{\partial A_i} = \mathbf{0} \ \text{ at } \ t = 0 \tag{2.86}$$

It is seen from (2.84) and (2.86) that the sensitivity coefficients are computed using conventional time-history analysis, where $\partial \mathbf{U}/\partial A_i$ and $\mathbf{Q}$ are conceived as displacements and loads, respectively.

Consider the case where the representative response quantity $\widetilde{F}(\mathbf{A})$, for which the sensitivity coefficients are to be computed, is defined by the integral of a time-varying function $R(\mathbf{U}(\mathbf{A}, t), \mathbf{A})$ as

$$\widetilde{F}(\mathbf{A}) = \int_0^{t^{\mathrm{f}}} R(\mathbf{U}(\mathbf{A}, t), \mathbf{A}) \mathrm{d}t \qquad (2.87)$$

where t^{f} is the final time. For example, if $\widetilde{F}(\mathbf{A})$ represents the L_2-norm of the displacement vector over the specified time period $0 \leq t \leq t^{\mathrm{f}}$, R is given as

$$R(\mathbf{U}(\mathbf{A}, t), \mathbf{A}) = \sqrt{\mathbf{U}^\top(\mathbf{A}, t)\mathbf{U}(\mathbf{A}, t)} \qquad (2.88)$$

Other norms, such as L_1-norm and L_∞-norm, can also be used (see Appendix A.4.4 for definitions of norms (distances)). The displacement U_j at the specified time t_0 can also be formally defined using the Dirac delta function $\delta(t_0)$ as

$$R(\mathbf{U}(\mathbf{A}, t), \mathbf{A}) = U_j(\mathbf{A}, t)\delta(t_0) \qquad (2.89)$$

In the following, the arguments $\mathbf{A}$, $\mathbf{U}$, and t are omitted for brevity.

The sensitivity coefficients of the representative response $\widetilde{F}$ are computed from the following equation obtained by differentiating (2.87) with respect to A_i:

$$\frac{\partial \widetilde{F}(\mathbf{A})}{\partial A_i} = \int_0^{t^{\mathrm{f}}} \left(\frac{\partial R(\mathbf{U}, \mathbf{A})}{\partial A_i} + \sum_{j=1}^{n} \frac{\partial R(\mathbf{U}, \mathbf{A})}{\partial U_j} \frac{\partial U_j}{\partial A_i} \right) \mathrm{d}t \qquad (2.90)$$

This approach is classified as the direct differentiation method. Note that (2.90) is an integration of the known values, which can be computed by a simple algorithm of numerical integration, e.g., the trapezoidal rule, after $\partial U_j / \partial A_i$ ($i = 1, \ldots, n$) are known. Therefore, the computation of the time history of the design sensitivity vector $\partial \mathbf{U} / \partial A_i$ in (2.84) requires the most computational time; hence, the computational cost is proportional to the number of design variables.

2.5.2 Adjoint variable method

In the adjoint variable method, the adjoint variable vector $\boldsymbol{\lambda}(t)$ is defined as follows in the time interval $0 \leq t \leq t^{\mathrm{f}}$:

$$\ddot{\boldsymbol{\lambda}}^\top \mathbf{M} - \dot{\boldsymbol{\lambda}}^\top \mathbf{C} + \boldsymbol{\lambda}^\top \mathbf{K} = \frac{\partial R}{\partial \mathbf{U}} \qquad (2.91)$$

where

$$\frac{\partial R}{\partial \mathbf{U}} = \left(\frac{\partial R}{\partial U_1}, \ldots, \frac{\partial R}{\partial U_n} \right)^\top \qquad (2.92)$$

By premultiplying $\boldsymbol{\lambda}^\top$ on (2.84), and integrating with respect to t from 0 to t^{f}, we obtain

$$\int_0^{t^{\mathrm{f}}} \boldsymbol{\lambda}^\top \left(\mathbf{M}\frac{\partial \ddot{\mathbf{U}}}{\partial A_i} + \mathbf{C}\frac{\partial \dot{\mathbf{U}}}{\partial A_i} + \mathbf{K}\frac{\partial \mathbf{U}}{\partial A_i} - \mathbf{Q} \right) \mathrm{d}t = 0 \qquad (2.93)$$

Integrating (2.93) by parts, we have

$$\boldsymbol{\lambda}^\top \mathbf{M} \frac{\partial \dot{\mathbf{U}}}{\partial A_i}\bigg|_{t=0}^{t=t^f} + \boldsymbol{\lambda}^\top \mathbf{C} \frac{\partial \mathbf{U}}{\partial A_i}\bigg|_{t=0}^{t=t^f}$$

$$+ \int_0^{t^f} \left(-\dot{\boldsymbol{\lambda}}^\top \mathbf{M} \frac{\partial \dot{\mathbf{U}}}{\partial A_i} - \dot{\boldsymbol{\lambda}}^\top \mathbf{C} \frac{\partial \mathbf{U}}{\partial A_i} + \boldsymbol{\lambda}^\top \mathbf{K} \frac{\partial \mathbf{U}}{\partial A_i} \right) dt \qquad (2.94)$$

$$= \int_0^{t^f} \boldsymbol{\lambda}^\top \mathbf{Q} dt$$

Further integration leads to

$$\boldsymbol{\lambda}^\top \mathbf{M} \frac{\partial \dot{\mathbf{U}}}{\partial A_i}\bigg|_{t=0}^{t=t^f} + \boldsymbol{\lambda}^\top \mathbf{C} \frac{\partial \mathbf{U}}{\partial A_i}\bigg|_{t=0}^{t=t^f} - \dot{\boldsymbol{\lambda}}^\top \mathbf{M} \frac{\partial \mathbf{U}}{\partial A_i}\bigg|_{t=0}^{t=t^f}$$

$$+ \int_0^{t^f} \left(\ddot{\boldsymbol{\lambda}}^\top \mathbf{M} - \dot{\boldsymbol{\lambda}}^\top \mathbf{C} + \boldsymbol{\lambda}^\top \mathbf{K} \right) \frac{\partial \mathbf{U}}{\partial A_i} dt = \int_0^{t^f} \boldsymbol{\lambda}^\top \mathbf{Q} dt \qquad (2.95)$$

The boundary terms in (2.95) vanish at $t = 0$, because the initial conditions for displacements and velocities are given as (2.86). The following conditions are to be satisfied for the adjoint variable vector so that the boundary terms in (2.95) vanish at $t = t^f$:

$$\boldsymbol{\lambda}(t^f) = \mathbf{0}, \quad \dot{\boldsymbol{\lambda}}(t^f) = \mathbf{0} \qquad (2.96)$$

By incorporating (2.91) into (2.95), and using (2.96), we can obtain the following relation:

$$\int_0^{t^f} \sum_{j=1}^n \frac{\partial R(\mathbf{U}, \mathbf{A})}{\partial U_j} \frac{\partial U_j}{\partial A_i} dt = \int_0^{t^f} \boldsymbol{\lambda}^\top \mathbf{Q} dt \qquad (2.97)$$

Therefore, from (2.90) and (2.97), we have

$$\frac{\partial \tilde{F}(\mathbf{A})}{\partial A_i} = \int_0^{t^f} \left(\frac{\partial R(\mathbf{U}, \mathbf{A})}{\partial A_i} + \boldsymbol{\lambda}^\top \mathbf{Q} \right) dt \qquad (2.98)$$

Note that (2.98) is an integration of the known values, which can be computed using a simple algorithm of numerical integration. Therefore, the computation of the time history of the adjoint variable vector in (2.91) requires the most computational time; hence, the computational cost is proportional to the number of response quantities for which the sensitivity coefficients are to be found.

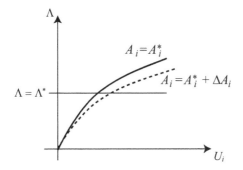

FIGURE 2.8: Geometrically nonlinear load-displacement relation.

2.6 Nonlinear responses

So far we considered the responses defined by geometrically linear analysis, for which the linear relation (2.7) is satisfied between the loads and displacements. However, for flexible structures, including shallow latticed domes and arch-type trusses in civil and architectural engineering, the effect of geometrical nonlinearity should be incorporated in the strain-displacement relation for computation of responses against external loads. Fig. 2.8 illustrates the relation between the representative displacement component U_j and the load factor Λ for a proportional loading $\mathbf{P} = \Lambda\mathbf{P}_0$ defined with load pattern vector $\mathbf{P}_0$. The equilibrium path plotted in the solid line may move to the dashed line due to a modification of the design variable A_i, e.g., from A_i^* to $A_i^* + \Delta A_i$. The purpose of sensitivity analysis here is to compute the rate of change of displacements at the specified load factor Λ^*.

Consider an elastic conservative system that has the total potential energy $\Pi(\mathbf{U}(\mathbf{A}), \mathbf{A})$ for the fixed value of Λ, which is a function of $\mathbf{U}(\mathbf{A})$ and $\mathbf{A}$. The equilibrium equations are obtained from the stationary condition of $\Pi(\mathbf{U}(\mathbf{A}), \mathbf{A})$ as

$$\frac{\partial \Pi(\mathbf{U}(\mathbf{A}), \mathbf{A})}{\partial \mathbf{U}} = \mathbf{0} \tag{2.99}$$

By differentiating (2.99) with respect to a design variable A_i, we can obtain the following equations for computing the sensitivity coefficients of $\mathbf{U}$ with respect to A_i:

$$\mathbf{K}_{\mathrm{T}}\frac{\partial \mathbf{U}}{\partial A_i} + \frac{\partial^2 \Pi}{\partial A_i \partial \mathbf{U}} = \mathbf{0} \tag{2.100}$$

where $\mathbf{K}_{\mathrm{T}}$ is called the tangent stiffness matrix, for which the (i,j)-component $K_{\mathrm{T},ij}$ is given as

$$K_{\mathrm{T},ij} = \frac{\partial^2 \Pi}{\partial U_i \partial U_j} \tag{2.101}$$

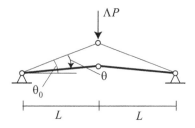

FIGURE 2.9: A two-bar shallow truss.

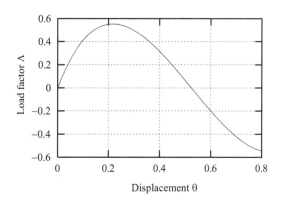

FIGURE 2.10: Equilibrium path of the two-bar shallow truss.

Hence, the tangent stiffness matrix is symmetric for a conservative system.

For the case of proportional loading, $\Pi(\mathbf{U}(\mathbf{A}), \mathbf{A})$ is defined with the strain energy $S(\mathbf{U}(\mathbf{A}), \mathbf{A})$ and the external work $\Lambda^*\mathbf{P}_0^\top \mathbf{U}$ at $\Lambda = \Lambda^*$ as

$$\Pi(\mathbf{U}(\mathbf{A}), \mathbf{A}) = S(\mathbf{U}(\mathbf{A}), \mathbf{A}) - \Lambda^*\mathbf{P}_0^\top \mathbf{U} \qquad (2.102)$$

and (2.100) is reduced to

$$\mathbf{K}_{\mathrm{T}}\frac{\partial \mathbf{U}}{\partial A_i} + \frac{\partial^2 S}{\partial A_i \partial \mathbf{U}} = \Lambda^*\frac{\partial \mathbf{P}_0}{\partial A_i} \qquad (2.103)$$

which has a similar form as the geometrically linear case in (2.8). Indeed, if we use

$$S(\mathbf{U}(\mathbf{A}), \mathbf{A}) = \frac{1}{2}\mathbf{U}^\top \mathbf{K}\mathbf{U} \qquad (2.104)$$

then (2.103) reduces to (2.8).

Example 2.5

Consider a two-bar shallow truss, as shown in Fig. 2.9, subjected to a vertical load ΛP, where the thin and thick lines represent the undeformed and deformed configurations, respectively.

The cross-sectional areas and the elastic modulus are A and E, respectively, for both members. Therefore, the truss is symmetric with respect to the center vertical axis. The initial angle of each member to the horizontal axis is denoted by θ_0, and the rotation θ is taken as the generalized displacement.

Since the deformed configuration is also symmetric with respect to the center axis, the strain $\varepsilon(\theta)$ of both members is given as

$$\varepsilon(\theta) = \frac{L/\cos(\theta_0 - \theta)}{L/\cos\theta_0} - 1 \tag{2.105}$$

Then the total potential energy $\Pi(\theta, A)$ is written as

$$\Pi(\theta, A) = EA\varepsilon^2(\theta)\frac{L}{\cos\theta_0} - \Lambda PL[\tan\theta_0 - \tan(\theta_0 - \theta)] \tag{2.106}$$

By differentiating $\Pi(\theta, A)$ with respect to θ, we obtain the following equilibrium equation:

$$2EAL\varepsilon(\theta)\frac{\sin(\theta_0 - \theta)}{\cos^2\theta_0} - \Lambda PL\frac{1}{\cos^2(\theta_0 - \theta)} = 0 \tag{2.107}$$

From (2.107), we have the load-displacement relation

$$\Lambda P = 2EA\frac{\sin(\theta_0 - \theta)[\cos(\theta_0 - \theta) - \cos\theta_0]}{\cos(\theta_0 - \theta)} \tag{2.108}$$

which is plotted in Fig. 2.10 with $P = 1$, $\theta_0 = \pi/6$, $E = 10$, and $A = 1$. Note that Λ reaches the maximum at the *limit point*, as θ is increased, and attains 0 at $\theta = \theta_0 = \pi/6$, when the two bars are colinear.

Differentiation of (2.107) with respect to A at the fixed value of Λ at Λ^* leads to the following equation for computing the sensitivity coefficient of θ with respect to A:

$$\begin{aligned} &\left\{ 2EAL\frac{\sin^2(\theta_0 - \theta)}{\cos^3\theta_0} - 2EAL\varepsilon(\theta)\frac{\cos(\theta_0 - \theta)}{\cos^2\theta_0} \right. \\ &\left. + 2\Lambda^*PL\frac{\sin(\theta_0 - \theta)}{\cos^3(\theta_0 - \theta)} \right\}\frac{\partial\theta}{\partial A} + 2EL\varepsilon(\theta)\frac{\sin(\theta_0 - \theta)}{\cos^2(\theta_0 - \theta)} = 0 \end{aligned} \tag{2.109}$$

2.7 Shape sensitivity analysis of trusses

So far, we considered cross-sectional areas as design variables for application to sizing optimization using a gradient-based approach. However, the nodal coordinates are to be modified in the process of shape optimization. Therefore, in this section, we derive the basic equations of shape sensitivity analysis of trusses with respect to nodal locations (Twu and Choi 1992). A three-dimensional truss member, as shown in Fig. 2.3, is also used in this section.

The 6×6 member stiffness matrix $\mathbf{k}_i$ with respect to the global coordinates is given in (2.6). Because L_i and the unit vector $\mathbf{t}_i$ of member direction depend on the nodal coordinates, the sensitivity coefficients of these values are needed to compute the shape sensitivity coefficients of $\mathbf{k}_i$.

Let (X_1^i, Y_1^i, Z_1^i) and (X_2^i, Y_2^i, Z_2^i) denote the global coordinates of nodes 1 and 2, respectively, of member i. Then L_i and $\mathbf{t}_i$ are given as

$$L_i = \sqrt{(X_2^i - X_1^i)^2 + (Y_2^i - Y_1^i)^2 + (Z_2^i - Z_1^i)^2} \tag{2.110}$$

$$\mathbf{t}_i = \frac{1}{L_i}(X_2^i - X_1^i, Y_2^i - Y_1^i, Z_2^i - Z_1^i)^\top \tag{2.111}$$

By differentiating (2.110) with respect to X_1^i and X_2^i, respectively, we obtain

$$\frac{\partial L_i}{\partial X_1^i} = -\frac{1}{L_i}(X_2^i - X_1^i), \quad \frac{\partial L_i}{\partial X_2^i} = \frac{1}{L_i}(X_2^i - X_1^i) \tag{2.112}$$

Hence, using (2.111) with the notation $\mathbf{t}_i = (t_{i,1}, t_{i,2}, t_{i,3})^\top$, simpler expressions

$$\frac{\partial L_i}{\partial X_1^i} = -t_{i,1}, \quad \frac{\partial L_i}{\partial X_2^i} = t_{i,1} \tag{2.113}$$

are obtained. The sensitivity coefficients of $t_{i,1}$ with respect to X_1 and X_2 are derived as

$$\frac{\partial t_{i,1}}{\partial X_1^i} = \frac{1}{L_i}\left(-1 - t_{i,1}\frac{\partial L_i}{\partial X_1^i}\right) = -\frac{1 - t_{i,1}^2}{L_i}$$
$$\frac{\partial t_{i,1}}{\partial X_2^i} = \frac{1}{L_i}\left(1 - t_{i,1}\frac{\partial L_i}{\partial X_2^i}\right) = \frac{1 - t_{i,1}^2}{L_i} \tag{2.114}$$

The sensitivity coefficients with respect to the Y- and Z-coordinates are derived in a similar manner.

Then the sensitivity coefficients of $\mathbf{d}$ are readily obtained from (2.4), and those of $\mathbf{k}_i$ with respect to the nodal coordinates, e.g., X_1^i, are computed by differentiating (2.6) as

$$\frac{\partial \mathbf{k}_i}{\partial X_1^i} = -\frac{A_i E}{L_i^2}\frac{\partial L_i}{\partial X_1^i}\mathbf{d}_i\mathbf{d}_i^\top + \frac{A_i E}{L_i}\frac{\partial \mathbf{d}}{\partial X_1^i}\mathbf{d}^\top + \frac{A_i E}{L_i}\mathbf{d}\left(\frac{\partial \mathbf{d}}{\partial X_1^i}\right)^\top \tag{2.115}$$

which is incorporated into (2.10) and (2.11) to compute the shape sensitivity coefficients of the displacements. Therefore, the difference between design sensitivity and shape sensitivity exists only in the differentiation process of the local properties, and the formulations of the final set of linear equations are the same.

Chapter 3

Topology Optimization of Trusses

In this chapter, various problem formulations and methods of topology optimization are presented for pin-jointed trusses. A brief introduction and historical review are given in Sec. 3.1. The traditional concept of the Michell truss is presented in Sec. 3.2. Some typical problem formulations and methodologies are summarized in Secs. 3.3 and 3.4. The problem under stress constraints is discussed for continuous variables in Sec. 3.5 and discrete variables in Sec. 3.6. A genetic algorithm is applied in Sec. 3.7 for the problem considering nodal costs. A random search method is presented in Sec. 3.8. An approach based on semidefinite programming for eigenvalue constraints is presented in Sec. 3.9. Finally, application of data mining is discussed in Sec. 3.10.

3.1 Introduction

Configuration optimization of framed structures such as trusses and frames is categorized into two areas, i.e., optimization of nodal locations and optimization of connectivity of the nodes by the members. The former process is referred to as *geometry optimization*, and the latter is called *topology optimization*. Both processes usually include *sizing optimization* or traditional *optimum design* where the cross-sectional properties are optimized. *Simultaneous optimization of topology and geometry* may be referred to as *layout optimization*, or simply *configuration optimization*; see Sec. 1.1 for details of the classification and Sec. 1.10 for an illustrative example of geometry and topology optimization of a truss.

The history of configuration optimization of framed structures may be divided into three periods. Culmann (1875) seems to be the first to have formulated the geometry optimization of trusses (Prager 1974b) to find the fully-stressed design. However, papers by Michell (1904) and Maxwell (1890) are usually cited as the first work in the field of layout optimization of trusses. Although the study by Michell is important in view of a theoretical basis (Hemp 1973), it can be applied to limited types of structures and constraints, and the so-called Michell truss is not practically acceptable because it has an infinite number of members.

In the 1960s and 1970s, when computers were available only to a limited number of researchers and engineers, some computational approaches under stress constraints were presented and small-scale toy problems were solved. In this period, many important theoretical results for numerical methods, including optimality criteria approaches, were presented, and difficulties in topology optimization were extensively studied. The theoretical works, e.g., the explicit optimality criteria approach and the method based on so-called *grillage*, are summarized in review papers by Kirsch (1989a) and books by Rozvany (1976, 1989, 1997).

Recently, owing to the dramatic progress of computer technology, many numerical methods of topology optimization have been developed and applied to large-scale real-world structures. In this chapter, we concentrate on the numerical methods for topology optimization. Trusses are mainly considered here because rigidly jointed frames and pin-jointed trusses are modeled in similar formulations of the finite element method, and optimizing topologies of trusses is more difficult than optimizing frames because of instability at the joints.

In the widely used numerical approach to topology optimization, unnecessary members are removed from a highly connected *ground structure* while the nodal locations are fixed (Kirsch 1989a; Topping 1992); see Sec. 3.3 for details. Many methods have been presented using this ground structure approach; therefore, finding practically acceptable solutions seems to be a matter of computational capacity. However, difficulties still exist for problems with the constraints on stresses, local buckling, and eigenvalues of vibration. Development of efficient computational algorithms for optimizing large-scale trusses is also an important subject. Another difficulty in topology optimization is that the solution often turns out to be an unrealistic design due to the existence of unstable nodes, intersection of members, and the existence of extremely slender members.

Nakamura and Ohsaki (1992) investigated the characteristics of optimal topologies under eigenvalue constraints and showed that local instability and multiplicity of eigenvalues lead to serious difficulties in finding optimal topologies. Kim, Jang, and Kim (2008) optimized topology of a frame with flexible joints, where the joint stiffness is considered as a continuous design variable, while the sections of the beams are fixed. Frame optimization with flexible joints called 2-joints was also investigated for application to the framework of an automobile (Fredricson, Johansen, Klarbring, and Petersson 2003). Fredricson (2005) developed a material interpolation method, which is similar to continuum topology optimization, to optimize the flexibility of joints. Various optimization results of frames are presented in Chap. 5. The methodologies of truss topology optimization can be applied to other fields of engineering, including flow networks (Klarbring, Petersson, Torstenfelt, and Karlsson 2003; Evgrafov 2006).

3.2 Michell truss

In his pioneering paper, Michell (1904) derived a condition for the optimal layout of a framework to have the least volume of material under stress constraints against static loads (Hemp 1973). Let σ^t (> 0) and σ^c (< 0) denote the bounds of stress in tension and in compression, respectively. Using $\sigma_0 = (\sigma^t - \sigma^c)/2$ and a positive constant ε_0, the condition for optimality by Michell may be stated using the current terminologies as follows:

- A truss has minimum material volume, if

    - the equilibrium conditions are satisfied against the specified static loads,
    - the stress of each member is equal to the upper bound σ^t or the lower bound σ^c, and
    - there exists a deformation compatible with the strains that are equal to $\sigma_0 \varepsilon_0 / \sigma^t$ for members in tension and $\sigma_0 \varepsilon_0 / \sigma^c$ for those in compression.

Accordingly, the bars of the optimal pin-jointed framework are fully stressed and arranged in the directions of the principal strains of the displacement field. However, these conditions do not uniquely determine the displacement field. Therefore, the optimal layout should be searched using a trial-and-error process.

The structure obtained this way is called the *Michell structure* or the *Michell truss*. Hegemier and Prager (1969) derived explicit optimality conditions for the Michell truss and presented an approach without trial and error. They also showed that the truss with maximum stiffness is fully stressed and extended the concept of the Michell truss to optimization under frequency constraints and for creep problems. Schmidt (1962) extended Michell's theory to a statically determinate structure under multiple loading conditions. He showed that the condition of fully-stressed design is a necessary, but not sufficient, condition of optimality. Exact analytical benchmark solutions were presented by Lewiński and Rozvany (2007).

It is derived from the condition of the Michell truss that the infinitely dense members exist in the two perpendicular directions of the principal strains of the displacement field. To overcome the practical difficulty due to the existence of too many members, Prager (1974b) introduced the cost of node, or connection. Assuming that the solution is fully stressed, the cost coefficient for the cross-sectional area A_i is increased from the member length L_i by a certain specified amount b_0 representing the cost of the connections; i.e., the total structural cost V is formulated as

$$V = \sum_{i=1}^{m} A_i (L_i + b_0) \tag{3.1}$$

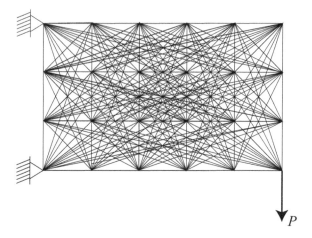

FIGURE 3.1:   A highly connected ground structure.

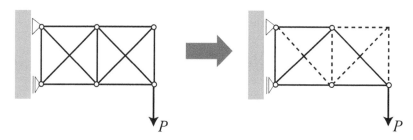

FIGURE 3.2:   Illustration of topology optimization.

where m is the number of members. In this case, the conditions for optimality are obtained as

$$L_i|\varepsilon_i| = (L_i + b_0)\varepsilon^* \quad \text{for} \quad A_i > 0, \quad (i = 1, \ldots, m) \tag{3.2}$$

where ε^* is a positive constant, and ε_i represents the strain of the ith member. Note that ε_i may represent the strain rate for a plastic limit design problem. This way, the number of members is successfully reduced to obtain a practically acceptable layout.

3.3   Topology optimization problem

Consider a problem of finding optimal locations of members, or connectivity of the nodes by members, from a set of existable nodes and members under constraints on the responses against specified static loads. The structure with all the candidate nodes and members is called the *ground structure*, which includes many nodes and members, as illustrated in Fig. 3.1; preferably a member should exist between every pair of nodes including supports.

The approach that removes unnecessary members from the ground structure by optimization is called the *ground structure approach*, and the paper by Dorn, Gomory, and Greenberg (1964) is usually cited as the first work on this approach. Fig. 3.2 illustrates the process of topology optimization of a plane truss with six nodes including two supports, where the dotted lines indicate the members to be removed after optimization. There have been extensive studies on this approach since the 1960s, e.g., Dobbs and Felton (1969); see, e.g., the review articles by Kirsch (1989a), Topping (1984), and the book by Rozvany (1997).

Let A_i and L_i denote the cross-sectional area and the length of the ith member, respectively. The number of members in the initial ground structure is denoted by m. Only inequality constraints $H_j \leq 0$ $(j = 1, \cdots, n^{\mathrm{I}})$ are considered, for simplicity, because the design requirements in various fields of engineering are generally given with inequality. If the responses represented by displacements and stresses are considered to be implicit functions of the continuous design variables $\mathbf{A} = (A_1, \ldots, A_m)^{\top}$, then H_j is conceived as a function of $\mathbf{A}$ only, as presented in Sec. 1.3. Therefore, the problem of minimizing the total structural volume $V(\mathbf{A})$ turns out to be a nonlinear programming (NLP) problem as

$$\text{Minimize} \quad V(\mathbf{A}) = \sum_{i=1}^{m} A_i L_i \tag{3.3a}$$

$$\text{subject to} \quad H_j(\mathbf{A}) \leq 0, \quad (j = 1, \cdots, n^{\mathrm{I}}) \tag{3.3b}$$

$$A_i \geq 0, \quad (i = 1, \cdots, m) \tag{3.3c}$$

where the upper bound for A_i is not given, for simplicity. The members with null cross-sectional areas are removed after optimization to obtain the optimal topology.

Although the cross-sectional areas are assumed to be continuous variables in the conventional ground structure approach, the topology optimization problem under stress constraints is virtually a combinatorial optimization problem with 0–1 variables defining the existence of members even for the case where A_i is a continuous variable, as demonstrated in Sec. 3.5. If the state variables defined by nodal displacements are regarded as independent continuous variables, the problem becomes a mixed integer linear or nonlinear programming problem (Floudas 1995).

Suppose A_i is selected from a discrete set of available sections. Then, the truss topology optimization problem for minimizing the total structural volume is formally written as

$$\text{Minimize} \quad V(\mathbf{A}) = \sum_{i=1}^{m} A_i L_i \tag{3.4a}$$

$$\text{subject to} \quad H_j(\mathbf{A}) \le 0, \quad (j = 1, \cdots, n^{\mathrm{I}}) \tag{3.4b}$$

$$A_i \in \mathcal{A}_i, \quad (i = 1, \cdots, m) \tag{3.4c}$$

where $\mathcal{A}_i$ represents the feasible set of A_i; e.g., $\mathcal{A}_i = \{0, A^0\}$ if the cross-sectional area of an existing member can take only the specified value A^0, and $\mathcal{A}_i = \{0, A_1^0, \ldots, A_r^0\}$ if A_i can have one of the pre-defined positive values $A_1^0, \ldots, A_r^0$ for an existing member and 0 for a nonexistent member.

3.4   Optimization methods

For the case in which constraints on displacements or compliance against static loads is considered, the topology optimization problem is formulated as an NLP problem based on the ground structure approach, and optimal solutions may be easily found by using an appropriate NLP algorithm or an optimality criteria approach in Sec. 1.7. Among many methods of NLP, sequential quadratic programming and the method of the moving asymptote are often used for structural optimization (see Appendix A.2.2 for details of NLP). Recently, new algorithms, including the interior-point method (Ben-tal and Nemirovski 1994; Kočvara 1997) and semidefinite programming (Ohsaki, Fujisawa, Katoh, and Kanno 1999), have been shown to be very effective for special problems under constraints on eigenvalues and/or compliance.

In the process of removing unnecessary members, the truss may become unstable due to the existence of a node without any connecting member or the existence of a hinge that connects only two colinear members. A lower bound A_i^{L} is therefore given as follows to prevent instability:

$$A_i \ge A_i^{\mathrm{L}}, \quad (i = 1, \cdots, m) \tag{3.5}$$

Note that A_i^{L} has a very small positive value, and the member with $A_i = A_i^{\mathrm{L}}$ should be removed from the optimal solution; i.e., A_i^{L} is introduced to avoid numerical instability in the optimization process, and not to avoid the unstable optimal solution. Petersson (2001) showed that the optimal solution and the corresponding displacements for the limit $A_i^{\mathrm{L}} \to 0$ converge to the values for $A_i^{\mathrm{L}} = 0$ if the strain energy of the solution with $A_i^{\mathrm{L}} = 0$ is bounded. Svanberg and Werme (2009) discussed the validity of assigning small lower bounds for a problem with 0–1 variables.

The possibility of obtaining an unstable solution may be reduced by considering multiple loading conditions, as demonstrated in Sec. 1.5. The method based on boundary cycle to restrict the feasible solution to the truss consisting of stable units only may also be useful (Kaveh 1986; Nakanishi and Nakagiri 1996a, 1996b). Including constraints on the shape and topology of the triangular unit is very effective to obtain practically acceptable designs (Nakanishi and Nakagiri 1997; Ohsaki and Kato 1999; Kawamura, Ohmori, and Koto 2002).

Ohsaki and Katoh (2005) assigned a lower bound for the number of members connected to an existing node to ensure stability (see Sec. 3.5.6 for details). Ohsaki and Watada (2008) and Watada and Ohsaki (2009b, 2009c) presented a mixed integer programming approach to select the layout of each unit of a regular truss from the predefined standard layouts; see Sec. 3.6 for details. The stable topologies of bar-joint structures are extensively investigated in the field of structural rigidity in applied mathematics utilizing graph theoretical approaches (Kaveh 1991; Graver, Servatius, and Servatius 1993; Kaveh 2004; Avis, Katoh, Ohsaki, Streinu, and Tanigawa 2007).

Although an optimal solution is found by simply applying an NLP algorithm, it is very difficult to find the *global* optimal solution even for a simple problem with stress and/or displacement constraints, as demonstrated in Sec. 3.5. Minimization of the compliance under a single loading condition without an upper bound for the cross-sectional area is considered as a min-max problem for minimizing the compliance with respect to the design variable, simultaneously maximizing the total potential energy with respect to the displacements (Ben-Tal and Bendsøe 1993; Bendsøe, Ben-Tal, and Zowe 1994). A strictly convex dual problem is then formulated and is efficiently solved by a primal-dual approach (Beckers and Fleury 1997) (see Sec. 1.8.3 for several reformulations of the truss optimization problem under compliance constraints).

Heuristic approaches, e.g., genetic algorithm (GA), can also be applied to truss topology optimization problems (Jenkins 1991; Ohsaki 1995); see Sec. 3.7. Recently, some evolutionary approaches have been presented (Dasgupta and Michalewicz 1997; Hajela 1997; Kwan 1998; Yang, Xie, Steven, and Querin 1999a; Lagaros, Papadrakakis, and Kokossalakis 2002; Greiner, Emperador, and Winter 2004; Ebenau, Rottschäfer, and Thierauf 2005; Kicinger, Arciszewski, and de Jong 2005). The term *evolutionary* is very confusing because it is used with several different meanings. A GA with evolving parameters may be called evolutionary, or even a simple GA may be an evolutionary approach (Kwan 1998). The algorithm with local rules that is similar to the optimality criteria approaches and many variants of growing processes are also *evolutionary*.

Simulated annealing (SA) is also useful for topology optimization (Topping, Khan, and Leite 1996; Tagawa and Ohsaki 1999). The advantage of using SA is that both continuous and discrete variables can be included without any difficulty. Tabu search may also be very effective for truss topology optimization (Bennage and Dhingra 1995). Luh and Lin (2008) used the *ant system* for truss topology optimization. Because the computational cost of analysis

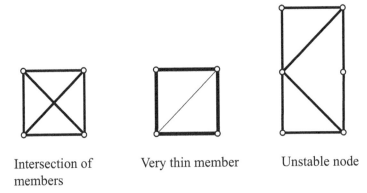

Intersection of members   Very thin member   Unstable node

FIGURE 3.3: Infeasible topologies of trusses.

is very large for these heuristic methods, an efficient method of reanalysis presented in Sec. 3.8.2 may be utilized for reducing the computational cost (Kirsch 1993, 1995; Kirsch and Liu 1995).

The ground structure approach is regarded as the standard procedure for topology optimization, although there are many difficulties, which are summarized as follows:

1. Too many members and nodes are needed in the initial ground structure, because only removal of members is possible, and addition of members and nodes is very difficult without resort to a heuristic approach.

2. The optimal topology strongly depends on the initial design, and an infinite number of nodes and members is needed if the nodal locations are also to be optimized.

3. Unrealistic optimal solutions with very long members, intersection and/or overlap of members, etc., are often obtained.

4. The truss becomes unstable due to the existence of a node connected by two colinear members only, if too many members are removed.

The unrealistic or infeasible topologies are illustrated in Fig. 3.3, including intersecting members, very thin members, and unstable nodes. Note that intersection of members is sometimes allowed, e.g., for a truss or a frame with moderately thin braces; see Sec 5.1.4.

The nodal locations can also be considered as variables in order to optimize the geometry of a truss (Ohsaki and Nishiwaki 2005). This process is sometimes referred to as the *extended ground structure approach*.

3.5   Stress constraints

3.5.1   Introduction

One of the main difficulties in topology optimization under stress constraints is that the constraints need not be satisfied by the member with vanishing cross-sectional area; i.e., the constraint obviously does not exist for a member that does not exist, as noted by Sved and Ginos (1968), Sheu and Schmit (1972), Nagtegaal (1973), and Kirsch (1989a, 1990). Therefore, the problem becomes discrete in the formulation of constraints with a 0–1 variable indicating existence/nonexistence of a member, and it is very difficult to find the *global* optimal solution even for a simple truss topology optimization problem with a few members. The constraint of this type is called the *design dependent constraint*. If we consider a process of continuously decreasing the cross-sectional area of a member from a finite positive value to 0, the constraint suddenly disappears at the final state with null cross-sectional area. As a result, the feasible region is not convex in general; hence, the optimal solution is often located at a cusp or an isolated point of the feasible region, as illustrated in Sec. 3.5.3.

The optimal topology under stress constraints considering a single loading condition can be derived as a solution of a linear programming (LP) problem with member forces as design variables (Dorn, Gomory, and Greenberg 1964; Ringertz 1985). The solution is globally optimal if the optimal truss is statically determinate (Sved 1954). For the case in which the optimal topology is unstable due to the existence of colinear members connected by a hinge, as shown in Fig. 3.3, some members should be added or the unstable joints should be fixed.

To overcome the difficulty due to discontinuity in the problem formulation, several relaxation approaches as well as branch-and-bound-type iterative methods have been presented. Ringertz (1985) presented a method for problems with stress and displacement constraints, where a compatible set of strains and displacements is first calculated for specified cross-sectional areas. An NLP problem is then solved under stress constraints while fixing the displacements. Ringertz (1986) proposed an approach to obtaining the lower-bound solution by solving an NLP problem under displacement constraints only. The stress constraints are successively given for members that violate the constraints to obtain an upper-bound solution. The efficiency of the solution may be evaluated through comparison of the objective value with a lower-bound value that can be found by solving an LP problem neglecting the compatibility conditions (Kirsch 1989b).

A relaxation method has been presented by Cheng (1995) and Cheng and Guo (1997) for obtaining a good approximate solution. In their approach, called *ε-relaxed approach*, the stress constraint is relaxed for a member with a small cross-sectional area. It is very difficult, however, to determine an appropriate value of the parameter for relaxing the constraints and to assign the initial solution that can reach the global optimal solution, as noted by Stolpe and Svanberg (2001b). The convergence properties of this approach were discussed by Bruggi (2008). Guo and Cheng (2000) extended their method to incorporate an extrapolation approach (Nakamura and Ohsaki 1992; Ohsaki and Nakamura 1996).

Duysinx and Sigmund (1998) presented a relaxation approach to a continuum topology optimization problem. Evgrafov and Patriksson (2003) presented a stochastic relaxation approach to stress constrained problems. Stolpe and Svanberg (2003b) showed that a stress constrained problem can be solved by a traditional NLP approach. Sui, Du, and Guo (2006) presented a method for topology optimization of frames using different filter functions for weight, stiffness, and stress bound of the frame element.

3.5.2   Governing equations

Consider an elastic truss subjected to multiple static loading conditions. The problem is to obtain an optimal topology and the optimal member cross-sectional areas that minimize the total structural volume under constraints on stresses of members, where the conventional ground structure approach is used.

Let $\mathbf{P}^k$ denote the vector of the kth set of nodal loads. The vector of axial forces corresponding to $\mathbf{P}^k$ is denoted by $\mathbf{N}^k = (N_1^k, \ldots, N_m^k)^\top$, where m is the number of members in the ground structure. The equilibrium equations are given as

$$\mathbf{D}\mathbf{N}^k = \mathbf{P}^k, \quad (k = 1, \ldots, n^{\mathrm{P}}) \tag{3.6}$$

where n^{P} is the number of loading conditions, and the $n \times m$ matrix $\mathbf{D}$ is called the equilibrium matrix with n being the number of degrees of freedom.

Let $\mathbf{U}^k$ denote the vector of nodal displacements against $\mathbf{P}^k$. The elongation of the ith member corresponding to $\mathbf{U}^k$ is denoted by δ_i^k. The compatibility condition between $\mathbf{U}^k$ and δ_i^k is written as

$$\mathbf{d}_i^\top \mathbf{U}^k = \delta_i^k, \quad (i = 1, \ldots, m; \ k = 1, \ldots, n^{\mathrm{P}}) \tag{3.7}$$

where $\mathbf{d}_i$ is the ith column of $\mathbf{D}$. Let ε_i^k and σ_i^k denote the strain and stress, respectively, of the ith member for the kth loading condition. Then, the strain, stress, and axial force of the ith member against the kth loading condition are obtained from $\mathbf{U}^k$ as

$$\varepsilon_i^k = \frac{\delta_i^k}{L_i}, \quad \sigma_i^k = E\varepsilon_i^k, \quad N_i^k = A_i\sigma_i^k \tag{3.8}$$

where A_i and L_i are the cross-sectional area and length of the ith member, respectively, and E is the elastic modulus. Eqs. (3.6)–(3.8) are combined into

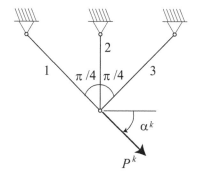

FIGURE 3.4: A three-bar truss subjected to multiple loading conditions.

the following stiffness equation:

$$\mathbf{K}\mathbf{U}^k = \mathbf{P}^k, \quad (k = 1, \ldots, n^{\mathrm{P}}) \tag{3.9}$$

where $\mathbf{K}$ is the stiffness matrix.

3.5.3 Discontinuity in stress constraint

The difficulties in topology optimization under stress constraints considering multiple loading conditions were first discussed by Sved and Ginos (1968) using a simple three-bar truss as follows:

Example 3.1
Consider a three-bar truss as shown in Fig. 3.4. The three loading conditions are defined as $(P^k, \alpha^k) = (40, \pi/4), (30, \pi/2), (20, 3\pi/4)$, where P^k and α^k denote the magnitude and direction of the kth load, respectively, as defined in Fig. 3.4. The lower and upper bounds $(\sigma_i^{\mathrm{L}}, \sigma_i^{\mathrm{U}})$ for the stress are $(-5, 5)$, $(-20, 20)$, and $(-5, 5)$, for members 1, 2, and 3, respectively. Suppose the length of member 2 is equal to 1, for simplicity. The objective function is the total structural volume that is given as

$$V = A_2 + \sqrt{2}(A_1 + A_3) \tag{3.10}$$

The optimal solution that minimizes V under stress constraints for three loading conditions is $A_1 = 8.0$, $A_2 = 1.5$, and $A_3 = 0$, where $V = 12.812$. The stresses of three members are $(5, 0, -5)$, $(0, 20, 20)$, and $(-2.5, 18.856, 21.356)$ for the three loading conditions. Note that the stress of member 3 for P^3 is 21.356, which is greater than the upper bound 5. The stress constraint, however, need not be satisfied, because $A_3 = 0$ and member 3 does not exist. Note that the stress of member 3 can be computed from the nodal displacements as (3.7) and (3.8).

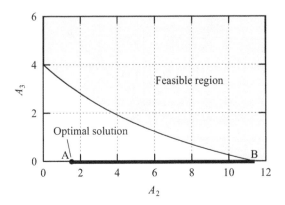

FIGURE 3.5:   Feasible region of the three-bar truss.

If we assign the stress constraint even for the nonexistent member, the optimal solution is found to be $A_1 = 7.099$, $A_2 = 1.849$, and $A_3 = 2.897$, where $V = 15.986$, which is greater than the objective value 12.812 of the global optimal solution. This result suggests that the optimal solution exists at a singular point 'A' in the feasible region, which is plotted in Fig. 3.5 in the A_2–A_3 plane for $A_1 = 8.0$. Note that the feasible region includes the line AB in addition to the shaded region.

Incorporating the discontinuity into the constraints, the topology optimization problem with stress constraints is formulated as

$$\text{Minimize} \quad V = \sum_{i=1}^{m} A_i L_i \tag{3.11a}$$

$$\text{subject to} \quad \sigma_i^{\mathrm{L}} \le \sigma_i^k \le \sigma_i^{\mathrm{U}} \text{ for } A_i > 0,$$

$$(i = 1, \dots, m; \ k = 1, \dots, n^{\mathrm{P}}) \tag{3.11b}$$

$$A_i \ge 0, \quad (i = 1, \dots, m) \tag{3.11c}$$

The stress may be calculated from axial force N_i^k as $\sigma_i^k = N_i^k / A_i$; however, this relation cannot be used for a removed member because both N_i^k and A_i vanish. On the other hand, the strain ε_i^k can be calculated from the elongation δ_i^k, which is easily found from the displacements of the nodes connected by the removed member, if they exist, and the stress is calculated using the relations (3.8) (Cheng and Jiang 1992) as we have done in Example 3.1. Therefore, there is no discontinuity in σ_i^k itself at $A_i = 0$ if the truss after removal of the ith member is stable, and if the stress is obtained from the strain, not from the axial force; i.e., only the definition of the stress constraints (3.11b) is discontinuous. A constraint of this type is said to be a *design dependent constraint* or a *vanishing constraint* (Rozvany 2001; Izmailov and

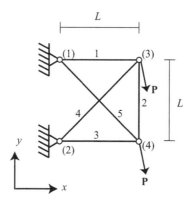

FIGURE 3.6: A five-bar truss.

Solodov 2008). The optimal topology with violating the stress constraint in the vanishing members is called *singular optimal topology*.

Constraint (3.11b) suggests that the stress constraints should be relaxed in the vicinity of $A_i = 0$ if the optimal topologies are to be found by the ground structure approach. Cheng and Guo (1997) presented the *ε-relaxed approach* that reformulates the stress constraint as

$$(\sigma_i^{\mathrm{L}} - \sigma_i^k)A_i \le e \tag{3.12a}$$
$$(\sigma_i^k - \sigma_i^{\mathrm{U}})A_i \le e \tag{3.12b}$$
$$A_i \ge e^2 \tag{3.12c}$$

where the parameter e has a sufficiently small positive value. The optimal topology may be found by successively reducing e to 0.

Example 3.2

As a small example, consider a five-bar truss, as shown in Fig. 3.6, that was solved by Cheng and Guo (1997). The units for force and length are omitted for brevity. Two loading conditions are considered, where the loads $\mathbf{P} = (P_x, P_y)^{\top} = (5, -50)^{\top}$ are applied at nodes 3 and 4, respectively. The parameters are $E = 1.0$ and $e = 0.0001$. The upper and lower bounds for stresses are ± 20 for member 2 and ± 5 for the remaining members. The optimization software package IDESIGN Ver. 3.5 (Arora and Tseng 1987), which utilizes sequential quadratic programming, is used.

If the initial design is selected as $(A_1, \ldots, A_5) = (5.0, 8.0, 15.0, 1.0, 20.0)$, then the global optimal topology is successfully found by the ε-relaxed approach with $(A_1, \ldots, A_5) = (1.0000, 2.5000, 10.0000, 0.0, 14.1421)$ and $V = 35.520$. Note that the truss consists of four members, and the stress of the nonexistent member 4 for the load at node 3 is -15.0, which violates the constraint.

However, if the initial solution is given as $A_i = 10.0$ for all the members, a non-optimal solution $(A_1, \ldots, A_5) = (7.2194, 4.7319, 4.5318, 8.7956, 7.8310)$ with $V = 39.986$ is reached. In order to investigate the convergence property, optimal solutions are found starting with cross-sectional areas with uniform random distribution between 0 and 20.0. Among the ten trials, the global optimal solution was found twice. The numbers of trials that lead to the non-optimal solution with five bars, four bars without members 3 and 5, respectively, are 4, 1, and 3.

To prevent convergence to a local optimal solution, it is natural to use a continuation method, where the parameter e in (3.12) is successively reduced to 0, while the optimal solution for a certain value of e is chosen as the initial solution for optimization corresponding to a slightly reduced value of e. However, even using this approach, convergence to the global optimal solution is not guaranteed (Petersson 2001; Stolpe and Svanberg 2001b), because the trajectory of the solution with respect to the parameters e may be discontinuous near the global optimal solution.

3.5.4   Discontinuity due to member buckling

For pin-jointed trusses, the lower-bound stress σ_i^{L} (< 0) should be defined by the Euler buckling stress if local buckling is considered (Cheng and Guo 1997). It is known that the optimal topology strongly depends on the values of stress bounds in tension and compression especially for bridge-type structures (Achtziger 1996; Oberndorfer, Achtziger, and Hörnlein 1996).

As a simple example, consider a four-bar truss, as shown in Fig. 3.7, subjected to a compressive load to the horizontal members 1 and 2. The lengths of members 1 and 2 are L, and the radius of gyration is r irrespective of the cross-sectional area; i.e., we assume a sandwich section, a rectangular section with constant height, or a pipe section with an appropriate relation between the radius and thickness. If we assume $P > 0$, members 1 and 2 are in a compressive state, and the lower-bound stress is defined with the Euler buckling stress $-\pi^2 E/(r/L)^2$, of which the absolute value is smaller than the yield stress if the slenderness ratio is larger than the critical slenderness ratio. This dependence of the lower-bound stress on the member length leads to more difficulties in topology optimization of pin-jointed trusses. In the practical design process, the lower-bound stress should be defined by dividing the Euler buckling stress by a safety factor, which is not considered here for simplicity.

Since the stresses of vertical members 3 and 4 are 0 for the truss in Fig. 3.7, these members are removed as the result of optimization. Hence, the optimal solution has two colinear members 1 and 2 connected by pin joint 2, and the optimal truss is unstable. The unstable truss can be stabilized by fixing pin joint 2. Consequently, a long member with length $2L$ emerges, and the

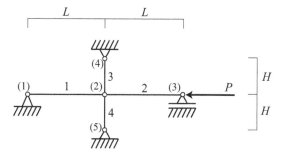

FIGURE 3.7: A four-bar truss under a compressive load.

lower-bound stress should be replaced by $-\pi^2 E/(r/2L)^2$, which is 1/4 of the initially assigned value. Therefore, the stress constraint will be violated by the optimal solution that satisfies the stress constraints with equality in the short two members 1 and 2.

One possible strategy to avoid the existence of unstable optimal topology may be to include constraints on global buckling or imperfection of nodal location in the problem formulation (Zhou 1996) (see Sec. 3.9.4 for topology optimization under linear buckling constraints). In this case, unrealistically very slender members exist in the optimal topology, as pointed out by Nakamura and Ohsaki (1992) for a problem with frequency constraints; see Sec. 3.9.2. For the four-bar truss in Fig. 3.7, members 3 and 4 will have very small cross-sectional areas to add vertical stiffness at node 2. The solution with thin bracing members obtained this way usually has a smaller objective value than the solution with a long member after fixing the unstable nodes and increasing the cross-sectional areas to satisfy the local buckling constraints.

However, in some cases the solution with thin bracing members has a larger objective value than that with long members in compression (Rozvany 1996). For example, suppose $H \gg L$ in Fig. 3.7, although this is a very unrealistic situation. Let $\tilde{A}_1$ and $\tilde{A}_2$ denote the optimal cross-sectional areas before fixing node 2; i.e., $\tilde{A}_1 = \tilde{A}_2 = \tilde{A} = -P/\sigma^{\mathrm{L}}$, where $\sigma^{\mathrm{L}} = -\pi^2 E/(r/L)^2$ and $\tilde{A}_3 = \tilde{A}_4 = 0$. After fixing node 2, the cross-sectional areas of members 1 and 2 should be increased to $4\tilde{A}$ to satisfy the Euler buckling constraint; hence the total structural volume increases to $\tilde{V} = 8L\tilde{A}$. If the global buckling constraint is introduced in the optimization problem of the four-bar truss with an appropriately small lower-bound linear buckling load, members 3 and 4 have a small cross-sectional area denoted by A^{L}, and the cross-sectional areas of members 1 and 2 are $\tilde{A}$ at the optimal solution. Then the total structural volume is $2L\tilde{A} + 2HA^{\mathrm{L}}$, which is larger than $\tilde{V}$ if $H > 3L\tilde{A}/A^{\mathrm{L}}$. The value of A^{L} may be given by incorporating the buckling constraint also for the bracing members, as discussed by Takagi and Ohsaki (2004) for the design of column-type frames with lateral braces. Achtziger (1999a, 1999b) defined *chain* as a

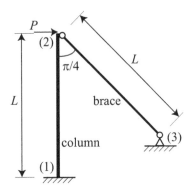

FIGURE 3.8: Cantilever column supported by a brace.

set of colinear members without connecting members at the unstable nodes, and assigned the Euler buckling constraints using the buckling length of the *active chain*. He also discussed the possibility of obtaining a stable truss by introducing constraints on the slenderness ratio.

Guo, Cheng, and Yamazaki (2001) used a quadratic extended interior penalty function (Haftka and Starnes 1976) for approximating the local buckling constraints of members with small cross-sectional areas, where the section is assumed to be a circular solid. Evgrafov (2005) noted, for the problem of minimizing the total structural volume under a linear buckling constraint, that the solution with a small lower-bound cross-sectional area has far larger objective value than that with the vanishing lower bound, because the existence of the nodes connected by slender members only may have local (nodal) buckling under very small loads.

Note that local instability can be avoided if rigidly jointed frame elements are used. Ohsaki (1997b) used frame elements for configuration optimization; see Chap. 4. A semi-rigid connection can also be used for preventing local instability (Ramrakhyami, Frecker, and Lesieutre 2009). Instability of the optimal truss is closely related to nonuniqueness of the displacement (Kočvara and Outrata 2006), which may lead to the nonuniqueness of optimal cross-sectional areas. Therefore, a unique solution can be found by minimizing a strictly convex function with respect to the cross-sectional areas and displacements among the set of nonunique solutions. For example, a *physical* displacement can be obtained by minimizing the norm of transverse displacements of the members related to Green's strain. Uniqueness of the optimal solutions of building frames is discussed in Sec. 5.2.

Example 3.3
Discontinuity due to member buckling also exists in the topology optimization problem of a braced frame. Consider, for example, a cantilever column, as shown in Fig. 3.8, subjected to a lateral load P and supported by a brace

modeled by a truss member (Hagishita and Ohsaki 2007). Let A_c and I_c denote the cross-sectional area and the second moment of inertia of the column, respectively. The cross-sectional area of the brace is denoted by A_b. The column and brace have the same length L. The stress σ_b of the brace is derived as

$$\sigma_b = -\frac{\sqrt{2}PL^3 A_c}{6I_c A_c + 3I_c A_b + A_b A_c L^2} \tag{3.13}$$

First, we assume that the lower-bound stress of the brace is independent of A_b. Then, from (3.13), the value $\widetilde{A}_b$ of A_b at which σ_b is equal to the lower bound σ^L (< 0) is derived as

$$\widetilde{A}_b = -\frac{A_c(\sqrt{2}PL^3 A_c + 6\sigma^L I_c)}{\sigma^L(3I_c + A_c L^2)} \tag{3.14}$$

Consider a problem of minimizing the structural volume of the brace under constraint $\sigma_b \geq \sigma^L$ of the stress of the brace, while fixing the cross-section of the column; i.e., the cross-sectional area of the brace is simply minimized in this small example. It is seen from (3.14) that $\widetilde{A}_b$ cannot have a positive value, i.e., the stress constraint is satisfied for $A_b \geq 0$ if I_c has a sufficiently large value so that $\sqrt{2}PL^3 A_c + 6\sigma^L I_c < 0$ is satisfied with $\sigma^L < 0$. Therefore, the brace is not needed if the column is sufficiently stiff. In contrast, if $\sqrt{2}PL^3 A_c + 6\sigma^L I_c > 0$, then there exists a positive value $\widetilde{A}_b$ satisfying (3.14), and the stress constraint is violated for $0 < A_b < \widetilde{A}_b$. However, no stress constraint is needed for the brace if $A_b = 0$. Therefore, the feasible region is discontinuous, and the optimal solution $A_b = 0$ exists at an isolated point of the feasible region.

Next, we consider the member buckling constraint and assume that the Euler buckling stress is proportional to A_b, which holds, for example, for a solid rectangular section with variable height and constant width. In this case, σ^L is defined to be proportional to A_b with a positive coefficient α as $\sigma^L = -\alpha A_b$, and (3.13) with $\sigma_b = \sigma^L$ reduces to

$$\alpha(3I_c + A_c L^2)A_b^2 + 6\alpha I_c A_c A_b - \sqrt{2}PL^3 A_c = 0 \tag{3.15}$$

Therefore, there always exists a positive value A_b satisfying the stress constraint with equality; hence, the feasible region is disjoint and the optimal solution exists at the isolated point $A_b = 0$.

3.5.5 Mathematical programming approach

3.5.5.1 Reformulation of the problem

As we have seen in Example 3.3, a singular optimal topology with violated stress constraints by vanishing members cannot be generally found by solving the original NLP problem (3.11) using a gradient-based NLP algorithm.

Therefore, in order to resolve the discontinuity of the problem, several methods have been presented for reformulation of the problem. Equivalence among the problems under constraints on compliance, stress, and plastic limit load has been extensively studied since the 1970s (Hemp 1973). A problem with a single loading condition can be written in terms of the member forces and the cross-sectional areas as (Kirsch 1989b; Muralidhar and Rao 1997; Stolpe and Svanberg 2003b)

$$\text{Minimize} \quad V(\mathbf{A}) = \sum_{i=1}^{m} A_i L_i \tag{3.16a}$$

$$\text{subject to} \quad \mathbf{DN} = \mathbf{P} \tag{3.16b}$$

$$A_i \sigma_i^{\mathrm{L}} \leq N_i \leq A_i \sigma_i^{\mathrm{U}}, \quad (i = 1, \ldots, m) \tag{3.16c}$$

$$A_i \geq 0, \quad (i = 1, \ldots, m) \tag{3.16d}$$

which is an LP problem with variables $\mathbf{A}$ and $\mathbf{N}$, and the superscript indicating load number is omitted. The redundant forces are often used as independent variables in Problem (3.16) of a statically indeterminate truss to reduce the number of variables using the equilibrium conditions (3.16b), and to investigate the properties of the optimal solutions. Sedaghati and Esmailzadeh (2003) used singular value decomposition (see Appendix A.1.3) for extracting the independent force components. Note that this approach is widely used for tension structures (Pellegrino 1993; Ohsaki, Zhang, and Ohishi 2008). However, we use the force vector of all members for a simple presentation of the formulations.

Because the existence of displacements compatible to $\mathbf{A}$ and $\mathbf{N}$ is not assured in Problem (3.16), the solution to this problem gives a lower bound for the objective value of the original NLP problem with compatibility conditions. However, there always exists a displacement vector compatible to $\mathbf{A}$ and $\mathbf{N}$ for the solution of Problem (3.16), because the solution of the LP problem exists at a vertex of the the feasible region that represents a statically determinate truss (Sheu and Schmit 1972). Hence, the optimal solution of the original NLP problem with compatibility conditions can be found by solving Problem (3.16).

For a degenerate case in which the solution of Problem (3.16) is statically indeterminate, i.e., the solution of the LP problem exists on a hyperplane defining the boundary of the feasible region, there exists a statically determinate truss at a vertex of the feasible region that has the same objective value as the degenerate optimal solution. Prager (1976) and Save (1983) investigated the characteristics of optimal solutions of a three-bar statically indeterminate truss.

It is seen from (3.16c) that the stress constraint is obviously satisfied by a vanishing member with $A_i = 0$ and $N_i = 0$. This way, the discontinuity in stress constraints can be avoided for a problem with a single loading condition by simply assigning constraints on member forces instead of stresses.

Note that the solution by Stolpe and Svanberg (2003b) satisfies all the stress constraints even for the vanishing members.

We introduce the vectors of slack variables $\mathbf{N}^+ = (N_1^+, \ldots, N_i^+)^\top$ and $\mathbf{N}^- = (N_1^-, \ldots, N_i^-)^\top$ for $\mathbf{N}$ as

$$\mathbf{N} = \mathbf{N}^+ - \mathbf{N}^-, \quad \mathbf{N}^+ \geq \mathbf{0}, \quad \mathbf{N}^- \geq \mathbf{0} \tag{3.17}$$

For the case with $\sigma_i^{\mathrm{L}} = -\sigma_i^{\mathrm{U}}$, Problem (3.16) can be reformulated as an LP problem, as follows, with respect to the member forces only (Achtziger, Bendsøe, Ben-Tal, and Zowe 1992):

$$\text{Minimize} \quad V(\mathbf{N}^+, \mathbf{N}^-) = \sum_{i=1}^{m} \frac{1}{\sigma_i^{\mathrm{U}}} (N_i^+ + N_i^-) L_i \tag{3.18a}$$

$$\text{subject to} \quad \mathbf{D}(\mathbf{N}^+ - \mathbf{N}^-) = \mathbf{P} \tag{3.18b}$$

$$N_i^+ \geq 0, \quad N_i^- \geq 0, \quad (i = 1, \ldots, m) \tag{3.18c}$$

By minimizing $V(\mathbf{N}^+, \mathbf{N}^-)$, one of the inequalities in (3.18c) is satisfied with equality in each member. Furthermore, the optimal truss is statically determinate, and $V(\mathbf{N}^+, \mathbf{N}^-)$ turns out to be the total structural volume with

$$A_i = \frac{1}{\sigma_i^{\mathrm{U}}} (N_i^+ + N_i^-) \tag{3.19}$$

The general problem including multiple loading conditions can be reformulated as a mathematical program with complementarity constraints (MPCC) (Izmailov and Solodov 2008). The stress constraints can be converted to a global constraint using a p-norm or Kreisselmeier–Steinhauser (KS) measure (Kreisselmeier and Steinhauser 1983; Wrenn 1998; Sobieszczanski-Sobieski 1992; Qiu and Li 2010). The p-norm σ^{P} of the stress is defined as

$$\sigma^{\mathrm{P}} = \left(\sum_{i=1}^{m} \sigma_i^p \right)^{\frac{1}{p}} \tag{3.20}$$

Note that σ^{P} gives an upper bound for the maximum value of σ_i and converges to the maximum value as p is increased to ∞. However, an appropriately large value has to be given for p to ensure smoothness of the function.

This approach has been extensively studied in the field of continuum shape and topology optimization (Allaire, Jouve, and Maliot 2004), where the density of the material of each finite element is chosen as a design variable. The so-called *solid isotropic microstructure with penalty* or *solid isotropic material with penalization* (SIMP) approach is used (Bendsøe 1989; Rozvany, Zhou, and Birker 1992). The nonexistence of the constraint for the vanishing material density of a plate element can be incorporated by rewriting the constraint using the p-mean in conjunction with the ε-relaxed approach as (Duysinx and

Bendsøe 1998; Bruggi 2008)

$$\left[\frac{1}{m} \sum_{i=1}^{m} \left(\max \left\{ 0, \frac{\sigma_i}{\rho_i^\eta \sigma^U} + e - \frac{e}{\rho_i} \right\} \right)^p \right]^{\frac{1}{p}} \leq 1 \qquad (3.21)$$

where ρ_i is the density of the ith element, m is the number of elements, σ^U is the upper bound for the von Mises stress σ_i, and η is a penalization parameter for intermediate density.

3.5.5.2   Branch-and-bound method

For general problems with stress and displacement constraints under multiple loading conditions, reformulation of the problem to an LP problem is not possible. Therefore, a global optimization algorithm, e.g., the branch-and-bound method, should be used for obtaining the optimal topology.

Consider the following NLP problem under n^P loading conditions:

$$\text{Minimize} \quad \sum_{i=1}^{m} A_i L_i \qquad (3.22a)$$

$$\text{subject to} \quad \mathbf{K}(\mathbf{A}) \mathbf{U}^k = \mathbf{P}^k, \quad (k = 1, \ldots, n^P) \qquad (3.22b)$$

$$\sigma_i^L \leq \sigma_i^k \leq \sigma_i^U, \quad (i = 1, \ldots, m; \; k = 1, \ldots, n^P) \qquad (3.22c)$$

$$U_i^L \leq U_i^k \leq U_i^U, \quad (i = 1, \ldots, n; \; k = 1, \ldots, n^P) \qquad (3.22d)$$

$$A_i \geq 0, \quad (i = 1, \ldots, m) \qquad (3.22e)$$

where U_i^U and U_i^L are the upper and lower bounds for U_i.

Ringertz (1986) proposed the following algorithm based on successive solution of the relaxed problem without stress constraints:

Step 1: Assign a sufficiently large value to the current optimal objective value V^*.

Step 2: Solve Problem (3.22) without stress constraints (3.22c) using an NLP algorithm. Stop if there is no violated stress constraint.

Step 3: Solve Problem (3.22) with stress constraints for all members.

Step 4: Generate a set of candidate trusses by removing each member with an active stress constraint from the optimal truss obtained in Step 3, and add them to the candidate list.

Step 5: Stop if the candidate list is empty; otherwise select the last candidate in the list, and remove it from the list.

Step 6: Go to Step 5 if equilibrium equations (3.22b) are not satisfied, i.e., if the truss is unstable.

Step 7: Solve Problem (3.22) for the candidate truss without stress constraints to obtain the optimal objective value V^{d}. If there is no violated stress constraint, then let $V^* = \min\{V^*, V^{\mathrm{d}}\}$, and go to Step 5. If $V^{\mathrm{d}} > V^*$, then go to Step 5.

Step 8: Solve Problem (3.22) with stress constraints to obtain the optimal objective value V^{s}. Let $V^* = \min\{V^*, V^{\mathrm{s}}\}$. Add new candidates to the list by removing each member with an active stress constraint from the optimal truss, and go to Step 5.

Since this algorithm is rather heuristic, obtaining the global optimal solution is not guaranteed. Furthermore, because Problem (3.22) is a nonconvex problem, the solution obtained using a gradient-based NLP algorithm at Steps 3, 7, and 8 may be a local optimal solution.

It is important for rapid termination of the branches of a branch-and-bound algorithm that strict valid constraints that are satisfied by the optimal solution are assigned to restrict the space of feasible solutions. Stolpe (2004) presented a branch-and-bound algorithm for the problem with stress, displacement, and local buckling constraints, where a formulation of simultaneous analysis and design (SAND), as shown in Sec. 1.13, is used, and the variables are the cross-sectional areas, member forces, stresses, and displacements, while the lower and upper bounds A_i^{L} and A_i^{U}, respectively, are given for A_i. Note that $A_i^{\mathrm{L}} = 0$ for a topology optimization problem. He introduced valid constraints on the structural volume as well as the compliance that is formulated using a matrix inequality. The quadratic equality constraint

$$N_i^k = A_i \sigma_i^k \tag{3.23}$$

of the ith member for the kth loading condition is relaxed to the following four linear inequalities to formulate the convex subproblems:

$$
\begin{aligned}
N_i^k &\geq A_i^{\mathrm{L}} \sigma_i^k + A_i \sigma_i^{\mathrm{L}} - A_i^{\mathrm{L}} \sigma_i^{\mathrm{L}}, \\
N_i^k &\geq A_i^{\mathrm{U}} \sigma_i^k + A_i \sigma_i^{\mathrm{U}} - A_i^{\mathrm{U}} \sigma_i^{\mathrm{U}}, \\
N_i^k &\leq A_i^{\mathrm{L}} \sigma_i^k + A_i \sigma_i^{\mathrm{U}} - A_i^{\mathrm{L}} \sigma_i^{\mathrm{U}}, \\
N_i^k &\geq A_i^{\mathrm{U}} \sigma_i^k + A_i \sigma_i^{\mathrm{L}} - A_i^{\mathrm{U}} \sigma_i^{\mathrm{L}}
\end{aligned}
\tag{3.24}
$$

For example, suppose $\sigma_i^{\mathrm{L}} = -1$, $\sigma_i^{\mathrm{U}} = 1$, $A_i^{\mathrm{L}} = 1$, and $A_i^{\mathrm{U}} = 2$. If $A_i = A_i^{\mathrm{L}} = 1$, the relation $N_i^k = A_i \sigma_i^k = \sigma_i^k$ is obtained from (3.24). This strict relation is also satisfied for $A_i = A_i^{\mathrm{U}} = 2$. However, if $A_i = 1.5$, the inequalities $1.5\sigma_i^k - 0.5 \leq N_i^k \leq 1.5\sigma_i^k + 0.5$ are derived. Therefore, (3.24) defines an approximate relation between force and stress.

3.5.6   Problem with stress and local constraints

3.5.6.1   Problem formulation

In this section, the topology optimization problem under stress constraints is first formulated as a mixed integer nonlinear programming (MINLP) problem (Kravanja, Kravanja, and Bedenik 1998). The local constraints on nodal instability and intersection of members are incorporated, and a moderately large lower bound is given for the cross-sectional area of an existing member (Ohsaki and Katoh 2005).

Let $y_i \in \{0, 1\}$ denote a variable indicating by $y_i = 1$ and $y_i = 0$, respectively, the existence and nonexistence of the ith member of the ground structure with many nodes and members. Stress constraints should be assigned only for members with $y_i = 1$. One of the drawbacks of the ground structure approach is that there often exist very thin members in the optimal topology, as discussed in Sec. 3.5.3, which can be prevented by assigning a moderately large lower bound A_i^L for the cross-sectional area of an existing member as

$$A_i^L y_i \leq A_i \leq A_i^U y_i, \quad (i = 1, \ldots, m) \tag{3.25}$$

where A_i^U is the upper bound for A_i. Note from (3.25) that $A_i = 0$ should be satisfied if $y_i = 0$.

Another drawback in topology optimization based on the ground structure approach is that an unstable optimal truss is often obtained. A node connecting only two colinear members, as shown in Fig. 3.3, is unstable to the transverse direction of the members. An unstable solution can be avoided by introducing the lower bound for the number of members connected to an existing node.

Let $x_r \in \{0, 1\}$ be the variable indicating nonexistence and existence of the rth node, respectively, by $x_r = 0$ and $x_r = 1$. The upper and lower bounds for the number of members connected to the rth node, if they exist, are denoted by C_r^U and C_r^L, respectively. Note that C_r^U is given to prevent the existence of a highly connected node. The set of indices of members connected to the rth node in the initial ground structure is denoted by J_r, and the following constraints are given:

$$x_r C_r^L \leq \sum_{i \in J_r} y_i \leq x_r C_r^U, \quad (r = 1, \ldots, s) \tag{3.26}$$

where s is the number of nodes including the supports. Note from (3.26) that $y_i = 0$ should be satisfied by all the members connected to a removed node with $x_r = 0$.

In a practical situation, intersection of members should also be avoided, although those members are needed in the initial ground structure so as not to restrict the design space. The ith pair of mutually intersecting members in the ground structure is denoted by S_i $(i = 1, \ldots, q)$. The following constraints are to be satisfied:

$$\sum_{j \in S_i} y_j \leq 1, \quad (i = 1, \ldots, q) \tag{3.27}$$

The topology optimization problem is then formulated as an MINLP problem:

$$\text{Minimize} \quad V(\mathbf{A}) = \sum_{i=1}^{m} A_i L_i \tag{3.28a}$$

$$\text{subject to} \quad \sigma_i^{\text{L}} y_i \leq \sigma_i^k y_i \leq \sigma_i^{\text{U}} y_i, \quad (i = 1, \ldots, m; \ k = 1, \ldots, n^{\text{P}}) \tag{3.28b}$$

$$\sigma_i^k = \frac{E}{L_i} \mathbf{d}_i^{\top} \mathbf{U}^k, \quad (i = 1, \ldots, m; \ k = 1, \ldots, n^{\text{P}}) \tag{3.28c}$$

$$N_i^k = A_i \sigma_i^k, (i = 1, \ldots, m; \ k = 1, \ldots, n^{\text{P}}) \tag{3.28d}$$

$$\mathbf{D} \mathbf{N}^k = \mathbf{P}^k, \quad (k = 1, \ldots, n^{\text{P}}) \tag{3.28e}$$

$$A_i^{\text{L}} y_i \leq A_i \leq A_i^{\text{U}} y_i, \quad (i = 1, \ldots, m) \tag{3.28f}$$

$$x_r C_r^{\text{L}} \leq \sum_{i \in J_r} y_i \leq x_r C_r^{\text{U}}, \quad (r = 1, \ldots, s) \tag{3.28g}$$

$$\sum_{j \in S_i} y_j \leq 1, \quad (i = 1, \ldots, q) \tag{3.28h}$$

$$y_i \in \{0, 1\}, \quad (i = 1, \ldots, m) \tag{3.28i}$$

$$x_r \in \{0, 1\}, \quad (r = 1, \ldots, s) \tag{3.28j}$$

where the variables are $\mathbf{A}$, $\mathbf{y}$, $\mathbf{x}$, $\mathbf{U}^k$, $\boldsymbol{\sigma}^k$, and $\mathbf{N}^k$ with $\mathbf{y} = (y_1, \ldots, y_m)^{\top}$, $\mathbf{x} = (x_1, \ldots, x_s)^{\top}$, and $\boldsymbol{\sigma}^k = (\sigma_1^k, \ldots, \sigma_m^k)^{\top}$ $(i = 1, \ldots, m; \ k = 1, \ldots, n^{\text{P}})$.

Note that constraint (3.28d) is nonlinear with respect to A_i and σ_i^k. If A_i takes 0 or 1, then (3.28d) can be converted to a pair of linear inequality constraints, as shown in Sec. 3.6. However, we assume that A_i has a real value. Furthermore, we cannot use the reformulation to (3.24), because the existence of singular (isolated) optimal topology is allowed here.

The objective value of Problem (3.28) is denoted by V^{MIP}. In the following, (3.28h) is called a constraint on *member intersection*, (3.28g) with (3.28f) is called a constraint on *nodal instability*, and (3.28f)–(3.28h) are referred to as *local constraints*.

3.5.6.2 Lower-bound and upper-bound solutions

A relaxed LP problem of Problem (3.28) is formulated by relaxing integer variables x_r and y_i to continuous ones and by neglecting compatibility constraints (3.28c) to find the lower bound of V^{MIP}. If $y_i = 1$, (3.28b) reads

$$\sigma_i^{\text{L}} \leq \sigma_i^k \leq \sigma_i^{\text{U}}, \quad (i = 1, \ldots, m; \ k = 1, \ldots, n^{\text{P}}) \tag{3.29}$$

and by multiplying A_i,

$$A_i \sigma_i^{\text{L}} \leq N_i^k \leq A_i \sigma_i^{\text{U}} \quad (i = 1, \ldots, m; \ k = 1, \ldots, n^{\text{P}}) \tag{3.30}$$

is derived. Note that (3.30) is satisfied also for $y_i = 0$ with $N_i = 0$ because constraint (3.28f) is imposed in the relaxed problem. Therefore, (3.30) is satisfied for $0 \leq y_i \leq 1$ if (3.28b) is satisfied.

Hence, the relaxed LP problem of MIP (3.28) is formulated as

$$\text{Minimize} \quad V(\mathbf{A}) = \sum_{i=1}^{m} A_i L_i \qquad (3.31\text{a})$$

$$\text{subject to} \quad \mathbf{DN}^k = \mathbf{P}^k, \quad (k = 1, \ldots, n^{\mathrm{P}}) \qquad (3.31\text{b})$$

$$A_i \sigma_i^{\mathrm{L}} \leq N_i^k \leq A_i \sigma_i^{\mathrm{U}}, \quad (i = 1, \ldots, m; \ k = 1, \ldots, n^{\mathrm{P}}) \qquad (3.31\text{c})$$

$$A_i^{\mathrm{L}} y_i \leq A_i \leq A_i^{\mathrm{U}} y_i, \quad (i = 1, \ldots, m) \qquad (3.31\text{d})$$

$$x_r C_r^{\mathrm{L}} \leq \sum_{i \in J_r} y_i \leq x_r C_r^{\mathrm{U}}, \quad (r = 1, \ldots, s) \qquad (3.31\text{e})$$

$$\sum_{j \in S_i} y_j \leq 1, \quad (i = 1, \ldots, q) \qquad (3.31\text{f})$$

$$0 \leq y_i \leq 1, \quad (i = 1, \ldots, m) \qquad (3.31\text{g})$$

$$0 \leq x_r \leq 1, \quad (r = 1, \ldots, s) \qquad (3.31\text{h})$$

where the variables are $\mathbf{A}$, $\mathbf{y}$, $\mathbf{x}$, and $\mathbf{N}^k$. Note again that Problem (3.31) is an LP problem, where the global optimality of the solution is guaranteed.

If a statically determinate truss satisfying the local constraints is obtained by solving LP (3.31), then the solution gives the global optimal topology of the original MIP problem. If a statically indeterminate truss is obtained, then the objective value V^{LP} of LP (3.31) is a lower bound of the true optimal objective value of MIP (3.28), because the solution of MIP (3.28) satisfies all the constraints of LP (3.31).

For a given set $\mathcal{I}$ of the indices of existing members of the optimal solution of Problem (3.31), i.e., $\mathcal{I} = \{j \mid y_j > 0\}$, the following NLP, called NLP$_\mathcal{I}$, is defined. If NLP$_\mathcal{I}$ is feasible, then its objective value gives an upper bound $V^{\mathrm{NLP}_\mathcal{I}}$ of V^{MIP}, because the solution of NLP$_\mathcal{I}$ satisfies all the constraints of MIP (3.28):

$$\text{Minimize} \quad V(\mathbf{A}) = \sum_{i \in \mathcal{I}} A_i L_i \qquad (3.32\text{a})$$

$$\text{subject to} \quad \sigma_i^{\mathrm{L}} \leq \sigma_i^k(\mathbf{A}) \leq \sigma_i^{\mathrm{U}}, \quad (i \in \mathcal{I}; \ k = 1, \ldots, n^{\mathrm{P}}) \qquad (3.32\text{b})$$

$$A_i^{\mathrm{L}} \leq A_i \leq A_i^{\mathrm{U}}, \quad (i \in \mathcal{I}) \qquad (3.32\text{c})$$

where the variables are $\mathbf{A}$ only.

3.5.6.3   Branch-and-bound method

Using the upper- and lower-bound solutions, the original MINLP problem (3.28) is solved using the standard branch-and-bound method, which is summarized as follows:

Step 0: Initialize the upper bound V^U as $V^U = \infty$. Let the set $\mathcal{A}$ of the active problems consist of the original MINLP problem (3.28).

Step 1: Select a problem P from $\mathcal{A}$ and remove it from $\mathcal{A}$.

Step 2: Solve the relaxed LP problem $\overline{P}$ of P, and select a member i from $\mathcal{I}$ with $0 < y_i < 1$ of the solution. Let P_0 and P_1 denote the subproblems of P by specifying $y_i = 0$ and 1, respectively. Solve the relaxed LPs (3.31) of P_0 and P_1, which are denoted by $\overline{P}_0$ and $\overline{P}_1$, respectively.

Step 3: Let V_0 and V_1 denote the optimal objective values of $\overline{P}_0$ and $\overline{P}_1$, respectively. If $V_0 > V^U$, set $y_i = 1$ and terminate P_0. If $V_1 > V^U$, set $y_i = 0$ and terminate P_1.

Step 4: If the topology conforming to the LP solution satisfies all the local constraints, then compute the objective value of NLP$_\mathcal{I}$ (3.32) denoted by $V^{\mathrm{NLP}_\mathcal{I}}$ after specifying the set $\mathcal{I}$ of the existing members.

Step 5: If $V^{\mathrm{NLP}_\mathcal{I}} < V^U$, then update V^U to $V^{\mathrm{NLP}_\mathcal{I}}$.

Step 6: If V_0 and/or V_1 is less than V^U, then add P_0 and/or P_1, respectively, to $\mathcal{A}$ and go to Step 1.

Step 7: If $\mathcal{A}$ is not empty, then go to Step 1.

Step 8: Output the best value of V^U. Compute the lower bound V^L backward from the bottom of the branching tree so that V^L of the parent problem P is updated by $\min\{V_0, V_1\}$ if $V^L < \min\{V_0, V_1\}$.

The basic branch-and-bound algorithm is presented in Appendix A.2.5.

At the first stage of the branching process, we can obtain an upper bound V^U by fixing y_i appropriately for all the members existing in the lower-bound solution so that the local constraints are satisfied. Furthermore, we can use the local constraints before solving the LP problem (3.31); i.e., x_r can be fixed at 1 if a member exists such that $y_i = 1$ ($i \in J_r$). Alternatively, x_r is equal to 0 if $y_i = 0$ for all the members in J_r. Furthermore, y_i ($i \in S_r$) can be fixed at 0 if there exists a member with $y_j = 1$ ($j \in S_r, j \neq i$).

Because Problem (3.32) is nonconvex, the global optimum of the solution obtained by this branch-and-bound algorithm cannot be guaranteed. However, in the following numerical examples, it is confirmed that the same optimal solutions can be found starting from different initial solutions. Therefore, the nonconvexity of Problem (3.32) is not very strong, and the global optimal solution can be found for almost all cases.

3.5.6.4 Numerical examples

Optimal topologies of plane trusses are found using the branch-and-bound algorithm. In the following examples, the units of force and length are omitted

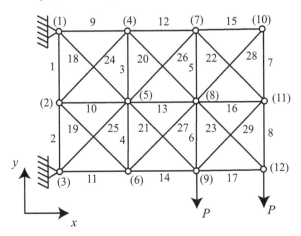

FIGURE 3.9: A 3×2 plane grid (from Ohsaki and Katoh, *Struct. Multidisc. Optimiz.*, 29, 190–197, 2005, Fig. 2, Copyright ©Springer Science+Business Media, reprinted with kind permission).

for brevity. Problem (3.31) is solved by HOPDM Ver. 2.13 (Gondzio 1995), which utilizes the higher-order primal-dual interior-point method. The solutions of NLP problems are found using NLPQL implemented as DNCONG in the IMSL library (Visual Numerics Inc. 1997), where sequential quadratic programming is used. Sensitivity coefficients of stresses and displacements with respect to the cross-sectional areas are computed using the direct differentiation method; see Sec. 2.2 for details.

As a simple example, consider the five-bar truss in Fig. 3.6. The bounds of the cross-sectional areas are $A_i^L = 0$ and $A_i^U = 20.0$. The local constraints are not given for comparison purposes with the results of Cheng and Guo (1997). However, from stability requirements, C_r^L should be 1 for the supports and 2 for the loaded nodes, and x_r should be 1 for all the nodes and supports. Other parameters are the same as in Example 3.2 in Sec. 3.5.3.

The initial LP solution has intersecting members 3 and 4, and member 4 is selected as the branching member. The optimal objective value of $\overline{P}_0$ is 33.500. Since the optimal truss of $\overline{P}_0$ is statically determinate, V^U is updated to 33.500. The optimal objective value of $\overline{P}_1$ is 32.500. All the members exist in the solution of $\overline{P}_1$, and the value of $V^{\mathrm{NLP}_\mathcal{I}}$ is 39.986. By solving LP problem (3.31) and NLP problem (3.32) 15 times and 4 times, respectively, we obtain the best upper bound solution $(A_1, A_2, A_3, A_4, A_5) = (1.0000, 2.5000, 10.0000, 0.0, 14.1421)$ with $V^U = 33.500$, which agrees with the result by Cheng and Guo (1997). The lower bound V^L is 32.500, which is slightly smaller than V^U.

Next we consider plane grid trusses with square units. A truss with a 3×2 grid is shown in Fig. 3.9. The lengths of the members in the x- and y-

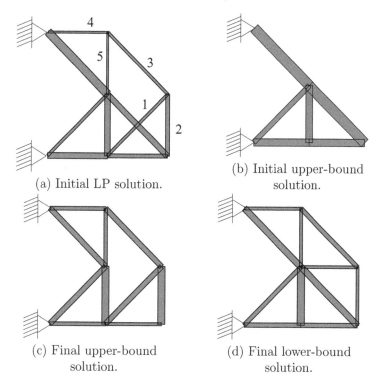

(a) Initial LP solution.

(b) Initial upper-bound solution.

(c) Final upper-bound solution.

(d) Final lower-bound solution.

FIGURE 3.10: Optimal solutions of 2×2 plane grid (from Ohsaki and Katoh, *Struct. Multidisc. Optimiz.*, 29, 190–197, 2005, Figs. 3–6, Copyright ©Springer Science+Business Media, reprinted with kind permission).

directions are 200. Irrespective of the number of grids, two loading conditions are considered, where the loads in the negative y-directions are applied at the node at the lowest end (node 12 in Fig. 3.9) and the node left of the lowest end (node 9 in Fig. 3.9), respectively. The magnitude of each load is 1000. The lower- and upper-bound cross-sectional areas are given as $A_i^L = 200$ and $A_i^U = 800$. The bounds for the stress are ± 2.0, and $C_r^U = 6$. The value of C_r^L is 1 for the supports, 2 for the node at the lowest end, and 3 for the remaining nodes. Note that these lower bounds are defined naturally from the requirements of stability and equilibrium, and they do not unnecessarily restrict the design space.

Optimal topology is first found for the 2×2 grid. The LP solution at the first step is as shown in Fig. 3.10(a), where the width of a member is proportional to its cross-sectional area. The objective value V^{LP} is 7.0000×10^6. To obtain an initial upper-bound solution, member 1 in Fig. 3.10(a) is removed because it has the smaller cross-sectional area in the pair of intersecting members. Note that this selection is heuristic; however, only a good upper bound is to

TABLE 3.1:   Optimization results of plane grid trusses (from Ohsaki and Katoh, *Struct. Multidisc. Optimiz.*, 29, 190–197, 2005, Table 1, Copyright ©Springer Science+Business Media, reprinted with kind permission).

No. of division	(m, n)	$(A_i^{\mathrm{L}}, A_i^{\mathrm{U}})$	No. of steps	No. of LP	No. of NLP	V^{U} $\times 10^7$	V^{L} $\times 10^7$
2×2	(20,14)	(200,800)	121	64	5	0.7900	0.7800
3×2	(29,20)	(200,800)	942	571	6	1.2900	1.2800
3×3	(42,28)	(200,800)	5874	3483	23	1.2467	1.2467
4×4	(72,46)	(200,800)	64890	42831	7	1.7067	1.7067
4×4	(72,46)	(200,600)	68656	42707	73	1.8373	1.7916
4×4	(72,46)	(400,800)	41001	26580	3	2.1507	2.1507

be found at this stage. After removing member 1, the node connected by members 2 and 3 is removed because it violates the local constraint (3.28g) with $C_r^{\mathrm{L}} = 3$, and, accordingly, members 4 and 5 are to be removed. NLP problem (3.32) is solved by fixing the values of y_i and x_r to find an upper bound solution, as shown in Fig. 3.10(b), where $V^{\mathrm{NLP}_\mathcal{I}}$ is 8.0000×10^6.

The branch-and-bound process is carried out to find the final upper-bound solution, as shown in Fig. 3.10(c), where $V^{\mathrm{U}} = 7.9000 \times 10^6$. The optimization results are listed in the first row of Table 3.1, where *No. of steps* means the number of different topologies that have been searched. In this example, only 121 topologies are searched out of the total of $2^{20} \simeq 10^6$ possible topologies. The final lower-bound solution is as shown in Fig. 3.10(d), where $V^{\mathrm{L}} = 7.8000 \times 10^6$.

Since this truss is statically indeterminate, the axial forces obtained by solving LP problem (3.31) are not correct. The maximum absolute value of the ratio of stress to the upper or lower bounds is 1.1111 if the compatibility conditions are considered; i.e., the solution does not satisfy stress constraints. Hence, V^{L} has a smaller value than V^{U}. However, the difference between V^{L} and V^{U} is very small and the good upper-bound solution has been found after solving NLP problem (3.32) only five times. If we do not incorporate the local constraints, the numbers of steps, LP, and NLP are 529, 350, and 58, respectively.

The optimization results for 3×2, 3×3, and 4×4 grids are also listed in Table 3.1. The final upper-bound solution for a 4×4 grid is shown in Fig. 3.11. Note that $V^{\mathrm{L}} = V^{\mathrm{U}}$ is satisfied for 3×3 and 4×4 grids, because the lower-bound solutions are statically determinate. The number of LP steps increases drastically as the size of the problem, such as the number of members, is increased. The number of NLP steps, however, is independent of the problem size, because it depends on the quality of the initial upper-bound solution. If A_i^{U} is decreased to 600 for the 4×4 grid, the number of NLP steps increases to 73, as shown in the fifth row of Table 3.1. If we do not consider the local constraints for the 3×2 truss, the number of steps, LP, and NLP are

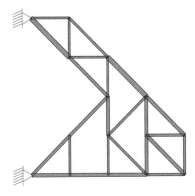

FIGURE 3.11: Final upper-bound solution for the 4 × 4 plane grid (from Ohsaki and Katoh, *Struct. Multidisc. Optimiz.*, 29, 190–197, 2005, Fig. 9, Copyright ©Springer Science+Business Media, reprinted with kind permission).

2364, 1468, and 391, respectively. We can observe from these results that the computational cost can be drastically reduced by using local constraints.

3.6 Mixed integer programming for topology optimization with discrete variables

3.6.1 Introduction

As we have seen in Secs. 3.3 and 3.5, the most popular approach to topology optimization of trusses is the ground structure approach, where the cross-sectional areas are regarded as continuous variables. However, in practical situations of civil and architectural engineering, the cross-sectional properties are selected from the discrete list of available standard sections. Furthermore, standard layouts are often used for plane and spatial trusses. Therefore, it is very important to investigate the optimal topology as an assembly of standard layouts, e.g., Schwedler dome and Lamella dome for spatial trusses, and Howe truss and K-truss for bridge-type plane trusses (Narayanan 2006).

In Sec. 3.5, we presented a mixed integer nonlinear programming (MINLP) approach to incorporate local constraints with continuous cross-sectional areas. In this section, we do not consider local constraints; however, we formulate the problem as an MINLP problem by considering the cross-sectional areas as discrete variables. An example of relaxation of the compliance minimization problem to a convex quadratic programming problem by Achtziger and Stolpe (2008, 2009) is first presented. Next, the problem under stress constraints is reformulated to an MIP problem following the approach of Stolpe and Svanberg (2003a), and we present the results of truss topology optimization as an assembly of standard layouts (Watada and Ohsaki 2009c).

3.6.2   Compliance minimization problem

Consider first a problem of minimizing the compliance under constraint on total structural volume. Let A_i and L_i denote the cross-sectional area and length of the ith member, respectively. A_i is selected from the set of available sections as $A_i \in \{0, a_i^1, \ldots, a_i^{r_i}\}$ $(i = 1, \ldots, m)$, where m is the number of members, and r_i is the number of available sections for the ith member if it exists. The ith member does not exist if $A_i = 0$. The truss is subjected to the nodal loads $\mathbf{P}$. The displacement vector and the stiffness matrix are denoted by $\mathbf{U}$ and $\mathbf{K(A)}$, respectively. Using the approach of simultaneous analysis and design (see Sec. 1.13), the optimization problem is formulated as

$$\text{Minimize} \quad \mathbf{P}^\top \mathbf{U} \tag{3.33a}$$

$$\text{subject to} \quad \mathbf{K(A)U} = \mathbf{P} \tag{3.33b}$$

$$\sum_{i=1}^{m} A_i L_i \leq V^{\mathrm{U}} \tag{3.33c}$$

$$A_i \in \{0, a_i^1, \ldots, a_i^{r_i}\}, \quad (i = 1, \ldots, m) \tag{3.33d}$$

where the variables are $\mathbf{A}$ and $\mathbf{U}$, and V^{U} is the specified upper bound of the total structural volume.

Problem (3.33) is a nonconvex quadratic programming (QP) problem with mixed continuous and discrete variables, which can be solved using a branch-and-bound method if the problem relaxing A_i to a continuous variable can be formulated as a convex programming problem. For this purpose, Achtziger and Stolpe (2008, 2009) presented two types of relaxed problems. Let λ denote an auxiliary variable. The lower and upper bounds for the continuous variable A_i are denoted simply by A_i^{L} and A_i^{U}, respectively; e.g., $A_i^{\mathrm{L}} = a_i^k$ and $A_i^{\mathrm{U}} = a_i^{k+1}$, $1 \leq k \leq r_i - 1$ with $a_i^0 = 0$ in the process of branch-and-bound. Using the obvious relation

$$\sum_{i=1}^{m} A_i^{\mathrm{L}} \leq \sum_{i=1}^{m} A_i \leq \sum_{i=1}^{m} A_i^{\mathrm{U}} \tag{3.34}$$

a relaxed problem can be formulated as

$$\text{Minimize} \quad -\mathbf{P}^\top \mathbf{U} + \lambda V^U - \sum_{i=1}^{m} \mu_i \tag{3.35a}$$

$$\text{subject to} \quad \left(\frac{1}{2L_i} \mathbf{U}^\top \mathbf{K}_i \mathbf{U} - \lambda \right) A_i^L + \mu_i \leq 0, \quad (i = 1, \ldots, m) \tag{3.35b}$$

$$\left(\frac{1}{2L_i} \mathbf{U}^\top \mathbf{K}_i \mathbf{U} - \lambda \right) A_i^U + \mu_i \leq 0, \quad (i = 1, \ldots, m) \tag{3.35c}$$

$$\lambda \geq 0 \tag{3.35d}$$

where the variables are $\mu_1, \ldots, \mu_m$, $\mathbf{U}$, and λ, and $\mathbf{K}_i$ is defined by

$$\mathbf{K} = \sum_{i=1}^{m} A_i \mathbf{K}_i \tag{3.36}$$

Here, notations consistent with those in this book have been used, while the volume of each member is taken as a design variable by Achtziger and Stolpe (2008).

Note that Problem (3.35) is a convex quadratic programming problem, for which the global optimal solution can be found using any method of nonlinear programming. However, the number of variables is very large if members exist between all pairs of nodes in the ground structure with many nodes. Therefore, Achtziger and Stolpe (2008, 2009) presented another *mildly convex* relaxed problem that can be solved efficiently with small computational cost.

3.6.3 Stress constraints

Suppose again the cross-sectional area A_i of the ith member is selected from the set of available sections $A_i \in \{0, a_i^1, \ldots, a_i^{r_i}\}$. Then the axial force vector $\mathbf{N} = (N_1, \ldots, N_m)^\top$ is given with respect to $\mathbf{U}$ using the constant vector $\mathbf{b}_i$, elastic modulus E, and the relations in (3.8) as

$$N_i = \frac{A_i E}{L_i} \mathbf{b}_i^\top \mathbf{U}, \quad (i = 1, \ldots, m) \tag{3.37}$$

Let σ_i^U and σ_i^L denote the upper and lower bounds for the stress of the ith member. The equilibrium matrix is denoted by $\mathbf{D}$, of which the ith column is equal to $\mathbf{b}_i$. Then the topology optimization problem for minimizing the total structural volume under stress constraints is formulated as

$$\text{Minimize} \quad V = \sum_{i=1}^{m} A_i L_i \tag{3.38a}$$

$$\text{subject to} \quad \mathbf{DN} = \mathbf{P} \tag{3.38b}$$

$$A_i \sigma_i^{\mathrm{L}} \le N_i \le A_i \sigma_i^{\mathrm{U}}, \quad (i = 1, \ldots, m) \tag{3.38c}$$

$$N_i = A_i \frac{E}{L_i} \mathbf{b}_i^{\mathsf{T}} \mathbf{U}, \quad (i = 1, \ldots, m) \tag{3.38d}$$

$$A_i \in \{0, a_i^1, \ldots, a_i^{r_i}\}, \quad (i = 1, \ldots, m) \tag{3.38e}$$

A 0–1 variable x_i^k is defined as

$$x_i^k = \begin{cases} 1 & \text{if } A_i = a_k^i \\ 0 & \text{otherwise} \end{cases}, \quad (i = 1, \ldots, m; \ k = 1, \ldots, r_i) \tag{3.39}$$

Then A_i is written with a_i^k and x_i^k as

$$A_i = \sum_{k=1}^{r_i} x_i^k a_i^k \tag{3.40}$$

with

$$\sum_{k=1}^{r_i} x_i^k \le 1 \tag{3.41}$$

Note that (3.40) and (3.41) indicate that at most only one of x_i^k ($k = 1, \ldots, r_i$) can take 1; if all of them are equal to 0, then $A_i = 0$ and member i is removed. Hence, a variable x_i, indicating existence and nonexistence of member i by $x_i = 1$ and 0, respectively, is defined as

$$x_i = \sum_{k=1}^{r_i} x_i^k \tag{3.42}$$

An auxiliary variable s_i^k is also used for the axial force as

$$s_i^k = x_i^k a_i^k \frac{E}{L_i} \mathbf{b}_i^{\mathsf{T}} \mathbf{U} \tag{3.43a}$$

$$N_i = \sum_{i=1}^{r_i} s_i^k \tag{3.43b}$$

Note that (3.43a) is nonlinear with respect to the variables x_i^k and $\mathbf{U}$.

Let $\mathbf{U}^{\min} = (U_1^{\min}, \ldots, U_n^{\min})^{\mathsf{T}}$ and $\mathbf{U}^{\max} = (U_1^{\max}, \ldots, U_n^{\max})^{\mathsf{T}}$ denote the vectors of lower and upper bounds of displacements, where n is the number of degrees of freedom. Sufficiently small lower bound $c_i^{\min}$ and large upper bound $c_i^{\max}$ are given for the elongation of member i corresponding to the displacements satisfying $\mathbf{U}^{\min} \leq \mathbf{U} \leq \mathbf{U}^{\max}$. Then the nonlinear relation (3.43a) is converted to the following pair of inequalities (Stolpe and Svanberg 2003a; Glover 1975):

$$(1 - x_i^k)c_i^{\min} \leq a_i^k \frac{E}{L_i}\mathbf{b}_i^\top \mathbf{U} - s_i^k \leq (1 - x_i^k)c_i^{\max} \qquad (3.44)$$

It is seen from (3.44) that if $x_i^k = 1$, then the left-hand side and right-hand side terms vanish, and, accordingly, relation (3.43a) is satisfied. In contrast, if $x_i^k = 0$, then the inequalities hold for any value of $\mathbf{U}$ satisfying $\mathbf{U}^{\min} \leq \mathbf{U} \leq \mathbf{U}^{\max}$; i.e., no constraint is given between $\mathbf{U}$ and s_i^k.

Additional requirements are to be assigned to obtain practical optimal solutions, for example (Ohsaki and Watada 2008; Watada and Ohsaki 2009b, 2009c):

- *Cond-1*: Intersecting members i and j in the ground structure cannot exist simultaneously in the optimal topology:

$$x_i + x_j \leq 1 \qquad (3.45)$$

- *Cond-2*: If member i exists, then member j exists:

$$x_j \geq x_i \qquad (3.46)$$

- *Cond-3*: If member i does not exist, then member j does not exist:

$$x_j \leq x_i \qquad (3.47)$$

- *Cond-4*: *Cond-2* and *Cond-3* are simultaneously satisfied:

$$x_j = x_i \qquad (3.48)$$

In the numerical examples, the members are classified into groups to represent the standard configurations. For this purpose, auxiliary variables $y_k \in \{0, 1\}$ $(k = 1, \ldots, n^y)$ are used for indicating by $y_k = 1$ and 0 the existence and nonexistence, respectively, of the kth group, where n^y is the number of groups.

Constraints for y_k are formulated as follows:

- *Cond-5*: Members i and j are included in group k; i.e., these members exist (do not exist) if group k exists (does not exist):

$$y_k = x_j = x_j \qquad (3.49)$$

- *Cond-6*: If group k exists, then group j exists:

$$y_j \geq y_k \qquad (3.50)$$

- *Cond-7*: If group k does not exist, then group j does not exist:

$$y_j \leq y_k \tag{3.51}$$

- *Cond-8*: If *Cond-6* and *Cond-7* are simultaneously satisfied, we make another group l containing groups j and k:

$$y_l = y_j = y_k \tag{3.52}$$

These constraints are symbolically written as follows using the vectors $\mathbf{x} = (x_1, \ldots, x_m)^\top$, $\mathbf{y} = (y_1, \ldots, y_{n^y})^\top$, and matrices $\mathbf{G}^\text{x}$, $\mathbf{G}^\text{y}$, $\mathbf{H}^\text{x}$, and $\mathbf{H}^\text{y}$:

$$\begin{aligned} \mathbf{G}^\text{x}\mathbf{x} + \mathbf{G}^\text{y}\mathbf{y} &= \mathbf{0}, \\ \mathbf{H}^\text{x}\mathbf{x} + \mathbf{H}^\text{y}\mathbf{y} &\leq \mathbf{0} \end{aligned} \tag{3.53}$$

The auxiliary variables and constraints can be systematically generated using a decision tree (Watada and Ohsaki 2009c).

Finally, the topology optimization problem for minimizing the total structural volume V is formulated as

$$\text{Minimize} \quad V = \sum_{i=1}^{m} \sum_{k=1}^{r_i} x_i^k a_i^k L_i \tag{3.54a}$$

$$\text{subject to} \quad \mathbf{DN} = \mathbf{P} \tag{3.54b}$$

$$N_i = \sum_{i=1}^{r_i} s_i^k, \quad (i = 1, \ldots, m) \tag{3.54c}$$

$$x_i^k a_i^k \sigma_i^\text{L} \leq s_i^k \leq x_i^k a_i^k \sigma_i^\text{U}, \quad (i = 1, \ldots, m; \ k = 1, \ldots, r_i) \tag{3.54d}$$

$$U_j^\text{min} \leq U_j \leq U_j^\text{max}, \quad (j = 1, \ldots, n) \tag{3.54e}$$

$$(1 - x_i^k)c_i^\text{min} \leq a_i^k \frac{E}{L_i} \mathbf{b}_i^\top \mathbf{U} - s_i^k \leq (1 - x_i^k)c_i^\text{max},$$

$$(i = 1, \ldots, m; \ k = 1, \ldots, r_i) \tag{3.54f}$$

$$\sum_{k=1}^{r_i} x_i^k \leq 1, \quad (i = 1, \ldots, m) \tag{3.54g}$$

$$\mathbf{G}^\text{x}\mathbf{x} + \mathbf{G}^\text{y}\mathbf{y} = \mathbf{0} \tag{3.54h}$$

$$\mathbf{H}^\text{x}\mathbf{x} + \mathbf{H}^\text{y}\mathbf{y} \leq \mathbf{0} \tag{3.54i}$$

$$x_i^k \in \{0, 1\}, \quad (i = 1, \ldots, m; \ k = 1, \ldots, r_i) \tag{3.54j}$$

$$y_i \in \{0, 1\}, \quad (i = 1, \ldots, n^y) \tag{3.54k}$$

which is an MIP problem with variables x_i^k, s_i^k $(i = 1, \ldots, m; \ k = 1, \ldots, r_i)$, U_j $(j = 1, \ldots, n)$, $\mathbf{x}$, $\mathbf{y}$, and $\mathbf{N}$.

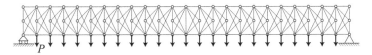

FIGURE 3.12: A bridge-type plane truss.

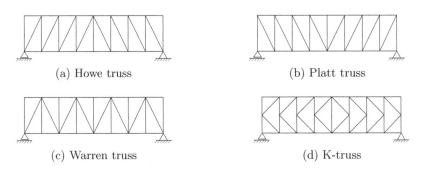

(a) Howe truss (b) Platt truss

(c) Warren truss (d) K-truss

FIGURE 3.13: Four standard layouts of plane trusses.

3.6.4 Numerical examples

Consider a bridge-type plane truss, as shown in Fig. 3.12, which represents the ground structure that has many members as an assemblage of the four standard layouts, namely, Howe truss, Platt truss, Warren truss, and K-truss, as shown in Figs. 3.13(a)–(d), respectively. Uniform vertical loads P are applied at the lower nodes. Incorporating the symmetry conditions of the loads and topology of the ground structure, the optimal topology of the truss is also assumed to be symmetric; hence, one of the half parts is to be optimized. We use the optimization library CPLEX 10.2 (ILOG 2007) for solving the MIP. Computation is carried out on a PC with Intel Xeon CPU 3.40GHz, 2.00GB RAM. The units of force and length are omitted for brevity.

The members are appropriately divided into groups so that one of the four types is selected for each of the 12 units of the half part. For this purpose, the members in a unit are classified into three groups, as shown in Fig. 3.14. The parameters are $H = 2.0$, $W = 12.0$, $E = 2.0$, and $\sigma_i^{\mathrm{U}} = 0.003$ for all members. The lower-bound stress is given as follows to incorporate the effect of member buckling:

$$\sigma_i^{\mathrm{L}} = -\frac{L_{\min}^2}{L_i^2}\sigma_i^{\mathrm{U}} \tag{3.55}$$

where $L_{\min}$ is the length of the shortest member in the ground structure. The cross-sectional areas are chosen from the list $\{0.0, 1.0, 1.8\}$; i.e., $a_i^1 = 1.0$, $a_i^2 = 1.8$, and $r_i = 2$ for all the members. In order to improve computational

FIGURE 3.14:   Classification of members in a unit.

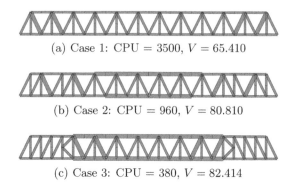

(a) Case 1: CPU = 3500, $V = 65.410$

(b) Case 2: CPU = 960, $V = 80.810$

(c) Case 3: CPU = 380, $V = 82.414$

FIGURE 3.15:   Optimal topologies and cross-sectional areas of the plane truss.

efficiency, the variable s_i^k is scaled to s_i^{k*} as

$$s_i^{k*} = \frac{L_i}{Ea_j^k} s_i^k \tag{3.56}$$

The optimal solutions are found for Cases 1–3 with three different load magnitudes $P = 0.00007$, 0.00012, and 0.00014, respectively. The optimal topologies are as shown in Figs. 3.15(a)–(c), where the width of each member is proportional to its cross-sectional area, and the CPU time (s) and the optimal objective value are shown in each figure. As is seen, the Warren truss tends to be selected for a smaller load, while the Howe truss is selected for a larger load. Because vertical members are allowed to be removed in the Warren truss, it has the smallest member density among the four types, and, accordingly, is selected for the case of a smaller load. Note that the CPU time strongly depends on the load magnitude; i.e., computational cost is smaller for a larger load, because the number of admissible sets of cross-sectional areas decreases as the load is increased.

We next optimize a dome truss as an assemblage of the Schwedler dome and Lamella dome, as shown in Figs. 3.16(a) and (b), respectively, subjected to the vertical load P at the top node. The cross-sectional areas are selected from the list $\{0.0, 1.0, 3.0, 5.0\}$, and we consider Cases 1–3 with loads $P = 0.001$, 0.002, and 0.0025, respectively. The whole structure is solved without considering the symmetry condition. The optimal topologies are as shown in Figs. 3.17(a)–(c). As is seen, the optimal truss is a Lamella dome for Case 1 with the

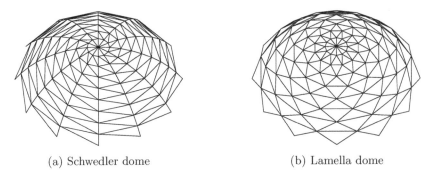

(a) Schwedler dome   (b) Lamella dome

FIGURE 3.16: Two standard layouts of dome trusses.

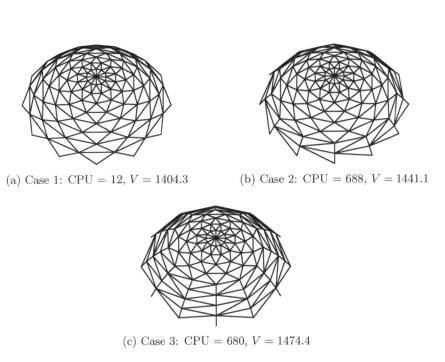

(a) Case 1: CPU = 12, $V = 1404.3$ (b) Case 2: CPU = 688, $V = 1441.1$

(c) Case 3: CPU = 680, $V = 1474.4$

FIGURE 3.17: Optimal topologies of the dome truss.

smallest load, while the Schwedler dome dominates for larger load magnitudes. Note that the cross-sectional areas are 1.0 for all the existing members except those connected to the center nodes for Cases 2 and 3, which are equal to 3.0 and 5.0, respectively. In this example, there is no correlation between the load magnitude and the CPU time.

3.7 Genetic algorithm for truss topology optimization

3.7.1 Introduction

The genetic algorithm (GA) has been successfully applied to optimization problems in various fields, including computer science, social science, and operations research. Although the GA was developed by Holland (1975), its concept have said to been proposed by Rechenberg (1965). GAs are very useful for problems with discontinuous cost and/or constraint functions because the gradients of the functions are not needed for finding approximate optimal solutions (Goldberg 1989).

Since the 1990s, GAs have been widely applied to structural optimization problems (Jenkins 1991; Hajela and Lin 1992; Rajeev and Krishnamoorthy 1997; Hayalioglu 2000; António 2002). Grierson and Pak (1993) applied a GA to a frame topology optimization problem, where the lists of topologies and cross-sectional properties as well as the nodal locations are encoded into binary strings. The null cross-sectional area is also included in the list to enable the removal of members. Hajela and Lee (1995) presented a two-level approach to ensure the stability of the truss. They also presented a scaling approach for constrained optimization problems, where the increase of the penalty term is limited to avoid too much penalty compared with the objective value.

GAs are extensively used for optimization of composite laminas. For instance, as an early development, Hajela and Lin (1991) optimized the stacking sequence of a composite beam with dampers. Marcelin, Trompette, and Dornberger (1995) developed a multiobjective GA for design of composite beams. Kogiso, Watson, Gürdal, and Haftka (1994) presented a binary tree approach for checking duplicate solutions as well as an operator for local improvement of composite laminates.

Kameshki and Saka (2003) optimized a frame with a semi-rigid connection using GA. Kocer and Arora (1999) compared the performances of GA and simulated annealing for optimization of transmission towers. GA is very useful for design under constraints on inelastic responses, for which the sensitivity coefficients of responses are difficult to obtain (Yun and Kim 2005). There are several variants of GAs, such as the *evolutionary algorithm* (EA). Greiner, Emperador, and Winter (2004) applied an improved EA with a rebirth operator to single and multiobjective optimization of frames. Giger and Ermanni (2006) presented a graph-based approach, where the operators of mutation and crossover are introduced for the adjacency matrix defining the topology of the truss.

In this section, an example of a GA is presented for topology optimization of trusses with stress constraints, considering the nodal costs (Ohsaki 1995).

3.7.2   Optimization considering nodal cost

3.7.2.1   Problem formulation

In the practical design process of frames and trusses, the cost of the nodes connecting the members may be sometimes equivalent to or greater than that of the members. Therefore, the nodal cost should be included in the cost (objective) function to obtain more realistic optimal topology.

A standard ground structure approach is used here. A node should exist if at least one member connected to the node has a positive cross-sectional area, and the size of the node is selected from a list of predefined values. Therefore, the nodal cost is a discontinuous function of the cross-sectional areas, which are considered as design variables, and no algorithm utilizing design sensitivity analysis can be applied even if the cross-sectional areas are continuous variables.

The length and the cost per unit volume of the ith member are denoted by L_i and c_i, respectively. Let $a_i(\mathbf{A})$ denote the maximum value of the cross-sectional areas among the members connected to the ith node. It is assumed here, for a simple presentation of the algorithm, that only one type of node is used; i.e., the cost $g_i(\mathbf{A})$ of the ith node is defined as

$$\begin{cases} g_i(\mathbf{A}) = \bar{g} \ \text{ for } \ a_i(\mathbf{A}) > 0 \\ g_i(\mathbf{A}) = 0 \ \text{ for } \ a_i(\mathbf{A}) = 0 \end{cases} \tag{3.57}$$

where $\bar{g}$ is the prescribed cost of a node.

Let n and n^{S} denote the number of degrees of freedom and the number of nodes, respectively. The truss is subjected to n^{P} sets of static loads, and the value corresponding to the kth load set is denoted by the superscript $(\cdot)^k$. The upper and lower bounds for the jth displacement component U_j^k are denoted by U_j^{U} and U_j^{L}, respectively. The bounds for the stress σ_i^k of the ith member are defined similarly. The topology optimization problem for minimizing the total cost $C(\mathbf{A})$ is formulated as

$$\text{Minimize} \quad C(\mathbf{A}) = \sum_{i=1}^{m} c_i A_i L_i + \sum_{i=1}^{n^{\mathrm{S}}} g_i(\mathbf{A}) \tag{3.58a}$$

$$\text{subject to} \quad U_j^{\mathrm{L}} \leq U_j^k \leq U_j^{\mathrm{U}}, \quad (j = 1, \dots, n; \ k = 1, \dots, n^{\mathrm{P}}) \tag{3.58b}$$

$$\sigma_i^{\mathrm{L}} \leq \sigma_i^k \leq \sigma_i^{\mathrm{U}} \ \text{for } A_i > 0,$$

$$(i = 1, \dots, m; \ k = 1, \dots, n^{\mathrm{P}}) \tag{3.58c}$$

Note from (3.58c) that the stress constraints are assigned only for the existing members.

3.7.3   Topological bit and fitness function

In order to solve Problem (3.58) using a GA, the cross-sectional area of each member is represented by a string of l^{e} binary data. Let A^0 denote the unit value of the cross-sectional area. For the case $l^e = 4$, a string 1010 for the ith member, for instance, indicates $A_i = (1 \times 2^3 + 0 \times 2^2 + 1 \times 2^1 + 0 \times 2^0) \times A_0 = 10A_0$. A *topological bit* G_i^e is added to the left of the string representing A_i to indicate the existence/nonexistence of member i. A member is to be removed if $A_i = 0$ or $G_i^e = 0$. For example, members with string 01001 ($A_i = 9A_0$, $G_i^e = 0$) and 10000 ($A_i = 0, G_i^e = 1$) are to be removed.

Although the GA is a heuristic approach without any strong theoretical basis, the *schema theorem* is usually used for explaining the convergence to an optimal solution by the GA. A pattern that has a certain meaning in a one-dimensional bit string is called a schema or building block. For example, a schema $H = 01**1**$ means that the first bit is fixed at 0, the second and fifth bits should be equal to 1, and the other bits can be either 0 or 1 as indicated by the wildcard $*$. Let $\delta(H)$ denote the *defining length* of the schema H that is the distance between the first and the last bits with fixed values. The number of bits with fixed values in H is called the *order*, which is denoted as $o(H)$; e.g., $\delta(H) = 4$ and $o(H) = 3$ for the schema $H = 01**1**$.

Let $M(H,t)$ denote the number of individuals that contain the schema H at the generation t. The mean fitness value of the individuals containing H is denoted by $f(H)$, whereas the mean fitness value of all individuals is denoted by $\overline{f}$. If we do not consider mutation or crossover, then the expected number of individuals containing H at the generation $t + 1$ is given as

$$M(H, t+1) = M(H,t)\frac{f(H)}{\overline{f}} \tag{3.59}$$

Let l denote the total length of the string. The probabilities of crossover and mutation are denoted by p^c and p^m, respectively. Then the probabilities of the schema H to be destroyed by mutation and crossover are given as $o(H)p^m$ and $p^c\delta(H)/(l-1)$, respectively. Because H may be destroyed by crossover and mutation simultaneously, the expected number of individuals containing H at the generation $t + 1$ is given with inequality as

$$M(H, t+1) \geq M(H,t)\frac{f(H)}{\overline{f}}\left(1 - p^c\frac{\delta(H)}{l-1} - o(H)p^m\right) \tag{3.60}$$

It is seen from (3.60) that the schema with the larger fitness value $f(H)$, i.e., $f(H)/\overline{f} > 1$, will dominate exponentially with the progress of generation.

For example, for our problem, consider a case where the nonexistence of member 1 leads to a larger fitness value. If the topological bit G_i^e is not incorporated, the solutions without member 1 are represented by a schema $H_a = 0000***\ldots$, which leads to $\delta(H_a) = 3$ and $o(H_a) = 4$. If G_i^e is incorporated at the left of the string representing the cross-sectional area, the schema corresponding to $G_1^e = 0$ is $H_b = 0***\ldots$, i.e., $\delta(H_b) = 0$ and $o(H_b) = 1$; hence, the probability of destroying the building block that leads to a larger fitness value is reduced. Therefore, from the schema theorem, incorporation of the topological bit leads to rapid convergence to optimal topology. If the nodal cost is moderately large, a truss with small number of members has a large fitness value, and such optimal topology can easily be found by incorporating the topological bit.

In order to prevent numerical difficulties for computing the displacements and stresses of the truss, a small value A_i^L is assigned for the cross-sectional area of a member with $A_i = 0$ or $G_i^e = 0$. The member with $A_i = A_i^L$ is to be removed from the final optimal solution. Hence, $a_j(\mathbf{A})$ in the conditions (3.57) is redefined as the maximum value of $A_i - A_i^L$ among the members connected to node j in the ground structure. The constraint (3.58c) is also modified as

$$\sigma_i^L \leq \sigma_i^k \leq \sigma_i^U \text{ for } A_i > A_i^L, \quad (i = 1, \ldots, m; \ k = 1, \ldots, n^S) \tag{3.61}$$

A random number $0 \leq \tau < 0$ is generated to assign the value of each topological bit at the initial generation. If τ is less than the prescribed probability p^e of existence, then $G_i^e = 1$; otherwise $G_i^e = 0$.

The fitness function is defined as

$$C^* = C_0 - \sum_{i=1}^{m} c_i A_i L_i - \sum_{i=1}^{n^S} g_i - \sum_{k=1}^{n^P} \left(\sum_{i=1}^{n} d_{kj}^D + \sum_{i=1}^{m} d_{ki}^S \right) \tag{3.62}$$

where C_0 is a large positive value to prevent the existence of an individual with a negative fitness value. Note that the value of C_0 does not have any effect on the optimal solution if a ranking strategy is used. The last term in (3.62) is the penalty for violating the stress and displacement constraints, where d_{kj}^D and d_{ki}^S are defined with parameters C^D, C^S, e_D^U, e_D^L, e_S^U, and e_S^L as

$$d_{kj}^D = C^D + e_D^U \left(1 - \frac{U_j^k}{U_j^U} \right)^2 + e_D^L \left(1 - \frac{U_j^k}{U_j^L} \right)^2 \tag{3.63a}$$

$$d_{kj}^S = C^S + e_S^U \left(1 - \frac{\sigma_i^k}{\sigma_i^U} \right)^2 + e_S^L \left(1 - \frac{\sigma_i^k}{\sigma_i^L} \right)^2 \tag{3.63b}$$

Note that e_D^U, e_D^L, e_S^U, and e_S^L have positive values, respectively, if the corresponding constraints on the responses are violated; otherwise they are 0. Furthermore, $C^D > 0$ only if $e_D^U > 0$ or $e_D^L > 0$; otherwise $C^D = 0$. C^S is defined similarly.

The ranking strategy is used here to prevent the dominance of overly fit designs in the early generation, and to ensure moderate convergence in the final stage. Let R_j denote the rank of the ith individual (solution), and define the fitness function V_j as

$$V_j = (R_j - n^S)^2 \tag{3.64}$$

where $R_j = 1$ for the individual with the smallest value of C^*. Then the following simple selection strategy is used to calculate the probability p_j of

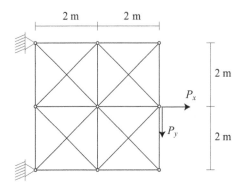

FIGURE 3.18: A 20-bar plane truss.

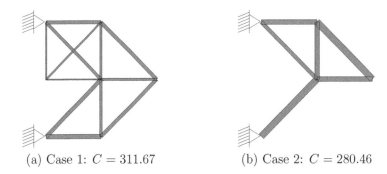

(a) Case 1: $C = 311.67$ (b) Case 2: $C = 280.46$

FIGURE 3.19: Optimal topologies of the 20-bar plane truss.

reproduction:

$$p_j = \frac{V_j}{\overline{V}}, \quad \overline{V} = \sum_{k=1}^{s} V_k \tag{3.65}$$

where s is the number of individuals in a generation (size of the population).

3.7.4   Numerical examples

Optimal topologies are found for a 20-bar plane truss, as shown in Fig. 3.18, under stress constraints to discuss the effectiveness of the use of the topological bit. The truss is subjected to the two loading conditions $(P_x, P_y) = (0, 10000 \text{ N})$ and $(7071.07 \text{ N}, -7071.07 \text{ N})$. The parameters are $A^0 = 50.0 \text{ mm}^2$, $s = 150$, $c_i = 5.0 \times 10^{-6} \text{ mm}^{-3}$ for all members, $\overline{g} = 25.0$, $p^c = 1$, and $p^m = 0.01$. It is assumed here that the cost of a support is the same as that of a node. The bounds for stress are $\sigma_i^U = -\sigma_i^L = 98.0 \text{ N/mm}^2$ for all members. A very small value A_i^L is given for the cross-sectional area of a removed

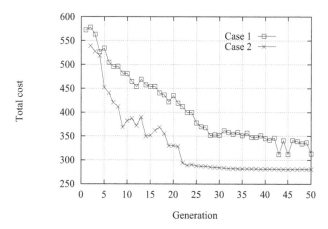

FIGURE 3.20: History of total cost C of the most fit design of the 20-bar plane truss.

member with $A_i = 0$ or $G_i^e = 0$. Therefore, there may exist nearly unstable design $A_i = A_i^L$ for all members in the optimization process if only the stress constraints are considered. Note that such a design has a large fitness value, because the member cost is extremely small and the stress constraint is allowed to be violated in all members. Furthermore, the nodal cost for the design $A_i = A_i^L$ for all members is 0. Because the displacements of this solution are very large, the convergence to this trivial optimal solution can be successfully prevented by imposing constraints on displacements. Hence, an artificial upper bound $U_j^U = 10.0$ mm is given for each displacement component.

Optimal topologies are found for two cases with $p^e = 1.0$ and 0.5, which are denoted as Cases 1 and 2, respectively. Note that $p^e = 1.0$ in Case 1 means that the topological bit is not used. Fig. 3.19(a) shows the optimal topology for Case 1. The truss has six nodes and nine members, and the total cost C is 311.67 with member cost 161.67. The optimal topology for Case 2 is as shown in Fig. 3.19(b). Five nodes and six members exist in this truss, and the total cost C and the total member cost are 280.46 and 155.46, respectively. As is seen, optimal topologies with a small number of members can be found if the topological bit and the probability of existence are utilized. The history of the total cost of the most fit solution in each generation is plotted in Fig. 3.20 for Cases 1 and 2; it shows that the use of the topological bit leads to rapid convergence of the total cost to the optimal value.

3.8   Random search method using exact reanalysis

3.8.1   Introduction

As we have discussed in previous sections, the cross-sectional areas of frames and trusses are often selected from a catalogue of available values; therefore, a search method that requires a large computational cost for structural analysis should be carried out for finding the optimal solution with discrete variables. Hence, heuristic approaches can be effectively used for finding approximate optimal solutions within a practically acceptable computational cost. Among many heuristic approaches, single-point-search methods, e.g., simulated annealing and tabu search, can obtain approximate solutions with a very small number of response evaluations; see Appendix A.3 for details. Since these approaches utilize local search that evaluates the performances of the solutions in the neighborhood of the current solution, the computational cost can be further reduced if the method called *exact reanalysis* is employed for structural analysis of a slightly modified structure.

In this section, we first introduce an exact reanalysis method of trusses subjected to static loads and summarize the random search algorithm in Ohsaki (2001a) for topology optimization of trusses with a discrete list of cross-sectional areas.

3.8.2   Exact reanalysis

3.8.2.1   Historical review

The method for computing the exact responses after modification of the variable by the specified amount is called *exact reanalysis*. There are several approaches to exact reanalysis of static responses of structures, e.g., the expansion method, the reduced basis method, and a method based on inversion of a modified matrix. Melosh and Luik (1968) presented a simple approach to reanalysis of trusses. Argyris and Roy (1972) presented a more general approach allowing changes in the number of degrees of freedom and support conditions. Kirsch and Rubinstein (1972) discussed convergence properties of the iterative reanalysis process. Kirsch (1994, 1996) and Kirsch and Liu (1995) presented an efficient method combining the reduced basis method and the expansion method. Exact reanalysis of eigenvalues can be found in Chen, Yang, and Lian (2000), Chen, Wu, and Yang (2006), and Kirsch (2000). There are other methods called the virtual distortion method (Putresza and Kolakowski 2001), the pseudoforce method (Deng and Ghosn 2001), and the pseudodistortion method (Makode, Corotis, and Ramirez 1999).

Calculation of the inverse of a modified matrix has been discussed in many fields of engineering and mathematics; developments in general methodologies are summarized in the review article by Henderson and Searle (1981). Kavlie,

Graham, and Powell (1971) developed a stiffness-based method for computing the displacements of the modified design based on the Sherman-Morrison-Woodbury (SMW) formula (Sherman and Morrison 1950; Woodbury 1950). Akgün, Garcelon, and Haftka (2001) extended the SMW formula to the nonlinear reanalysis problem. However, it is very important in the application to the static analysis problem of structures that the inverse of the stiffness matrix of the initial structure is not usually known; i.e. the matrix is only decomposed into triangular matrices (Ohsaki 2001a).

3.8.2.2 Inverse of the modified matrix

The formulation of the inverse of the modified matrix using the inverse of the original matrix was studied independently in various areas of mathematics and engineering (Henderson and Searle 1981). In the early 20th century, some formulations were derived indirectly from the inverse of a partitioned matrix. Let $\mathbf{K}$, $\mathbf{B}$, $\mathbf{C}$, and $\mathbf{D}$ denote matrices with appropriate size, and $\mathbf{I}$ and $\mathbf{O}$ be the identity matrix and null matrix, respectively. Schur (1917) derived the following formula:

$$\begin{pmatrix} \mathbf{K}^{-1} & \mathbf{O} \\ -\mathbf{C}\mathbf{K}^{-1} & \mathbf{I} \end{pmatrix} \begin{pmatrix} \mathbf{K} & \mathbf{B} \\ \mathbf{C} & \mathbf{D} \end{pmatrix} = \begin{pmatrix} \mathbf{I} & \mathbf{K}^{-1}\mathbf{B} \\ \mathbf{O} & \mathbf{D} - \mathbf{C}\mathbf{K}^{-1}\mathbf{B} \end{pmatrix} \tag{3.66}$$

from which the relations of the determinants are obtained as

$$\begin{vmatrix} \mathbf{K} & \mathbf{B} \\ \mathbf{C} & \mathbf{D} \end{vmatrix} = |\mathbf{K}||\mathbf{D} - \mathbf{C}\mathbf{K}^{-1}\mathbf{B}| = |\mathbf{D}||\mathbf{K} - \mathbf{B}\mathbf{D}^{-1}\mathbf{C}| \tag{3.67}$$

The following formulation of the inverse of a partitioned matrix was explicitly derived by Banachiewicz (1937), although it is sometimes attributed to Schur:

$$\begin{pmatrix} \mathbf{K} & \mathbf{B} \\ \mathbf{C} & \mathbf{D} \end{pmatrix}^{-1}$$
$$= \begin{pmatrix} \mathbf{K}^{-1} + \mathbf{K}^{-1}\mathbf{B}(\mathbf{D} - \mathbf{C}\mathbf{K}^{-1}\mathbf{B})^{-1}\mathbf{C}\mathbf{K}^{-1} & -\mathbf{K}^{-1}\mathbf{B}(\mathbf{D} - \mathbf{C}\mathbf{K}^{-1}\mathbf{B})^{-1} \\ -(\mathbf{D} - \mathbf{C}\mathbf{K}^{-1}\mathbf{B})^{-1}\mathbf{C}\mathbf{K}^{-1} & (\mathbf{D} - \mathbf{C}\mathbf{K}^{-1}\mathbf{B})^{-1} \end{pmatrix} \tag{3.68}$$

Hotelling (1943) derived a similar formula as

$$\begin{pmatrix} \mathbf{K} & \mathbf{B} \\ \mathbf{C} & \mathbf{D} \end{pmatrix}^{-1} = \begin{pmatrix} (\mathbf{K} - \mathbf{B}\mathbf{D}^{-1}\mathbf{C})^{-1} & -\mathbf{K}^{-1}\mathbf{B}(\mathbf{D} - \mathbf{C}\mathbf{K}^{-1}\mathbf{B})^{-1} \\ -\mathbf{D}^{-1}\mathbf{V}(\mathbf{D} - \mathbf{C}\mathbf{K}^{-1}\mathbf{B})^{-1} & (\mathbf{D} - \mathbf{C}\mathbf{K}^{-1}\mathbf{B})^{-1} \end{pmatrix} \tag{3.69}$$

By comparing the leading terms of (3.68) and (3.69), we obtain the following formula for the inverse of a sum of a matrix:

$$(\mathbf{K} - \mathbf{B}\mathbf{D}^{-1}\mathbf{C})^{-1} = \mathbf{K}^{-1} + \mathbf{K}^{-1}\mathbf{B}(\mathbf{D} - \mathbf{C}\mathbf{K}^{-1}\mathbf{B})^{-1}\mathbf{C}\mathbf{K}^{-1} \tag{3.70}$$

Since the 1950s, more general formulas have been investigated directly for finding the inverse of a sum of matrices. In particular, if $\mathbf{K}$ and $\mathbf{D}$ are symmetric and $\mathbf{C} = \mathbf{B}^{\top}$, the following relation holds (Henderson, Kempthorne,

Searle, and von Krosigk 1959):

$$(\mathbf{K} + \mathbf{BDB}^\top)^{-1} = \mathbf{K}^{-1} - \mathbf{K}^{-1}\mathbf{B}(\mathbf{D}^{-1} + \mathbf{B}^\top\mathbf{K}^{-1}\mathbf{B})^{-1}\mathbf{B}^\top\mathbf{K}^{-1} \qquad (3.71)$$

which demands $\mathbf{D}$ be nonsingular. If $\mathbf{D}$ is singular, the following formula can be used (Harville 1976):

$$(\mathbf{K} + \mathbf{BDB}^\top)^{-1} = \mathbf{K}^{-1} - \mathbf{K}^{-1}\mathbf{BD}(\mathbf{I} + \mathbf{B}^\top\mathbf{K}^{-1}\mathbf{BD})^{-1}\mathbf{B}^\top\mathbf{K}^{-1} \qquad (3.72)$$

3.8.2.3   Exact reanalysis of trusses

Consider a truss subjected to static nodal loads $\mathbf{P}$. Let n denote the number of degrees of freedom, and $\mathbf{K}$ denote the $n \times n$ stiffness matrix. The nodal displacement vector $\mathbf{U}^0$ of the initial (reference) truss before modification is obtained from

$$\mathbf{KU}^0 = \mathbf{P} \qquad (3.73)$$

The matrix $\mathbf{K}$ is a function of the vector $\mathbf{A} = (A_1, \ldots, A_m)^\top$ of cross-sectional areas, where m is the number of members. Because the components of $\mathbf{K}$ are proportional to A_i, $\mathbf{K}$ is written as the sum of $A_i\mathbf{K}_i$ with $\mathbf{K}_i$ being the $n \times n$ stiffness matrix for unit cross-sectional area of the ith member. Note that the member properties are defined by the matrices and vectors of size n for simple presentation of equations. Computation is carried out, however, using element-size matrices and vectors.

Suppose the cross-sectional area of the kth member is increased by ΔA_k. Eq. (3.73) for the modified truss is formulated as

$$(\mathbf{K} + \Delta A_k\mathbf{K}_k)\mathbf{U}^1 = \mathbf{P} \qquad (3.74)$$

The objective here is to find the displacement vector $\mathbf{U}^1$ of the modified truss for the given load vector $\mathbf{P}$. In the following, the superscripts 0 and 1 denote the values of initial and modified trusses, respectively. It is important to note here that the inverse of $\mathbf{K}$ is not usually known; i.e., $\mathbf{K}$ is only decomposed into triangular matrices for computing $\mathbf{U}^0$. Therefore, it is very important to derive a more explicit formula without the inverse of the initial stiffness matrix for computing $\mathbf{U}^1$ of the modified truss (Ohsaki 2001a).

Let N_i^0 denote the axial force of the ith member of the initial truss. The relation between N_i^0 and $\mathbf{U}^0$ is written by using a vector $\mathbf{c}_i$ as

$$N_i^0 = A_i\mathbf{c}_i^\top\mathbf{U}^0 \qquad (3.75)$$

Let E and L_i denote the elastic modulus and the length of the ith member, and define vector $\mathbf{b}_i$ as

$$\mathbf{b}_i = \frac{L_i}{E}\mathbf{c}_i \qquad (3.76)$$

Then the relation between N_i^0 and the n-vector of equivalent nodal load vector $\mathbf{F}_i^0$ of the ith member is written as $\mathbf{F}_i^0 = N_i^0\mathbf{b}_i$. Hence, $\mathbf{b}_i$ denotes the nodal load vector corresponding to the unit axial force of the ith member. The matrix $\mathbf{K}_i$ is defined by using $\mathbf{c}_i$ and $\mathbf{b}_i$ as

$$\mathbf{K}_i = \mathbf{b}_i \mathbf{c}_i^\top \tag{3.77}$$

The inverse of the matrix $\mathbf{K} + \Delta A_k \mathbf{K}_k$ is obtained from (Henderson and Searle 1981)

$$\begin{aligned}
(\mathbf{K} + \Delta A_k \mathbf{K}_k)^{-1} &= \left(\mathbf{K} + \Delta A_k \mathbf{b}_k \mathbf{c}_k^\top\right)^{-1} \\
&= \mathbf{K}^{-1} - \frac{\Delta A_k \mathbf{K}^{-1} \mathbf{b}_k \mathbf{c}_k^\top \mathbf{K}^{-1}}{1 + \Delta A_k \mathbf{c}_k^\top \mathbf{K}^{-1} \mathbf{b}_k}
\end{aligned} \tag{3.78}$$

By post-multiplying $\mathbf{P}$ to (3.78), the following relation is derived:

$$\mathbf{U}^1 = \mathbf{U}^0 - \frac{\Delta A_k \mathbf{K}^{-1} \mathbf{b}_k \mathbf{c}_k^\top \mathbf{U}^0}{1 + \Delta A_k \mathbf{c}_k^\top \mathbf{K}^{-1} \mathbf{b}_k} \tag{3.79}$$

where (3.73) and (3.74) have been used.

Let superscript $k*$ denote the values of the initial truss subjected to the load vector $\mathbf{b}_k$. Then (3.79) is rewritten as

$$\begin{aligned}
\mathbf{U}^1 &= \mathbf{U}^0 - \frac{\Delta A_k \mathbf{U}^{k*} \mathbf{c}_k^\top \mathbf{U}^0}{1 + \Delta A_k \mathbf{c}_k^\top \mathbf{U}^{k*}} \\
&= \mathbf{U}^0 - \frac{\Delta A_k E \varepsilon_k^0}{1 + \Delta A_k E \varepsilon_k^{k*}} \mathbf{U}^0
\end{aligned} \tag{3.80}$$

where (3.75) has been used, and ε_k is the strain of the kth member. This way, $\mathbf{U}^1$ can be obtained without using $\mathbf{K}^{-1}$. The strain of the member with a null cross-sectional area may be computed from the displacements of the nodes connected to the member. Therefore, (3.80) can be used for addition or removal of a member between two existing nodes.

Next, we consider a case where the cross-sectional areas A_k of members ($k = 1, \ldots, q$) are modified simultaneously. The stiffness matrix of the modified truss is formulated as

$$\begin{aligned}
\mathbf{K} + \sum_{k=1}^{q} \Delta A_k \mathbf{K}_k &= \mathbf{K} + \sum_{k=1}^{q} \frac{\Delta A_k E}{L_k} \mathbf{b}_k \mathbf{b}_k^\top \\
&= \mathbf{K} + \mathbf{B} \mathbf{D} \mathbf{B}^\top
\end{aligned} \tag{3.81}$$

where the ith column of the $n \times q$ matrix $\mathbf{B}$ is $\mathbf{b}_i$, and $\mathbf{D}$ is a diagonal matrix defined as

$$\mathbf{D} = \operatorname{diag}(\Delta A_1 E / L_1, \ldots, \Delta A_q E / L_q) \tag{3.82}$$

Then, the inverse of the modified stiffness matrix is written as (3.71). By post-multiplying $\mathbf{P}$ on both sides of (3.71), the following relation is derived:

$$\begin{aligned}
\mathbf{U}^1 &= \mathbf{U}^0 - \mathbf{K}^{-1} \mathbf{B} (\mathbf{D}^{-1} + \mathbf{B}^\top \mathbf{K}^{-1} \mathbf{B})^{-1} \mathbf{D} \mathbf{B}^\top \mathbf{U}^0 \\
&= \mathbf{U}^0 - \mathbf{C}^* (\mathbf{D}^{-1} + \mathbf{B}^\top \mathbf{K}^{-1} \mathbf{B})^{-1} \mathbf{y}
\end{aligned} \tag{3.83}$$

TABLE 3.2: Results of reanalysis of displacements (mm).

	u_a	u_b	u_c
Initial	-2.23732	-2.43724	-3.05510
Case 1	-2.11225	-2.30694	-2.90569
Case 2	-2.20397	-2.38722	-3.02175
Case 3	-2.10492	-2.29644	-2.89815

where the ith component of the q-vector $\mathbf{y}$ is $\varepsilon_i^0 L_i$, and the ith column of the $n \times q$ matrix $\mathbf{C}^*$ is $\mathbf{U}^{i*}$.

A temporary variable vector $\mathbf{z}$ is defined as the solution of

$$(\mathbf{D}^{-1} + \mathbf{B}^\mathsf{T}\mathbf{K}^{-1}\mathbf{B})\mathbf{z} = \mathbf{y} \tag{3.84}$$

Eq. (3.84) is simply rewritten as

$$(\mathbf{D}^{-1} + \mathbf{Y}^*)\mathbf{z} = \mathbf{y} \tag{3.85}$$

where the (i,j)-component of $\mathbf{Y}^*$ is equal to $\varepsilon_j^{i*}L_j$; i.e., the inverse matrix $\mathbf{K}^{-1}$ need not be explicitly computed also for this case. Then $\mathbf{U}^1$ is obtained from

$$\mathbf{U}^1 = \mathbf{U}^0 - \mathbf{U}^*\mathbf{z} \tag{3.86}$$

For the case of the modification of a single variable, i.e., $q = 1$, the first component of $\mathbf{y}$ is $\varepsilon_1^0 L_1$, and the $(1,1)$-components of $\mathbf{D}$ and $\mathbf{Y}^*$ are $\Delta A_1 E/L$ and $\varepsilon_1^{1*}L_1$, respectively. Hence, the first component z_1 of $\mathbf{z}$ is obtained as

$$z_1 = \frac{\Delta A_1 E \varepsilon_1^0}{1 + \Delta A_1 E \varepsilon_1^{1*}} \tag{3.87}$$

Therefore, (3.80) for modification of a single variable is a special case of (3.86) for modification of multiple variables.

Example 3.4

As an example, the nodal displacements after modification of cross-sectional areas are computed using exact reanalysis for a 2×2 plane grid truss, as shown in Fig. 3.21. The length of the members in the x- and y-directions are 2.0 m, and $P = 100.0$ kN. The elastic modulus is 205.8 kN/mm^2. The displacements u_a, u_b, and u_c in the y-direction of nodes 'a', 'b', and 'c', respectively, for the initial truss with $A_i = 1000.0$ mm^2 for $i \neq 1$ and $A_1 = 0$ are as shown in the first row of Table 3.2, where members 1, 2, and 3 are defined in Fig. 3.21. Reanalysis has been carried out for the following three cases:

Case 1: Increase A_2 from 1000.0 mm^2 to 1500.0 mm^2 and use (3.80) for reanalysis.

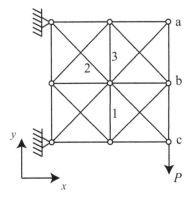

FIGURE 3.21: A 2×2 plane grid truss.

Case 2: Increase A_1 from 0 to 500.0 mm^2 and use (3.80) for reanalysis.

Case 3: Increase A_2 and A_3 to 1500.0 mm^2 simultaneously and use (3.85) and (3.86).

The results are listed in Table 3.2. It is confirmed that exact responses have been found for all the cases, including Case 2, corresponding to the addition of a member.

3.8.3   Random search for topology optimization of trusses

3.8.3.1   General algorithm

We consider truss optimization with inequality constraints for the response quantities $\widehat{H}_j(\mathbf{A})$, e.g., nodal displacements and member stresses, which are given as

$$\widehat{H}_j(\mathbf{A}) \le H_j^{\mathrm{U}}, \quad (j = 1, \cdots, n^{\mathrm{I}}) \tag{3.88}$$

where H_j^{U} is the upper bound for $\widehat{H}_j(\mathbf{A})$, and n^{I} is the number of constraints. Note that the lower-bound constraints are also converted into the standard form and included in (3.88). Then the constraints are normalized assuming $H_j^{\mathrm{U}} > 0$ as

$$H_j(\mathbf{A}) = \frac{\widehat{H}_j(\mathbf{A})}{H_j^{\mathrm{U}}} - 1 \le 0, \quad (j = 1, \dots, n^{\mathrm{I}}) \tag{3.89}$$

The objective function $V(\mathbf{A})$ is the total structural volume. The value of A_i is to be selected from the list $\mathcal{A} = \{a_1, \dots, a_r\}$, where r is the number of available sections. Then, the optimization problem is formulated as

$$\text{Minimize} \quad V(\mathbf{A}) \tag{3.90a}$$

$$\text{subject to} \quad H_j(\mathbf{A}) \leq 0, \quad (j = 1, \cdots, n^{\mathrm{I}}) \tag{3.90b}$$

$$A_i \in \mathcal{A} \tag{3.90c}$$

In order to show the effectiveness of the exact reanalysis method applied to topology optimization of a truss, a random search algorithm is used with a technique for improving convergence to the global optimal solution; see Appendix A.3 for the basic algorithm of random search. The simplest approach of utilizing a random search for a constrained optimization problem may be to add a penalty term of the constraints to the objective function. In this case, however, it is very difficult to assign an appropriate value for the penalty coefficient so that the magnitude of the objective function is equivalent to that of the penalty term. Therefore, the penalty for violating the constraints is implicitly given here by the threshold level for defining the increase or decrease of the design variables.

Assuming that $H_j(\mathbf{A})$ is a decreasing function of $\mathbf{A}$, which usually holds for static responses, the optimization algorithm is presented as follows:

Step 1: Assign the initial values of $A_1, \ldots, A_m$ from the given list $\mathcal{A}$, and set parameters $\overline{H}$, τ, and d.

Step 2: Evaluate the objective function $V(\mathbf{A})$ and the constraints $H_j(\mathbf{A})$, and find the maximum value H^{max} among $H_j(\mathbf{A})$ $(j = 1, \ldots, n^{\mathrm{I}})$. At this step, the stiffness matrix $\mathbf{K}$ is reconstructed and its Cholesky decomposition is carried out.

Step 3: Assign the parameter $\lambda = 0.5 + \tau H^{\mathrm{max}}$, where τ (> 0) adjusts the threshold level for increase or decrease of the variable. The parameter λ is replaced with 0 or 1 if $\lambda < 0$ or $\lambda > 1$, respectively. As is seen in Step 4, λ defines the probability of increase of the cross-sectional area of a member. If $H^{\mathrm{max}} < 0$, i.e., if all the constraints are satisfied, then $\lambda < 0.5$ and the cross-sectional areas have a small probability of increase.

Step 4: Generate a uniform random number between 0 and 1 to select the member k for which the cross-sectional area is to be modified. Suppose r allowable values $a_1, \ldots, a_r$ in $\mathcal{A}$ are listed in increasing order, and $A_k = a_i$ at the current step. Generate another random number $0 \leq p < 1$ to decide an increase or decrease of A_k. Calculate the smallest integer q, which defines the amount of increase of A_k, as an integer that is not less than $|\lambda - p|/d$ with d being the scaling parameter. Modify A_k to a_j if $p < \lambda$, where $j = \min\{r, i + q\}$; otherwise decrease A_k to a_j with $j = \max\{1, i - q\}$. Note that a smaller d leads to a larger modification of A_k.

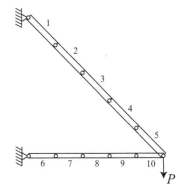

FIGURE 3.22: Optimal solution for the 5 × 5 plane grid.

Step 5: Compute constraint functions for the modified design using the exact reanalysis method. Accept the design if $H^{\max} < \overline{H}$, and go to Step 2 if the number of accepted designs is less than the specified limit.

Step 6: Go to Step 4 if $H^{\max} \geq \overline{H}$ and the number of trials at this stage is less than the specified limit.

Step 7: Select the best design satisfying $H^{\max} < 0$ from the list of accepted designs and terminate the process.

3.8.3.2 Numerical examples

Optimal topologies are found for plane square grids under stress constraints. The ground structure for a 2 × 2 grid is shown in Fig. 3.21. The length of the members in the horizontal and vertical directions are 2.0 m, and $P = 200.0$ kN. The upper bound for the absolute value of stress is 40.0 N/mm². The elastic modulus is 2.058×10^5 N/mm². The parameters are $H^{\mathrm{U}} = 0.3$, $\tau = 10.0$, and $d = 0.1$. In the following, the units of force and length are kN and mm, respectively.

The list of available cross-sectional areas is given, including the small value 1.0 as $\mathcal{A} = \{1.0, 2000.0, 4000.0, 6000.0, 8000.0\}$. Note that the member with $A_i = 1.0$ is to be removed after optimization and the stress constraint need not be satisfied by such a member. An artificial upper bound U^{U} is given for the nodal displacements to prevent convergence to the obvious and meaningless solution such that $A_i = 1.0$ for all the members (see Sec. 3.7). The initial solution is given as $A_i = 8000.0$ for all members.

Optimization has been carried out for 2 × 2, 3 × 3, 4 × 4, and 5 × 5 grids starting with 100 different random initial solutions for each case. The limit for the number of modifications of the cross-sectional area is 5000, and the limit for the trial for generating an acceptable neighborhood solution at each step is

TABLE 3.3: Result of optimization with 100 different random seeds.

	U^{U}	s^{ave}	t^{ave}	V^{ave} ($\times 10^8$)	V^{min} ($\times 10^8$)	V^{max} ($\times 10^8$)	n^{opt}
2×2	2.5	469	8278	0.072970	0.069292	0.89280	77
3×3	5.0	288	4303	1.3094	1.0397	1.7994	26
4×4	7.0	564	7933	1.8443	1.3866	2.5491	8
5×5	10.0	1026	14,297	2.5671	1.7227	3.6490	3

2000. Let s^{ave} and t^{ave} denote the average numbers of steps and trial analyses before reaching the minimum objective value within 5000 steps. Average, minimum, and maximum values of the minimum objective value within 5000 steps among 100 different random initial solutions are denoted by V^{ave}, V^{min}, and V^{max}, respectively.

Table 3.3 shows the values of U^{U}, s^{ave}, t^{ave}, V^{ave}, V^{min}, and V^{max} for each type of grid, where n^{opt} is the number of trials that reached the solution with $V = V^{\mathrm{min}}$; i.e., the possibility of obtaining the optimal solution decreases as the problem size is increased. It has been confirmed that an appropriate value of U^{U} leads to rapid convergence to nearly optimal solutions. The value of U^{U} is a little greater than the maximum displacement of the truss with optimal topology for each case; i.e., the displacement constraint is not active.

The optimal solution of the 5×5 grid is shown in Fig. 3.22. The cross-sectional areas are 6000.0 for members 6–10 and 8000.0 for the remaining existing members. The value of V^{min} for each case in Table 3.3 corresponds to the objective value of the apparent optimal solution similar to the truss in Fig. 3.22. It may be observed from Table 3.3 that the average number s^{ave} of analysis that needs decomposition of the stiffness matrix is less than 10% of the average number t^{ave} of trial analysis that can be done using formulas (3.85) and (3.86) of exact reanalysis. The ratio of CPU time for obtaining optimal topology of the 5×5 grid by using the reanalysis method to that without reanalysis is 0.2250. Therefore, computational cost may be drastically reduced utilizing the method of exact reanalysis.

3.9 Multiple eigenvalue constraints

3.9.1 Introduction

The eigenvalues of free vibration as well as the linear buckling load factor, which is obtained by solving an eigenvalue problem, are important performance measures of trusses and frames. Therefore, there have been many studies of optimization under eigenvalue constraints. If the fundamental eigenvalue of the optimum design is simple, the optimization problem may easily be solved using a nonlinear programming or an optimality criteria approach (Venkayya and Tishler 1983; Sadek 1989), because there is no difficulty in calculating the sensitivity coefficients of the eigenvalue with respect to the design variable, as presented in Sec. 2.3.

However, it is well known that the optimum designs for a specified fundamental eigenvalue often have multiple (repeated) values. Such an optimal structure was presented by Olhoff and Rasmussen (1977), where necessary conditions for optimality of a column with a variable cross-section are discussed in a variational form, and an optimal column under buckling constraint is found using an optimality criteria approach. Masur (1984) showed that the necessary conditions of Olhoff and Rasmussen (1977) are also sufficient conditions in the case of the bimodal optimal solution of symmetric structures. Early developments in this field are summarized by Olhoff (1980).

Difficulties in optimizing distributed parameter structures (continuum) for specified multiple eigenvalues are discussed extensively in Haug and Cea (1981). It has been shown that multiple eigenvalues are not differentiable in the ordinary sense, and only directional derivatives or subgradients with respect to the design variables may be calculated. Bochenek and Gajewski (1986) found optimal cross-sectional areas of arches that have at most three-fold eigenvalues corresponding to in-plane and out-of-plane buckling modes. For finite dimensional structures with a moderately large number of design variables, however, it is very difficult to find optimal structures with multiple eigenvalues using conventional optimality criteria approaches or nonlinear programming algorithms.

Several computational approaches have been developed for sensitivity analysis of multiple eigenvalues of finite dimensional structures (Haug and Choi 1982; Haug, Choi, and Komkov 1986; Seyranian 1993) (see Sec. 2.3 for sensitivity analysis of double eigenvalue of vibration). Khot and Kamat (1985) presented an optimality criteria approach for trusses with multiple frequency constraints. Rodorigues, Guedes, and Bendsøe (1995) developed necessary conditions for optimality for problems under constraints on the linear buckling load factor and presented an adjoint variable formulation for sensitivity analysis. Developments in this field are summarized in Seyranian, Lund, and Olhoff (1994).

In spite of the theoretical developments of sensitivity analysis and optimization methods for problems under multiple eigenvalue constraints, it is very difficult to guarantee global convergence for large-scale structures. Sergeyev and Mróz (2000) presented a gradient-based approach that can be applied to the bimodal case and investigated the properties of optimal solution for the problem under stress and frequency constraints. Nakamura and Ohsaki (1988) presented a parametric programming approach to trace a set of optimal solutions under multiple eigenvalue constraints and extended it to truss topology optimization (Nakamura and Ohsaki 1992). Their method has been shown to be applicable to topology optimization of frames (Ohsaki and Naka-mura 1993). Although their method is effective for the bimodal case, it is very difficult to extend it to obtain optimal solutions with a larger multiplicity of eigenvalues.

As an alternative approach, semidefinite programming (SDP) has been shown to be effective for structural optimization problems with multiple eigen-values because it does not need explicit derivatives of eigenvalues (see Appendix A.2.4 for details of SDP). SDP can also be applicable to problems with compliance constraints (de Klerk, Roos, and Terlaky 1995) and robust design (Ben-Tal and Nemirovski 1997). Achtziger and Kočvara (2007a, 2007b) presented a nonlinear SDP approach for the problem of minimizing the lowest eigenvalue under constraints on the structural volume and compliance considering multiple load cases. Sequential application of SDP is also effective for the problem with linear buckling constraints (Kanno, Ohsaki, and Katoh 2001).

3.9.2 Multiple eigenvalues in optimal topology

Let $\mathbf{M}^{s}(\mathbf{A})$ and $\mathbf{M}^{0}$ denote the $n \times n$ mass matrices corresponding to the structural and nonstructural masses, respectively, where n is the number of degrees of freedom. Note that $\mathbf{M}^{s}$ is a function of the vector $\mathbf{A}$ of m design variables. The eigenvalue problem of free vibration is formulated as

$$\mathbf{K}\mathbf{\Phi}_r = \Omega_r(\mathbf{M}^{0} + \mathbf{M}^{s})\mathbf{\Phi}_r, \quad (r = 1, \ldots, n) \tag{3.91}$$

where Ω_r and $\mathbf{\Phi}_r$ are the rth eigenvalue and eigenmode, respectively, which are implicit functions of $\mathbf{A}$. The vector $\mathbf{\Phi}_r$ is ortho-normalized by

$$\mathbf{\Phi}_r^{\top}\mathbf{M}\mathbf{\Phi}_s = \delta_{rs}, \quad (r, s = 1, \ldots, n) \tag{3.92}$$

where δ_{rs} is the Kronecker delta.

The objective here is to find an optimal solution for minimizing the total structural volume $V(\mathbf{A})$ under constraint such that the fundamental eigenvalue Ω_1 is not less than the specified lower bound Ω^{L}. Considering the occurrence of multiple eigenvalues, the optimization problem is formulated as

$$\text{Minimize} \quad V(\mathbf{A}) = \sum_{i=1}^{m} A_i L_i \tag{3.93a}$$

$$\text{subject to} \quad \Omega_r(\mathbf{A}) \geq \Omega^{L}, \quad (r = 1, \cdots, n) \tag{3.93b}$$

$$A_i \geq A_i^{L} \quad (i = 1, \cdots, m) \tag{3.93c}$$

where L_i is the length of the ith member, and a small positive lower bound A_i^{L} is usually given for A_i in order to prevent instability of the truss. The optimal topology is found by removing the members with $A_i = A_i^{L}$ as the result of optimization.

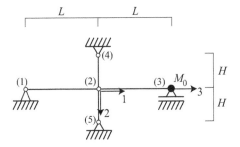

FIGURE 3.23: A simple four-bar truss with concentrated mass.

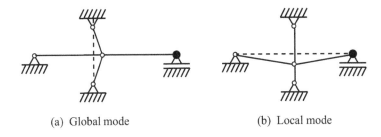

(a) Global mode (b) Local mode

FIGURE 3.24: Eigenmodes of the four-bar truss.

Example 3.5

Systematic occurrence of multiple eigenvalues is illustrated using a small example of a four-bar truss, as shown in Fig. 3.23, that has a nonstructural mass at the horizontal roller support. The cross-sectional areas of horizontal and vertical members are denoted by A_1 and A_2, respectively; i.e., we have two design variables. Suppose, for simplicity, the parameters are given without units as $L = 1$, $H = L/2 = 0.5$, and $M_0 = 1$. The elastic modulus is 1, and the mass density of the members is 0.01.

If the cross-sectional areas are given as $A_1 = A_2 = 1$, which leads to a very small structural mass compared with the nonstructural mass, the eigenvalues are 0.4957, 133.8, and 266.7; i.e., the lowest eigenvalue is much smaller than the second and third eigenvalues. The lowest eigenmode is such that the center node and roller support move horizontally, and the axial deformation of the vertical members is zero, as illustrated in Fig. 3.24(a). This type of mode associated with vibration of nonstructural masses is referred to as the *global mode*.

Suppose $A_1 = A_2 = 1$ is given as the initial solution for optimization with lower-bound eigenvalue $\Omega^L = 0.4957$. Since the vertical members do not deform in the lowest mode, A_2 decreases as the optimization process proceeds, while A_1 is almost constant. It is obvious, however, that this pin-jointed truss is unstable if the vertical members are removed. Consequently, A_2 should have an extremely small value, and the two lowest eigenvalues coincide even in the

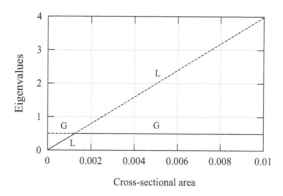

FIGURE 3.25: Relation between the eigenvalues and cross-sectional area A_2 of the four-bar truss; G: global mode, L: local mode.

optimal solution of this simple truss. One of the lowest eigenmodes is such that the center node vibrates vertically without vibration of the nonstructural mass, as shown in Fig. 3.24(b). This type of mode is referred to as the *local mode*. The two lowest eigenvalues are plotted in Fig. 3.25 for a range of small values of A_2 with $A_1 = 1$, where the solid and dotted lines indicate the lowest and second modes, respectively, and 'L' and 'G' indicate the local mode and global mode, respectively. From the practical point of view, however, optimal topology with extremely thin vertical members, which are conceived as *secondary members*, is not needed, and designers are not interested in the local mode that is simply suppressed by adding flexural stiffness at the center node.

3.9.3   Semidefinite programming for topology optimization

As noted earlier, the convergence of the nonlinear programming algorithm for Problem (3.93) deteriorates if the fundamental eigenvalue becomes multiple. In contrast, the optimal solution with multiple eigenvalues can be found without any difficulty by converting Problem (3.93) into an SDP problem (Wolkowicz, Saigal, and Vandenberghe 2000; Ohsaki, Fujisawa, Katoh, and Kanno 1999). As we have shown in Sec. 1.9, the eigenvalue constraint with the prescribed lower bound Ω^L is converted to the positive semidefiniteness of the $n \times n$ matrix $\mathbf{X}$ defined as

$$\mathbf{X} = \mathbf{K} - \Omega^L(\mathbf{M}^s + \mathbf{M}^0) \tag{3.94}$$

For a truss, we have

$$\mathbf{K} = \sum_{i=1}^{m} A_i \mathbf{K}_i, \quad \mathbf{M} = \sum_{i=1}^{m} A_i \mathbf{M}_i + \mathbf{M}^0 \tag{3.95}$$

where $\mathbf{K}_i$ and $\mathbf{M}_i$ are constant $n \times n$ matrices.

Let $\mathbf{X} \succeq \mathbf{O}$ indicate that the symmetric matrix $\mathbf{X}$ is positive semidefinite. The inner product $\mathbf{Y} \bullet \mathbf{Z}$ of the $n \times n$ matrices $\mathbf{Y}$ and $\mathbf{Z}$, for which their (i,j)-components are denoted by Y_{ij} and Z_{ij}, respectively, is defined as

$$\mathbf{Y} \bullet \mathbf{Z} = \sum_{i=1}^{n}\sum_{j=1}^{n} Y_{ij}Z_{ij} \qquad (3.96)$$

Then, using (3.94), (3.95), and $\mathbf{X} \succeq \mathbf{O}$, the primal problem of SDP for the topology optimization problem is formulated as

$$\text{Minimize} \quad V(\mathbf{A}) = \sum_{i=1}^{m} A_i L_i \qquad (3.97a)$$

$$\text{subject to} \quad \sum_{i=1}^{m}(\mathbf{K}_i - \Omega^L \mathbf{M}_i)A_i - \Omega^L \mathbf{M}^0 \succeq \mathbf{O} \qquad (3.97b)$$

$$A_i \geq A_i^{\mathrm{L}}, \quad (i = 1, \cdots, m) \qquad (3.97c)$$

where A_i^{L} is a small positive lower bound for A_i. The dual of the SDP problem (3.97) is formulated as

$$\text{Maximize} \quad \Omega^L \mathbf{M}_0 \bullet \mathbf{Y} \qquad (3.98a)$$

$$\text{subject to} \quad (\mathbf{K}_i - \Omega^L \mathbf{M}_i) \bullet \mathbf{Y} + \eta_i = L_i, \quad (i = 1, \ldots, m) \qquad (3.98b)$$

$$\eta_i \geq 0, \quad (i = 1, \ldots, m) \qquad (3.98c)$$

$$\mathbf{Y} \succeq \mathbf{O} \qquad (3.98d)$$

where η_i is a slack variable.

Notice here that Problems (3.97) and (3.98) are formally categorized as dual and primal forms of SDP problems, respectively. However, we regard Problem (3.97) as a primal problem, because it is derived from the original problem (3.93). Problems (3.97) and (3.98) may be solved simultaneously using the primal-dual interior-point method (Mehrotra 1992). It is important to note here that the sensitivity coefficients of eigenvalues with respect to the design variables $\mathbf{A}$ are not needed in the optimization process. Therefore, there is no difficulty, as will be shown in the examples, in finding the solutions with multiple fundamental eigenvalues. Furthermore, it is very important in engineering applications that a symmetric solution is always found using an interior-point method without explicit assignment of symmetry conditions, if the structure is symmetric, and a symmetric initial solution is assigned (Kanno, Ohsaki, Murota, and Katoh 2001; Kanno, Ohsaki, and Katoh 2002). Note that the eigenvalue optimization problem of a frame with a solid section turns out to be a nonlinear SDP problem, because $\mathbf{K}$ and/or $\mathbf{M}$ are nonlinear functions of the design variables representing the cross-sectional parameters (Kanno and Ohsaki 2007).

SDP has been shown to be effective for topology optimization of trusses considering compliance under given static loads. The problem under constraints on eigenvalues as well as compliances against multiple loading conditions $\mathbf{P}^1, \ldots, \mathbf{P}^{n^{\mathrm{P}}}$ is formulated as the following SDP problem (Achtziger and Kočvara 2007b):

$$\text{Minimize} \quad V \tag{3.99a}$$

$$\text{subject to} \quad \begin{pmatrix} W^{\mathrm{U}} & -\mathbf{P}^{k\top} \\ -\mathbf{P}^k & \mathbf{K}(\mathbf{A}) \end{pmatrix} \succeq \mathbf{O}, \quad (k = 1, \ldots, n^{\mathrm{P}}) \tag{3.99b}$$

$$\sum_{i=1}^{m} A_i L_i \leq V \tag{3.99c}$$

$$\mathbf{K}(\mathbf{A}) - \Omega^{\mathrm{L}}(\mathbf{M}^{\mathrm{s}}(\mathbf{A}) + \mathbf{M}^0) \succeq \mathbf{O} \tag{3.99d}$$

$$A_i \geq 0, \quad (i = 1, \cdots, m) \tag{3.99e}$$

where the variables are $\mathbf{A}$ and V, and W^{U} is the upper bound for the compliance. Note that (3.95) is to be used for explicit formulation of the constraints with respect to $\mathbf{A}$.

Many interior-point methods for linear and quadratic programming have been extended to solve SDPs (Kojima, Shindoh, and Hara 1997; Sturm 1999). Several free libraries, e.g., SDPA (Fujisawa, Kojima, and Nakata 1997), are available. In the following examples, we use SeDuMi Ver. 1.1 (Sturm 1999; Pólik 2005).

3.9.4   Linear buckling constraint

The linear buckling load factor is a basic performance measure of stability of structures mainly for thin-walled structures and frames with slender members. The multiple buckling load factors of the optimal solution were first presented for columns by Olhoff and Rasmussen (1977). Similar to the case of the eigenvalue of vibration, multiple buckling load factors are not continuously differentiable, and the optimization problem is effectively formulated using an SDP problem with matrix inequality constraints. However, contrary to the vibration problem, the geometrical stiffness matrix for formulating the linear buckling problem depends on internal forces that are usually nonlinear functions of the design variables. Therefore, in this section, an optimization method is presented for the nonlinear SDP problem based on the successive solution of linearized SDPs (Kanno, Ohsaki, and Katoh 2001).

The optimal cross-sectional areas of trusses subjected to proportional loads are to be found so as to minimize the total volume for the specified linear buckling load factor. The quasi-static proportional loads are defined by the load factor Λ and the specified vector $\mathbf{P}_0$ of the load pattern as $\mathbf{P} = \Lambda \mathbf{P}_0$. Let $\mathbf{K}_{\mathrm{G}}$ denote the geometrical stiffness matrix corresponding to the unit load factor. The rth linear buckling load factor Λ_r and the corresponding buckling mode $\boldsymbol{\Phi}_r$ are defined by

$$(\mathbf{K} + \Lambda_r \mathbf{K}_G)\boldsymbol{\Phi}_r = \mathbf{0}, \quad (r = 1, \ldots, n) \tag{3.100}$$

Note that $\mathbf{K}_G$ is generally indefinite and there exist positive and negative buckling load factors as the solution of (3.100). However, we are concerned with the smallest positive value of Λ_r, because the directions of the loads are usually fixed. Therefore, the reciprocals of load factors $1/\Lambda_r$ are numbered in nonincreasing order, i.e., Λ_1 is the smallest positive buckling load factor.

The positive lower bound Λ^L is specified for positive Λ_r; however, no constraint is given for the negative values of Λ_r, i.e., Λ_r should satisfy

$$\Lambda_r \geq \Lambda^L \text{ or } \Lambda_r < 0, \quad (r = 1, \ldots, n) \tag{3.101}$$

which is alternatively written as

$$\frac{1}{\Lambda_r} \leq \frac{1}{\Lambda^L}, \quad (r = 1, \ldots, n) \tag{3.102}$$

Hence, the topology optimization problem for minimizing the total structural volume under linear buckling constraints is formulated as

$$\text{Minimize} \quad V(\mathbf{A}) = \sum_{i=1}^m A_i L_i \tag{3.103a}$$

$$\text{subject to} \quad \frac{1}{\Lambda_r(\mathbf{A})} \leq \frac{1}{\Lambda^L}, \quad (r = 1, \ldots, n) \tag{3.103b}$$

$$A_i \geq A_i^L, \quad (i = 1, \ldots, m) \tag{3.103c}$$

where A_i^L is a small positive lower bound for A_i.

The generalized eigenvalue problem (3.100) is reformulated as

$$-\mathbf{K}_G \boldsymbol{\Phi}_r = \frac{1}{\Lambda_r} \mathbf{K} \boldsymbol{\Phi}_r, \quad (r = 1, \ldots, n) \tag{3.104}$$

Using Rayleigh's principle and the positive definiteness of $\mathbf{K}$, the condition (3.103b) is reduced to

$$-\frac{\boldsymbol{\Psi}^\top \mathbf{K}_G \boldsymbol{\Psi}}{\boldsymbol{\Psi}^\top \mathbf{K} \boldsymbol{\Psi}} \leq \frac{1}{\Lambda^L} \tag{3.105}$$

for any n-vector $\boldsymbol{\Psi}$; see Appendix A.1.2. Eq. (3.105) is then rewritten as the following condition of positive semidefiniteness of a matrix:

$$\frac{1}{\Lambda^L} \mathbf{K} + \mathbf{K}^G \succeq \mathbf{O} \tag{3.106}$$

Furthermore, $\mathbf{K}_G$ is written as follows as a linear function of the vector $\mathbf{N}(\mathbf{U}) = (N_1(\mathbf{U}), \ldots, N_m(\mathbf{U}))^\top$ of member forces against $\mathbf{P}_0$, which are functions of nodal displacement vector $\mathbf{U}$:

$$\mathbf{K}_G = \sum_{i=1}^m N_i(\mathbf{U}) \mathbf{K}_{Gi} \tag{3.107}$$

where $\mathbf{K}_{\mathrm{G}i}$ is an $n \times n$ symmetric constant matrix. Consequently, Problem (3.103) can be equivalently reformulated as

$$\text{Minimize} \quad V(\mathbf{A}) = \sum_{i=1}^{m} A_i L_i \tag{3.108a}$$

$$\text{subject to} \quad \frac{1}{\Lambda^{\mathrm{L}}} \sum_{i=1}^{m} A_i \mathbf{K}_i + \sum_{i=1}^{m} N_i(\mathbf{U}) \mathbf{K}_i^{\mathrm{G}} \succeq \mathbf{O} \tag{3.108b}$$

$$\sum_{i=1}^{m} A_i \mathbf{K}_i \mathbf{U} = \mathbf{P}_0 \tag{3.108c}$$

$$A_i \geq A_i^{\mathrm{L}}, \quad (i = 1, \ldots, m) \tag{3.108d}$$

Note that Problem (3.108) cannot be formulated as an SDP problem, because the equality constraints (3.108c), i.e., the equilibrium equations, are nonlinear with respect to $\mathbf{A}$ and $\mathbf{U}$ for a statically indeterminate structure. Ben-Tal, Jarre, Kočvara, Nemirovski, and Zowe (2000) proposed a successive SDP algorithm that solves Problem (3.108) using the semidefinite relaxation of (3.108c).

Kanno, Ohsaki, and Katoh (2001) proposed a method based on sequential linearization of the nonlinear SDP problem (3.108). Because $\mathbf{U}$ is an implicit function of $\mathbf{A}$ that is defined by (3.108c), $\mathbf{N}(\mathbf{U})$ is also regarded as a function of $\mathbf{A}$, which is denoted by $\widetilde{\mathbf{N}}(\mathbf{A})$. Let the superscript (k) indicate the iteration counter for the sequential SDP. The axial forces $\widetilde{N}_i(\mathbf{A})$ $(i = 1, \ldots, m)$ are approximated by $\widetilde{N}_i^{(k)}(\mathbf{A})$ defined by the values at $\mathbf{A} = \mathbf{A}^{(k)}$ using two methods as

$$\text{Method (a)}: \quad \widetilde{N}_i^{(k)}(\mathbf{A}) = N_i(\mathbf{A}^{(k)}) + \nabla N_i(\mathbf{A}^{(k)})^{\top}(\mathbf{A} - \mathbf{A}^{(k)}) \tag{3.109a}$$

$$\text{Method (b)}: \quad \widetilde{N}_i^{(k)}(\mathbf{A}) = N_i(\mathbf{A}^{(k)}) \tag{3.109b}$$

In Method (a), N_i is approximated as a linear function of the design variables $\mathbf{A}$ utilizing the sensitivity coefficients of $\mathbf{N}$. Method (b) simply fixes N_i at the current solution $\mathbf{A}^{(k)}$.

Note that the sensitivity coefficients of buckling load factors are not used in these formulations. Therefore, the optimal design with a large multiplicity of buckling load factors can be found without any difficulty. It should be emphasized that a solution obtained by Method (a) satisfies the first-order optimality conditions of Problem (3.108). Therefore, the solution is guaranteed to be a stationary point of Problem (3.108). On the contrary, the optimality conditions are not generally satisfied by the solution obtained using Method (b).

3.9.5 Numerical examples

Optimal topologies are found for plane and space trusses under constraints on eigenvalues of free vibration. The material of the members is steel with

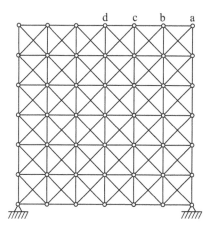

FIGURE 3.26: A 6 × 6 plane square grid.

the elastic modulus $E = 200 \text{ kN/mm}^2$ and the mass density $\rho = 7.86 \times 10^{-6} \text{ kg/mm}^3$.

SeDuMi Ver. 1.1 (Sturm 1999) is used for optimization. Note that Se-DuMi can find an optimal solution without explicit assignment of a positive lower bound for A_i. However, the minimum cross-sectional area $A_i^L = 1.0 \times 10^{-6} \text{ mm}^2$ is given to prevent the existence of a negative cross-sectional area at the optimal solution due to the nonzero tolerance of the constraints, and also to prevent inaccuracy of the eigenvalue analysis that is carried out for verification purposes. The computation was carried out on a PC (Xeon 3.8GHz with 2.0GB memory).

Consider first a 6 × 6 plane square grid, as shown in Fig. 3.26, where the lengths of the vertical and horizontal members are 200 mm. The optimal topologies are shown in Figs. 3.27(a)–(d), where the concentrated non-structural mass of 1000.0 kg is located at nodes 'a'–'d', respectively. The specified eigenvalue is 100.0 rad^2/s^2 for all cases. The width of each member in Fig. 3.27 is proportional to its cross-sectional area, and the members with $A_i = A_i^L$ have been removed. Therefore, the unstable nodes at which two colinear members are connected should be fixed after removal of the thin members. As is seen from Fig. 3.27(d), the symmetric optimal solution has been successfully found, for the case where the mass exists at the center node 'd', without explicit assignment of symmetry constraints. Especially in Fig. 3.27(a), a kind of net with small cross-sectional areas should exist to prevent instability of the longer members consisting of large cross-sectional areas.

The three lowest eigenvalues, total volume, number of iterations in the primal-dual interior-point algorithm, and CPU time are listed in Table 3.4. For the case with a mass at node 'a', the fundamental eigenvalue is duplicated as $\Omega_1 = \Omega_2 = 100.0 \text{ rad}^2/\text{s}^2$, where the corresponding eigenmodes are shown

TABLE 3.4: Eigenvalues (rad^2/s^2), total volume ($\times 10^7$mm^3), number of iterations, and CPU time (s) of the plane square 6 × 6 grid.

Location of mass	'a'	'b'	'c'	'd'
Ω_1 (rad^2/s^2)	100.00	100.00	100.00	100.00
Ω_2 (rad^2/s^2)	100.00	108.08	103.47	100.00
Ω_3 (rad^2/s^2)	115.98	131.12	125.92	102.66
Total volume	8.4811	7.8047	7.4250	7.3065
No. of iterations	29	21	20	22
CPU time	5.28	3.34	3.23	3.91

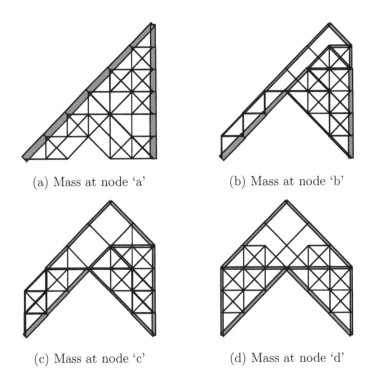

(a) Mass at node 'a'   (b) Mass at node 'b'

(c) Mass at node 'c'   (d) Mass at node 'd'

FIGURE 3.27: Optimal topologies of the 6 × 6 plane square grid.

in Fig. 3.28. Note that the mode in Fig. 3.28(a) is a global mode in which the node with nonstructural mass vibrates, whereas the mode in Fig. 3.28(b) is a local mode in which the node with mass does not move. As is seen from Table 3.4, the two lowest eigenvalues coincide if the mass exists at node 'a' or 'd'. For the case with a mass at node 'd', the two lowest eigenmodes are antisymmetric and symmetric as shown in Fig. 3.29. The number of iterations and the CPU time do not depend on the location of the mass or

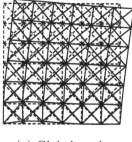

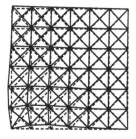

(a) Global mode   (b) Local mode

FIGURE 3.28: The two lowest eigenmodes of the optimal 6×6 plane square grid with mass at node 'a'.

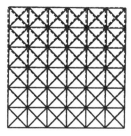

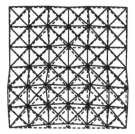

(a) Antisymmetric mode   (b) Symmetric mode

FIGURE 3.29: The two lowest eigenmodes of the optimal 6×6 grid with mass at node 'd'.

the multiplicity of the eigenvalues.

We next find optimal solutions for double-layer cylindrical grid trusses with different numbers of grids, where the configuration of the 7×7 grid is shown in Fig. 3.30. The truss has pin supports at the four lower corners and the lumped mass of 1000 kg at each node. Let n^{G} denote the number of units in both the longitudinal and span directions of the upper surface. The span length (m) between the supports is $2(n^{\mathrm{G}}-1)$ for the two directions. The open angle of the lower cylinder is 80 deg., and the distance between the lower and upper cylinders is 2 m. The lower and upper chords in each direction, and the diagonals have the same lengths, respectively. The specified fundamental eigenvalue is 1000 rad^2/s^2.

The optimal cross-sectional areas are shown in Figs. 3.31(a)–(d). As is seen, the lower chords and the diagonals near the supports as well as the upper chords near the center lines have large cross-sectional areas. The three eigenvalues coincide at the optimal solutions, as shown in the second column in Table 3.5, and the symmetry of the cross-sectional areas is attained without

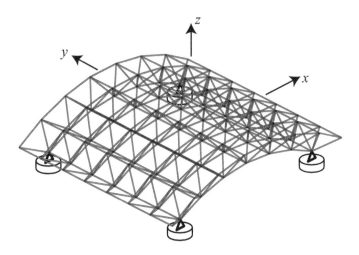

FIGURE 3.30: A 7×7 double-layer cylindrical grid.

TABLE 3.5:   Eigenvalues $(\mathrm{rad}^2/\mathrm{s}^2)$, total volume (m^3), number of iterations, and CPU time (s) of the cylindrical double-layer grid.

n^{G}	7	8	9	10
n	327	423	531	651
m	392	512	648	800
Ω_1	1000.0	1000.0	1000.0	1000.0
Ω_2	1000.0	1000.0	1000.0	1000.0
Ω_3	1000.0	1000.0	1000.0	1000.0
Ω_4	1134.0	1217.1	1287.2	1350.7
Total volume	1.9259	4.2570	8.9832	18.401
No. of iterations	32	35	34	37
CPU time	163.3	449.0	904.7	1829
$\mathrm{CPU}/(n+m)^3 \times 10^7$	4.393	5.493	5.520	5.987

assignment of explicit symmetry constraints. The three modes are as shown in Fig. 3.32. Mode 1 is symmetric with respect to the xz- and yz-planes, while Modes 2 and 3 are antisymmetric with respect to the xy- and yz-planes, respectively.

Optimization was also carried out for 8×8, 9×9, and 10×10 grids. The four lowest eigenvalues, total volume, number of iterations, and CPU time are listed in Table 3.5. Note that the algorithm did not converge for the 10×10 grid, and stopped with numerical difficulty; however, a feasible and nearly optimal solution was successfully found, because a primal-dual interior-point method that searches in the feasible region was used. As is seen from the last row of Table 3.5, the CPU time is almost proportional to $(n+m)^3$, as it is known that the computational cost of the interior-point method can be

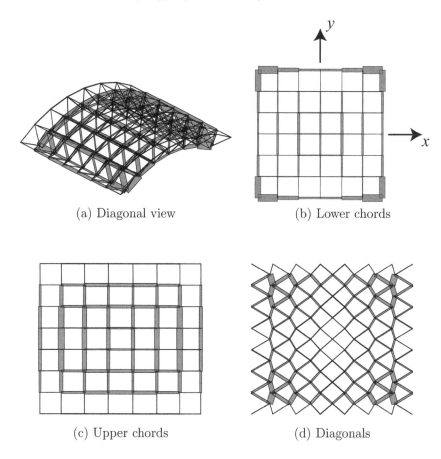

(a) Diagonal view (b) Lower chords

(c) Upper chords (d) Diagonals

FIGURE 3.31: Optimal solution of the 7×7 double-layer cylindrical grid.

estimated as a polynomial of the problem size.

3.10 Application of data mining

3.10.1 Frequent item set of decent solutions

Topology optimization may be conceived as a process of selecting the best solution from the set of the huge number of feasible solutions. For example, it is possible to select the optimal topology from the candidate list of trusses, as shown in Fig. 3.33, that represent substructures of a ground structure consisting of 10 bars that can exist. However, it is not possible to enumerate

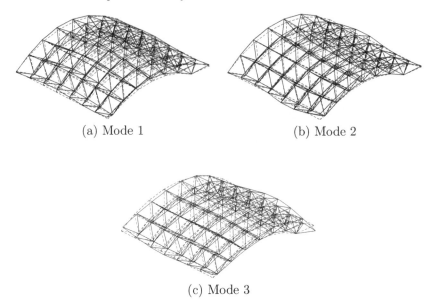

(a) Mode 1   (b) Mode 2

(c) Mode 3

FIGURE 3.32:   Three fundamental eigenmodes of the optimal solution of the 7×7 double-layer cylindrical grid.

all stable topologies from a ground structure with moderately large numbers of nodes and members. Therefore, methodologies developed in the field of data mining can be used for finding the set of existing/nonexisting members that leads to global optimal topology.

Data mining consists of the methods of *association rule, decision tree, clustering*, etc., for extracting the rules and knowledge from a huge database (Hand, Mannila, and Smyth 2001; Berry and Linoff 1997). The basic approaches of data mining for application to topology optimization are illustrated using a 10-bar truss in Fig. 3.34. Loads of 0.02 and -0.1 are applied in the x- and y-directions, respectively, at all the nodes, where units are omitted for brevity. The elastic modulus is 1, and the cross-sectional area of each member takes 0 or 1, for simplicity; i.e., the existing members have the same cross-sectional area, and, consequently, the total number of possible combinations of the members is 2^{10}. We consider a multiobjective optimization problem for minimizing the total structural volume and compliance.

All of the possible 2^{10} topologies have been enumerated to find only 40 stable topologies. The values of total structural volume and compliance of the 40 stable solutions are classified into four clusters in the objective function space using the data mining tool WEKA Ver. 3.5 (Witten and Frank 2000). The solutions in the four clusters are plotted with different marks in Fig. 3.35. The well-known method of K-means has been used for clustering, which is described as follows:

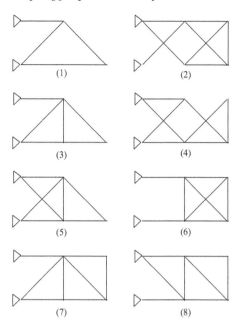

FIGURE 3.33: An example of a list of candidate topologies.

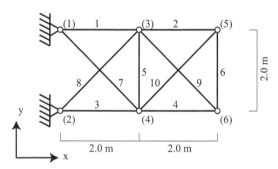

FIGURE 3.34: A 10-bar truss.

Step 1: Select four solutions that are preferably widely spaced in the objective function space, each of which is considered to be located at the center of each cluster.

Step 2: Assign each solution to the cluster that has the nearest center.

Step 3: Compute the center of each cluster in the objective function space, and go to Step 2 if not converged.

The 17 solutions indicated by '+' in Fig. 3.35 form a cluster of *decent solutions* that have smaller values of compliance and total structural volume.

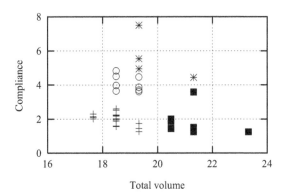

FIGURE 3.35: Clustering of the stable solutions of the 10-bar truss in the objective function space.

Each solution is designated by the set of existing members. Therefore, using the terminologies of data mining, the member and solution correspond to *item* and *transaction*, respectively. Let d denote the number of elements in the transaction database $\mathcal{D}$. In this example, $\mathcal{D}$ consists of 40 stable solutions. Let $\mathcal{G}_i$ denote the *cover* of the ith item (member) in $\mathcal{D}$; i.e., $\mathcal{G}_i$ is the set consisting of the solutions including member i. The number of elements (solutions) in $\mathcal{G}_i$ is denote by g_i. Then the *frequency* f_i of item (member) i is defined as $f_i = g_i/d$, and the set of items that has a frequency greater than the specified value α is called the *frequent item set*.

Next we define the set $\mathcal{Z}$ of the decent solutions that have a particular favorable property, e.g., small values of compliance and total structural volume for the 10-bar truss, and define the set $\mathcal{H}_i$ of solutions that have member i in $\mathcal{Z}$. The number of solutions in $\mathcal{H}_i$ is denoted by h_i. Consider a rule $\mathcal{G}_i \to \mathcal{Z}$, which states that the solution is a decent solution if it has member i. The *support* s_i and *confidence* c_i of this rule are defined by $s_i = h_i/d$ and $c_i = h_i/g_i$, respectively. This way, a reliable rule is characterized by large support and confidence.

Since the loads are applied at all nodes of the 10-bar truss, all nodes should exist so that the truss is stable. Therefore, the support of any rule $\mathcal{G}_i \to \mathcal{Z}$ has a moderately large value that is greater than 0.386. The confidences c_i of the rule $\mathcal{G}_i \to \mathcal{Z}$ are $c_1 = c_3 = 0.515$, $c_2 = c_4 = c_6 = c_9 = c_{10} = 0.412$, and $c_5 = c_7 = c_8 = 0.333$. Therefore, the stable solution with member 1 or 3 has a high possibility of being a decent solution. In contrast, let $\overline{\mathcal{G}}_i$ denote the set of solutions that do not include member i. The set of solutions that are not included in the set of decent solutions is denoted by $\overline{\mathcal{Z}}$. For the rule $\overline{\mathcal{G}}_i \to \overline{\mathcal{Z}}$, the members 1 and 3 have support 0.175 and confidence 1, which agrees with the results obtained by intuition; i.e., a truss has a large compliance if it does not have the horizontal members connected to the supports.

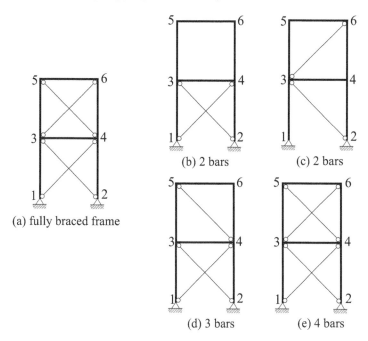

FIGURE 3.36: Predefined fully braced frame and ground structures as its subsets.

3.10.2 Topology mining of ground structures

One of the main difficulties in topology optimization of trusses and braces under stress constraints is that the local optimal solution obtained using an NLP algorithm strongly depends on the initial ground structure (GS), as demonstrated in Example 3.2, and the solution does not always improve as a result of increasing the number of members in the GS. Therefore, the methodologies developed in the field of data mining can be used for finding the *optimal ground structure* that has a moderately small number of members and leads to the global optimal topology. Hagishita and Ohsaki (2008b, 2008b) presented a method called *topology mining* for finding GSs that lead to the global optimal topology, where the *superior bar sets* are extracted using the *apriori algorithm* developed by Agrawal, Imielinski, and Swami (1998).

For example, consider a problem of optimizing the locations of the braces of a frame with fixed cross-sectional areas, such as the fully braced frame shown in Fig. 3.36(a). The total structural volume is minimized under stress constraints. The GSs shown in Figs. 3.36(b)–(e) are obtained as subsets of the fully braced frame. We assume that singular optimal topologies may exist for this problem; i.e., the stress constraints may not be satisfied by nonexistent braces.

TABLE 3.6:   Expression of topologies of the ground structures.

	Ground structure	Optimal objective value
(b)	$\{114, 123, 036, 045\}$	V_{b}
(c)	$\{014, 123, 136, 045\}$	V_{c}
(d)	$\{114, 123, 036, 145\}$	V_{d}
(e)	$\{114, 123, 136, 145\}$	V_{e}

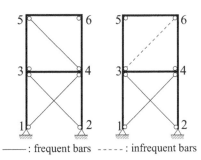

FIGURE 3.37:   Superior bar sets consisting of three braces in ground structures in Fig. 3.36(b)–(e).

The topologies of the braced frames in Figs. 3.36(b)–(e) are expressed by integer arrays, as shown in Table 3.6, where the existing bar between nodes i and j is represented by the three digits with '1' followed by the node numbers 'i' and 'j' as '$1ij$', while the nonexistent bar between nodes j and k is indicated by $0jk$ with '0' at the first digit. The GS defined by each array in the table is optimized using an NLP algorithm to find the optimal objective value denoted by $V_{\mathrm{b}}, \ldots, V_{\mathrm{e}}$. Let β denote the specified ratio of the number of *superior ground structures* to the total number of GSs. If $\beta = 0.25$ for 16 ($= 2^4$) GSs for the frames with four braces, then the four best GSs are regarded to be superior.

Suppose the frames in Figs. 3.36(b)–(e) are the four superior ground structures. The frequent bar set is defined as the set of items (braces) that has large frequency in the superior GSs. The infrequent bar set is also defined as the set of large frequency of nonexistence in the superior GSs. Note that infrequent bar does not mean a bar that exists in nonsuperior GSs. Then, we can obtain the superior bar sets, including the frequent and infrequent bar sets, by applying the *apriori* algorithm to the ground structures in Table 3.6.

The minimum frequency is assigned as 0.5; i.e., the set of braces that are existent/nonexistent in two or more ground structures is extracted. First, we can see from Figs. 3.36(b)–(e) that the frequent/infrequent bar sets, which

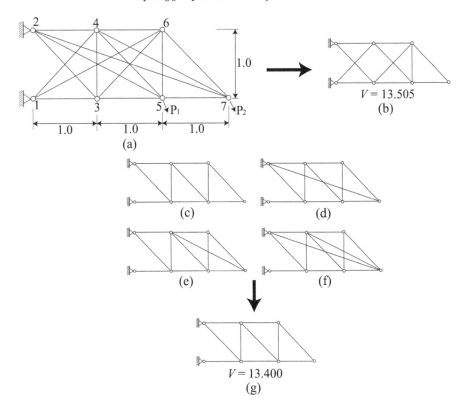

FIGURE 3.38: Optimization results of truss topology optimization: (a) fully connected ground structure, (b) optimal 11-bar truss from the fully connected ground structure, (c)–(f) ground structures that lead to the optimal 10-bar truss, (g) optimal 10-bar truss.

are called superior bar sets for brevity, consisting of single brace are given as

$$114, \quad 123, \quad 136, \quad 145, \quad 036, \quad 045 \qquad (3.110)$$

Then, utilizing the *apriori* algorithm, we should check the following sets to obtain the superior bar sets consisting of two braces:

$$\{114, 123\}, \ \{114, 136\}, \ \{114, 145\}, \ \{114, 036\}, \ \{114, 045\},$$
$$\{123, 136\}, \ \{123, 145\}, \ \{123, 036\}, \ \{123, 045\}, \ \{136, 145\}, \qquad (3.111)$$
$$\{136, 045\}, \ \{145, 036\}, \ \{036, 045\}$$

Note that a brace cannot be existent and nonexistent simultaneously; hence, sets such as $\{136, 036\}$ have been excluded. For example, the braces connecting nodes 1 and 4, and 2 and 3, respectively, exist simultaneously in three ground structures in Figs. 3.36(b)–(e); however, the braces connecting nodes

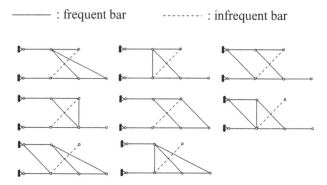

———— : frequent bar   -------- : infrequent bar

FIGURE 3.39:   An example of superior bar sets.

3 and 6, and 4 and 5, respectively, exist simultaneously only in one ground structure. By checking all the sets in (3.111), we have the following superior bar sets consisting of two braces:

$$\{114, 123\}, \ \{114, 145\}, \ \{114, 036\}, \ \{123, 136\},$$
$$\{123, 145\}, \ \{123, 036\}, \ \{123, 045\} \tag{3.112}$$

The superior bar sets consisting of three braces are the subsets of the union of the two sets in (3.112). Therefore, we should check

$$\{114, 123, 145\}, \ \{114, 123, 136\}, \ \{114, 123, 036\},$$
$$\{114, 123, 045\}, \ \{123, 136, 145\}, \ \{123, 136, 045\} \tag{3.113}$$

From (3.113), the following superior bar sets are obtained:

$$\{114, 123, 145\}, \ \{114, 123, 036\} \tag{3.114}$$

which are illustrated in Fig. 3.37. This way, the ground structures that contain superior bar sets can be obtained with moderately small computational cost. Utilizing this approach, Hagishita and Ohsaki (2008b) developed a method called *topology mining* for generating optimal ground structures that leads to global optimal topology.

As an example, optimal ground structures for a plane truss are found using topology mining. The fully connected ground structure has 16 members, as shown in Fig. 3.38(a). The truss is subjected to two loads, P_1 and P_2, independently at nodes 5 and 7, respectively; see Hagishita and Ohsaki (2008b) for details.

A simple application of the ground structure approach to the fully connected ground structure in Fig. 3.38(a) generates the 11-bar truss in Fig. 3.38(b), where $V = 13.505$. In order to generate the superior bar sets, ground structures are randomly generated 10^5 times and the optimal topologies are found from the 11,047 stable ground structures. As a result, we found four ground structures, shown in Figs. 3.38(c)–(f), that lead to the optimal 10-bar truss in Fig. 3.38(g) with the total structural volume $V = 13.400$, which is less than that of the 11-bar truss in Fig. 3.38(b). Topology mining is then carried out to find the superior bar sets, as shown in Fig. 3.39, which lead to the optimal 10-bar truss in Fig. 3.38(g). Note that these sets have been found with application of NLP for 140 stable solutions, which is very small compared with 11,047 for random enumeration.

Chapter 4

Configuration Optimization of Trusses

Configuration optimization of trusses is categorized into topology optimization and geometry optimization, as described in Secs. 1.1 and 1.10, and so far we have mainly considered topology optimization. In this chapter, some methods are presented for configuration optimization of trusses for optimizing topology, nodal locations, and cross-sectional areas simultaneously. A historical review is presented in Sec. 4.1 as an introduction to this chapter. Basic formulations and small examples are given in Sec. 4.2. An optimization method is presented for regular plane trusses with uniform cross-sectional area in Sec. 4.3. Link mechanisms are generated using the extended ground structure approach considering geometrical nonlinearity in Sec. 4.4.

4.1   Introduction

Geometry optimization of trusses and frames seems to be rather straightforward, because the nodal coordinates are considered as continuous design variables, and the optimal solutions may be found using appropriate methods of mathematical programming (Sadek 1986; Imai and Schmit 1982; Dems and Gatkowski 1995; Lin, Che, and Yu 1982; Saka 1980; Svanberg 1981). Early works on geometry optimization are found in Dobbs and Felton (1969) and Pedersen (1972).

In the process of geometry optimization, however, side constraints are usually given for the nodal coordinates to prevent unfavorable numerical instability due to the existence of extremely short members or closely spaced parallel members (Sadek 1986; Imai and Schmit 1982). Intersection of members should also be prevented in some cases. Therefore, it is very difficult to modify the topology of a truss or a frame through the process of geometrical optimization by continuously varying the nodes and removing the coalescent nodes and members. For these reasons, it is widely recognized that simultaneous optimization of topology and geometry of a truss is very difficult.

It is important to note that the topology optimization problem is a combinatorial optimization problem, whereas the geometry optimization problem has continuous variables and functions. Therefore, a hybrid approach may be used in which geometry optimization problems are solved successively for the truss obtained at each step of the upper-level topology optimization process (Hagishita and Ohsaki 2008b). However, computational effort seems to be very large if such two-level algorithms are applied to optimize large-scale trusses. Achtziger (2007) presented a mathematical programming approach based on implicit programming (Ben-Tal, Kočvara, and Zowe 1993).

Contrary to the ground structure approach to topology optimization, which allows only removal of members, some methods have been proposed for adding nodes and members from a simple *base truss* to generate an optimal topology and geometry (Rule 1994; McKeown 1998). The drawback of obtaining unrealistic or unstable optimal solutions for the ground structure approach may be avoided if a *growing process* is used. There is no theoretically clear criteria, however, for addition of nodes and members (Kirsch 1989a, 1996). The modification process of topology may be written in a formal shape grammar (Reddy and Cagan 1995a, 1995b). Shea, Cagan, and Fenves (1997) and Shea and Smith (2006) optimized transmission towers using shape annealing. Bojczuk and Mróz (1998b, 1999) used sensitivity information to determine whether the candidate modified topology is acceptable, and topology is modified by dividing a bar, or by adding a node in a triangular unit. Rychter and Musiuk (2007) defined the topological sensitivity of eigenvalues with respect to the flip of a diagonal member of a regular truss. Hagishita and Ohsaki (2009) presented a growing and removal process based on the mechanical properties of the modified structure. Yang and Soh (2002) used genetic programming for evolutionary optimization. Stadler (1999) developed an algorithm using the Steiner points of triangular units. Mróz and Bojczuk (2003) presented a method similar to a greedy method for finding the most efficient topology variation of a truss under compliance constraint.

4.2 General formulation and methodologies of configuration optimization

The problem of configuration optimization can be formulated using a so-called *extended* ground structure approach, where both the cross-sectional areas and nodal coordinates are considered as design variables (Kočvara and Zowe 1995). As noted in Sec. 4.1, the main difficulty of this problem is that some nodes may coincide during optimization, and the stiffness matrix becomes singular due to the existence of a member with zero length. Although these *coalescent nodes* (Ohsaki 1997b) or *melting nodes* (Achtziger 2007) should be avoided from a computational point of view, they are preferred because solutions with various topologies and geometries can be found

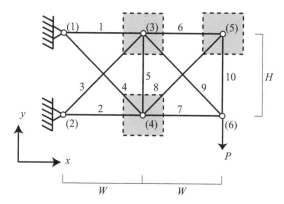

FIGURE 4.1: A 10-bar truss with variable nodal locations.

from the ground structure with relatively small numbers of members and nodes.

Let $\mathbf{A} = (A_1, \ldots, A_m)^\top$ denote the vector of cross-sectional areas, where m is the number of members in the initial ground structure. The vector consisting of variable nodal coordinates is denoted by $\mathbf{X} = (X_1, \ldots, X_{n^C})^\top$, where X_{n^C} is the number of variables of nodal coordinates. Note that $\mathbf{X}$ includes the coordinates in two directions for a plane truss and three directions for a spatial truss. The length of the ith member is denoted by $L_i(\mathbf{X})$. The lower bound for A_i is 0 in order to allow topology variation, while the upper bound is not given. If the upper bound X_i^{U} and the lower bound X_i^{L} are appropriately assigned for X_i to prevent the existence of coalescent nodes, the optimization problem for minimizing the total structural volume $V(\mathbf{A}, \mathbf{X})$ under general inequality constraints $H_j(\mathbf{A}, \mathbf{X}) \leq 0$ $(j = 1, \ldots, n^{\mathrm{I}})$ representing the constraints on stresses, displacements, etc., is formulated as

$$\text{Minimize} \quad V(\mathbf{A}, \mathbf{X}) = \sum_{i=1}^{m} A_i L_i \tag{4.1a}$$

$$\text{subject to} \quad H_j(\mathbf{A}, \mathbf{X}) \leq 0, \quad (j = 1, \ldots, n^{\mathrm{I}}) \tag{4.1b}$$

$$X_i^{\mathrm{L}} \leq X_i \leq X_i^{\mathrm{U}}, \quad (i = 1, \ldots, n^{\mathrm{C}}) \tag{4.1c}$$

$$A_i \geq 0, \quad (i = 1, \ldots, m) \tag{4.1d}$$

Example 4.1

As an illustrative example, consider a 10-bar truss, as shown in Fig. 4.1, subjected to a vertical load P, where $W = H = 2$ m. Optimal nodal locations are found for two cases with uniform and non-uniform cross-sectional areas, respectively. The total structural volume is minimized under constraint on the compliance. The locations of the loaded node and the supports are fixed. The gray regions in Fig. 4.1 are the feasible regions of the nodes with variable

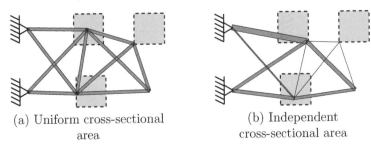

(a) Uniform cross-sectional
area

(b) Independent
cross-sectional area

FIGURE 4.2:   Optimal cross-sectional areas and nodal locations of the 10-bar truss.

coordinates, where the bounds of the coordinates are ±0.5 m from the initial values in the x- and y-directions. The parameters are the same as those for the 10-bar truss in Example 1.10 in Sec. 1.8. SNOPT Ver. 7.2 (Gill, Murray, and Saunders 2002), which utilizes sequential quadratic programming is used for optimization.

If the cross-sectional areas of all members have the same value, which is adjusted to have the specified compliance, the optimal shape is as shown in Fig. 4.2(a). Note that nodes 3, 4, and 5 exist, respectively, in the interior, along an edge, and at a vertex of the feasible region. If the cross-sectional areas of all members can vary independently, the optimal shape and cross-sectional areas are as shown in Fig. 4.2(b). Since small lower bounds are given for the cross-sectional areas, shape and topology can be simultaneously optimized by removing those members with lower-bound cross-sectional areas. However, the shape variation is limited within the feasible region of the nodal coordinates.

As an alternative formulation, a small lower bound L_i^{L} is given for $L_i(\mathbf{X})$ to prevent the existence of coalescent nodes as

$$\text{Minimize} \quad V(\mathbf{A},\mathbf{X}) = \sum_{i=1}^{m} A_i L_i \tag{4.2a}$$

$$\text{subject to} \quad H_j(\mathbf{A},\mathbf{X}) \leq 0, \quad (j=1,\ldots,n^{\mathrm{I}}) \tag{4.2b}$$

$$L_i(\mathbf{X}) \geq L_i^{\mathrm{L}}, \quad (i=1,\ldots,m) \tag{4.2c}$$

$$A_i \geq 0, \quad (i=1,\ldots,m) \tag{4.2d}$$

Problems (4.1) and (4.2) are standard nonlinear programming problems. However, it is not desirable to directly solve these problems for general large-scale trusses, because they are highly nonlinear, and the optimization procedure often terminates at a local optimal solution. One possible approach for improving the convergence to an approximate optimal solution is to fix the sizing variables $\mathbf{A}$ and the shape variables $\mathbf{X}$ alternatively, and find the optimal values of the free variables. Pedersen (1972) presented a method based on successive linearization.

Consider a case of a single loading condition, where the truss is subjected to a set of nodal loads $\mathbf{P} = (P_1, \ldots, P_n)^\top$ with n being the number of degrees of freedom. Achtziger (2007) presented a method based on the *implicit programming approach* (Ben-Tal, Kočvara, and Zowe 1993). Let $\mathbf{X} = (X_1, \ldots, X_{n^C})^\top$ denote the vector of variable nodal coordinates. Although the details are omitted here, the configuration optimization problem of minimizing the compliance under constraint on total structural volume is reformulated as

$$\text{Minimize} \quad \sum_{i=1}^{m} (\lambda_i + \mu_i) \mathbf{X}^\top \mathbf{C}_i \mathbf{X} \tag{4.3a}$$

$$\text{subject to} \quad \sum_{i=1}^{m} (\mu_i - \lambda_i) \sqrt{E_i} \mathbf{Q} \mathbf{C}_i \mathbf{X} + \mathbf{P} \tag{4.3b}$$

$$\lambda_i \geq 0, \quad \mu_i \geq 0, \quad (i = 1, \ldots, m) \tag{4.3c}$$

where λ_i, μ_i $(i = 1, \ldots, m)$, and $\mathbf{X}$ are the variables, $\mathbf{C}_i$ $(i = 1, \ldots, m)$ and $\mathbf{Q}$ are $n^C \times n^C$ constant matrices, and E_i is the elastic modulus of the ith member. As is seen, the objective and constraint functions are polynomials of the variables and the maximum order is three in the objective function.

Alternatively, an approximate solution can be obtained by modeling the truss using the frame element to prevent difficulties due to the existence of coalescent nodes. In the following example, we first verify that the displacements of trusses can be approximated by a frame member with an artificially small radius of gyration of area.

Example 4.2

Consider again the 10-bar truss in Fig. 4.1. Suppose the members have sandwich sections such that $I_i = h^2 A_i/4$, where I_i is the second moment of inertia of the ith member, and h is the distance between the flanges that is the same for all members. Other parameters are the same as those in Example 1.3 in Sec. 1.5. The optimal nodal locations and cross-sectional areas are found using SNOPT Ver. 7.2 (Gill, Murray, and Saunders 2002), where a very strict tolerance, 10^{-8}, is given for constraints and optimality conditions. First, the existence of coalescent nodes is not allowed, and the optimal solution is found under the same conditions as the truss in Fig. 4.1. Then, for $h = 1$, 10, and 100 (mm) the same topology and almost the same geometry as Fig. 4.2(b) have been obtained.

Table 4.1 shows the optimization results for various values of h as well as the result of the truss. As is seen, the optimal cross-sectional areas and the total structural volume converge to those of the truss as h is decreased. Although the bending deformation may be very large if h is small, the average axial deformation can be successfully approximated by the frame model, because

TABLE 4.1: Optimal cross-sectional areas and total structural volume of the 10-bar truss modeled by truss and frame elements with various values of h (mm).

Member number	A_i (mm^2)			
	Truss	$h = 1$	$h = 10$	$h = 100$
1	1383.54	1383.54	1383.53	1382.85
2	866.85	866.85	866.83	865.22
3	798.41	798.41	798.41	799.19
4	281.02	281.02	281.02	280.79
5	0.100	0.100	0.100	0.100
6	0.100	0.100	0.100	0.100
7	683.78	683.78	683.77	682.87
8	0.100	0.100	0.100	0.100
9	958.65	958.65	958.65	958.37
10	0.100	0.100	0.100	0.100
Total volume ($\times 10^7$mm^3)	1.43036	1.43036	1.43034	1.42875

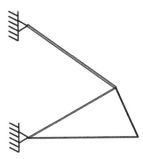

FIGURE 4.3: Optimal configuration and cross-sectional areas obtained by the frame model.

we consider only the infinitesimal deformation and each member consists of single element.

Optimal solutions are next found for the 10-bar truss without side constraints on the nodal locations. However, a small lower bound L^L is given for the length of each member to prevent the existence of too short members that leads to singularity of the stiffness matrix. The optimal configuration and cross-sectional areas obtained from the 10-bar truss with $H = 4.0$ m and $W = 2.0$ m in Fig. 4.1 are shown in Fig. 4.3, where $(h, L^L) = (1, 1)$. Note that the members with $A_i = A_i^L$ have been removed in Fig. 4.3. As is seen, a frame with four members and four nodes, including the supports, has been successfully generated. Table 4.2 shows the optimal nodal locations for various values of (h, L^L). Note that the optimization process does not converge strictly if $L^L = 0.0001$ mm. It is seen from Fig. 4.1 and Table 4.2 that nodes 4

TABLE 4.2: Optimal nodal locations and total structural volume of the 10-bar truss modeled by frame elements.

Node	Direction	(h, L^L)			
		$(1,1)$	$(1,0.1)$	$(0.1,1)$	$(0.0001,1)$
3	x	3203.19	3205.81	3205.38	3210.65
	y	1756.53	1753.85	1754.64	1749.05
4	x	3999.70	4000.06	3998.92	3999.00
	y	−0.953174	−0.0895076	−0.170447	−0.0844409
5	x	3205.39	3519.21	3385.73	3386.83
	y	1756.43	2839.75	5228.31	1133.13
Total volume		6.97819	6.96465	6.96896	6.97272
$(\times 10^6 \mathrm{mm}^3)$					

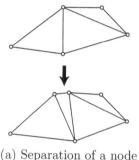

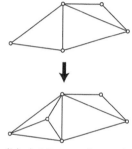

(a) Separation of a node   (b) Addition of a node

FIGURE 4.4: Topology variation of a truss.

and 6 are coalescent. Because the cross-sectional areas of members connected to node 5 are very small, the location of node 5 does not have any effect on the optimal solution and its objective value.

Finally, optimal nodal locations and cross-sectional areas are found for the truss that has the same topology as the frame in Fig. 4.3. The optimal objective value is 6.961021×10^6 mm^3, which is very close to the values in Table 4.2 obtained using the frame model. Therefore, optimal configuration of a truss can be successfully found after obtaining the optimal topology of the frame model.

So far, we have discussed nonlinear programming approaches for configuration optimization of trusses. Several heuristic approaches have been proposed utilizing topological variation in the optimization process. By allowing the addition of nodes and members, as well as variation of nodal coordinates, a complex optimal solution can be found starting with a simple ground structure (Bojczuk and Mróz 1998b).

Examples of topology variation are illustrated in Fig. 4.4. The location of a new node that most efficiently reduces the objective function is estimated using the sensitivity coefficients of the objective function with respect to the coordinates of the added or separated node. After addition of a node and the corresponding members, optimization of nodal location is carried out, and another node is to be added. A node may also be added at the intersection of the members (Martínez, Martí, and Querin 2007). Reddy and Cagan (1995b) presented an addition scheme of a triangular unit.

One of the main difficulties for application of the (extended) ground structure approach to configuration optimization is that a ground structure with many nodes and members is needed to obtain the optimal configuration with moderate complexity. However, in this case, optimization using nonlinear programming often reaches a local optimal solution especially for a problem with stress constraints. Therefore, a kind of *optimal* ground structure that can reach the global optimal solution from sufficiently small numbers of nodes and members should be found. For this purpose, Hagishita and Ohsaki (2009) presented an approach based on sequential topology optimization for the updated ground structure. In order to search solutions in wider design space, it is important that the new ground structure should be defined to increase the *complexity* of the design space. The algorithm is summarized as

Step 1: Define the initial ground structure that has sufficiently small numbers of nodes and members.

Step 2: Find the optimal topology considering the cross-sectional areas as design variables and remove the unnecessary nodes and members.

Step 3: Add nodes and members to the optimal topology to update the ground structure so as to have *maximum disturbance* to the optimal topology.

4.3   Optimization of a regular grid truss

4.3.1   Problem formulation

In this section, we summarize the difficulties in configuration optimization of trusses and present a systematic approach to optimize a regular truss (Ohsaki 1997b). Because the trusses in practical applications in civil engineering are usually regular, and the cross-sectional areas of the members are not allowed to have arbitrarily different positive values, it is practically desired to develop an optimization method for such trusses.

Consider a pin-jointed regular plane truss, as shown in Fig. 4.5, that consists of rectangular units with fixed topology. The size of a unit is allowed to vary under the condition that the regularity of the truss is preserved. The

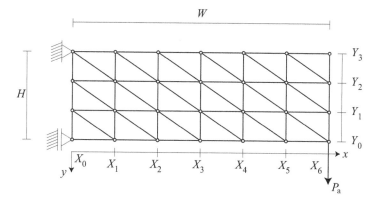

FIGURE 4.5: A plane rectangular truss ($N_x^{\mathrm{u}} = 6$, $N_y^{\mathrm{u}} = 3$).

numbers of units in the x- and y-directions are denoted, respectively, by N_x^{u} and N_y^{u}. The cross-sectional areas are assumed to be the same for all members. The problem considered here is to find the optimal topology, optimal nodal locations, and optimal cross-sectional area of a regular plane truss. The locations of the supports as well as the geometrical properties, e.g., the span length for defining the total dimension of the truss, are fixed in order to prevent the existence of an impractical optimal solution. The loads are applied at the limited number of nodes for which the locations are fixed.

Let n denote the number of degrees of freedom of the initial ground structure. The displacement vector is denoted by $\mathbf{U} = (U_1, \ldots, U_n)^\top$, and constraints are given as

$$|U_j| \leq U_j^{\mathrm{U}}, \quad (j = 1, \ldots, n) \tag{4.4}$$

where U_j^{U} is the upper bound for the absolute value of the jth displacement component. The design variables are the vector of variable nodal coordinates $\mathbf{X}$ and the cross-sectional area A that is the same for all the members. Note that the sizes of $\mathbf{X}$ and $\mathbf{U}$ depend on N_x^{u} and N_y^{u}. The set of vectors of nodal coordinates corresponding to feasible regular geometry is denoted by $\mathcal{X}$.

In the process of designing trusses, the cost of nodes is usually equivalent to that of the members. Therefore, from a practical point of view, it is very important to consider the nodal cost in a topology optimization problem as presented in Sec. 3.7.3. The cost of each node is denoted by B, which is the same for all the nodes. Let N^{n} denote the total number of nodes, which depends on N_x^{u} and N_y^{u}, and denote by c^{m} the cost coefficient for the unit volume of members. The length of the ith member is denoted by L_i, which is a function of $\mathbf{X}$. Then the configuration optimization problem of minimizing the total cost $C(\mathbf{X}, A)$ for fixed N_x^{u} and N_y^{u} is stated as

$$\text{Minimize} \quad C(\mathbf{X}, A) = c^{\mathrm{m}} \sum_{i=1}^{m} AL_i(\mathbf{X}) + N^{\mathrm{n}} B \tag{4.5a}$$

$$\text{subject to} \quad |U_j(\mathbf{X}, A)| \leq U_j^{\mathrm{U}}, \quad (j = 1, \ldots, n) \tag{4.5b}$$

$$\mathbf{X} \in \mathcal{X} \tag{4.5c}$$

Example 4.3

To illustrate the difficulties in configuration optimization of a truss with uniform cross-sectional area, optimum designs are found for a simple cantilever-type plane truss with fixed span length W and height H. Nodal cost is not considered here for a simple presentation of the results. A truss corresponding to $(N_x^{\mathrm{u}}, N_y^{\mathrm{u}}) = (3, 1)$ is shown in Fig. 4.6. The vertical member is assumed to remain vertical during design modification; i.e., the feasible trusses consist of rectangular units. The x-coordinates X_i $(0 \leq i \leq N_x^{\mathrm{u}})$ of the boundaries of the units are defined as shown in Fig. 4.6. The coordinates X_i $(1 \leq i \leq N_x^{\mathrm{u}} - 1)$ and the member cross-sectional area A are chosen as independent design variables. The load $P_{\mathrm{a}} = 49.0$ kN is applied in the y-direction at node 'a'. The parameters are $U^{\mathrm{U}} = 10.0$ mm, $c^{\mathrm{m}} = 1.0 \times 10^{-6}$ mm^{-3}, $W = 4000.0$ mm, $H = 2000.0$ mm, and the elastic modulus is $E = 205.8$ kN/mm^2. In the following, the units of force and length are kN and mm, respectively, and are omitted for brevity. Optimal trusses are found for $N_x^{\mathrm{u}} = 1$, 2, 3, and 4 using the optimization package IDESIGN (Arora and Tseng 1987), which utilizes sequential quadratic programming.

The values of X_i, A, and C of the optimum designs with fixed topologies are listed in Table 4.3. It is interesting to note that equally divided configurations are optimum for these statically determinate trusses. It is observed from Table 4.3 that the optimal truss for $N_x^{\mathrm{u}} = 2$ has the smallest value of the objective function; therefore, it is regarded as the global optimal solution if N_x^{u} is also considered as a variable.

———

As demonstrated in the example above, global optimal topology can be found by searching the smallest value of $C(\mathbf{X}, A)$ of the optimal solutions of Problem (4.5) corresponding to all the possible sets of N_x^{u} and N_y^{u}. Since this process is not practically acceptable for large-scale trusses, the computational cost may be reduced if the optimal solution with smaller numbers of N_x^{u} and N_y^{u} can be found from the ground structure with larger numbers of N_x^{u} and N_y^{u}. However, there exist serious difficulties due to discreteness in the design variable and the assumption of uniform cross-sectional area, as shown below.

4.3.1.1 Singularity and discontinuity of optimal solutions

For a cantilever-type truss with $N_y^{\mathrm{u}} = 1$, the vertical displacement U_{a} of the loaded node 'a' and the total cost C are computed for the trusses with

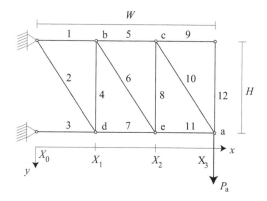

FIGURE 4.6: A plane cantilever-type truss ($N_x^u = 3, N_y^u = 1$).

TABLE 4.3: Optimal cross-sectional area, nodal location, and total cost for $N_x^u = 1, 2, 3,$ and 4.

N_x^u	1	2	3	4
A	913.35	602.71	670.95	611.32
X_0	0.0	0.0	0.0	0.0
X_1	4000.0	2000.0	1333.3	1000.0
X_2	—	4000.0	2666.7	2000.0
X_3	—	—	4000.0	3000.0
X_4	—	—	—	4000.0
C	13.218	10.642	12.967	16.736

$N_x^u = 2$ and 3 to investigate difficulties in finding optimal topologies by removing unnecessary nodes and members. The cross-sectional areas are 1.0 for all members of both trusses, and the nodal cost is not considered. Other parameters are the same as those in Example 4.3.

The total cost for the truss with $(N_x^u, X_1) = (2, 2000.0)$ is 1.7657, whereas that for $(N_x^u, X_1, X_2) = (3, 2000.0, 2001.0)$ is 2.1657. The difference observed here is due to the fact that members 4, 6, and 8 in Fig. 4.6 for $N_x^u = 3$ almost overlap with each other and the cross-sectional areas of the three members should be divided by 3 to be equivalent to the case of $N_x^u = 2$.

Next, the difficulty in optimizing pin-jointed trusses without rotational stiffness at the joints is discussed. The displacements U_a for $(N_x^u, X_1) = (2, 2000.0)$ and $(N_x^u, X_1, X_2) = (3, 2000.0, 2000.1)$ are 6.0271 and 6.9793, respectively, which are significantly different. Fig. 4.7 shows the deformation of the truss with $(N_x^u, X_2, X_3) = (3, 2000.0, 2000.1)$. As is seen, there exists discontinuity in displacement in the y-direction between $X = X_1$ and X_2, because the stiffness of member 6 in Fig. 4.6 is not infinite and the thin rectangular unit 'bdec' deforms into a parallelogram. To prevent this discontinuity,

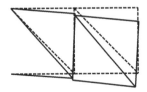

FIGURE 4.7: Deformation of a plane truss with $N_x^u = 3$, $X_1 = 2000.0$, and $X_2 = 2000.1$; solid line: deformed shape, dashed line: undeformed shape.

an additional constraint may be given such that the relative displacement between nodes 'b' and 'c' as well as 'd' and 'e' should be sufficiently small if the distance between X_1 and X_2 almost vanishes in the process of optimization.

The difficulties of configuration optimization of a regular truss with uniform cross-sectional area are summarized as follows:

1. The existence of extremely short members or elimination of some members leads to singularity of the stiffness matrix.

2. A member that has a larger cross-sectional area is generated if coalescent members are simply combined into a single member. In contrast, the global stiffness changes discontinuously if some among the coalescent members, leaving only one member, are removed.

3. Additional constraints are needed to restrict the relative displacement of the closely spaced nodes.

4. No node or member can be removed completely during the optimization process, because a node or a member removed once may turn out to be necessary in the final optimal solution.

4.3.1.2   Optimization algorithm

The third difficulty stated above is successfully avoided by modeling the truss as a rigidly jointed frame with a sufficiently small radius of gyration of area, as demonstrated in Sec. 4.2. To resolve the second and fourth difficulties, the cross-sectional area A_i of the ith member is considered to be a continuous function of $\mathbf{X}$. A sigmoid function is used to define the cross-sectional areas of the closely spaced members. For instance, A^* $(= A_6 = A_8)$ of the truss with $N_x^u = 3$ is defined as a function of $\Delta X_1 = X_2 - X_1$ as

$$A^* = A \tanh\left(\frac{\Delta X_1}{sX^*}\right) \tag{4.6}$$

where X^* is a parameter for defining the nondimensional value of the distance, and s is the shape parameter of the sigmoid function; a smaller value of s leads to a shape close to the step function. By using this function, A_6 and

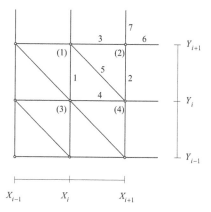

FIGURE 4.8: A unit of a rectangular truss.

A_8 are successfully decreased to a small value as ΔX_1 is decreased, while A_i for members 4, 5, and 7 remain equal to A. This way, continuous transition between the configurations corresponding to $N_x^u = 2$ and 3 is realized, and the second and fourth difficulties stated above are resolved.

Since the existence of a member with null cross-sectional area or with null length may lead to unfavorable singularity of the stiffness matrix, side constraint $\Delta X_i \geq D$ is given with a small positive value D. Thus, the nodes do not overlap completely, and the first difficulty is avoided.

For a regular plane truss as shown in Fig. 4.5, the configuration of the truss is defined in terms of the design variables X_i and Y_i, which are the coordinates of the nodes located at the intersection of the lines in the X- and Y-directions (see Tagawa and Ohsaki (1999) for a general algorithm for non-regular plane trusses).

The sigmoid functions S^x and S^y with respect to ΔX_i and ΔY_i are defined as

$$S^x(\Delta X_i) = \tanh\left(\frac{\Delta X_i}{sX^*}\right), \quad S^y(\Delta Y_i) = \tanh\left(\frac{\Delta Y_i}{sY^*}\right) \qquad (4.7)$$

where Y^* is a parameter for defining the nondimensional value of the distance in the Y-direction. Let B_k denote the cost of the kth node. For a rectangular unit, as shown in Fig. 4.8, B_k and A_i are calculated from

$$\begin{aligned}
&B_1 = BS^x(\Delta X_{i-1})S^y(\Delta Y_i), \quad B_2 = BS^x(\Delta X_i)S^y(\Delta Y_i), \\
&B_3 = BS^x(\Delta X_{i-1})S^y(\Delta Y_{i-1}), \quad B_4 = BS^x(\Delta X_i)S^y(\Delta Y_{i-1}), \\
&A_1 = AS^x(\Delta X_{i-1}), \quad A_2 = AS^x(\Delta X_i), \quad A_3 = AS^y(\Delta Y_i), \\
&A_4 = AS^y(\Delta Y_{i-1}), \quad A_5 = AS^x(\Delta X_i)S^y(\Delta Y_i), \quad A_6 = AS^y(\Delta Y_i), \\
&A_7 = AS^x(\Delta X_i)
\end{aligned} \qquad (4.8)$$

This way, optimization is carried out for the ground structure with sufficiently large N_x^u and N_y^u, and the nodes and members are removed to obtain the optimal configuration with a smaller number of units. However, for a truss that consists of a large number of units, the optimization process may converge to a local optimal solution with a larger number of units than that of the global optimal solution. In this case, an additional optimization should be carried out starting with the topology of the local optimal solution and randomly assigning the nodal coordinates. Hence, the optimization algorithm is summarized as follows:

Step 1: Define the dependence of A_j and B_k on ΔX_i and ΔY_i, and assign the parameters s, X^*, and Y^*. Set parameters c^m, U_j^U, B, and D as well as the radius of gyration of area of the members.

Step 2: Randomly generate the initial values of nodal coordinates, and find the optimal solution of the frame by using a nonlinear programming algorithm.

Step 3: If the initial and final topology as the result of optimization in Step 2 are the same, terminate the process.

Step 4: Update the topology by combining closely spaced nodes and members and go to Step 2.

4.3.2   Numerical examples

Optimal configurations are found for a regular plane truss, as shown in Fig. 4.5, where $E = 205.8$, $P_a = 98.0$, $W = 3000.0$, $H = 1000.0$, $X^* = Y^* = 500.0$, $D = 1.0$, $c^m = 1.0 \times 10^{-8}$, and $U_j^U = 10.0$ for all displacement components. The frame consists of a sandwich section with the flange distance $h = 100.0$. The cost of the supports is not included because the number of supports is fixed during optimization and the optimal topology does not depend on the cost of supports.

Note that a local optimal topology with large numbers of nodes and members is obtained if a small value is assigned for the parameter s in (4.6). In contrast, if s has a large value, the distribution of A_i and B_k are far from uniform. Therefore s is reduced linearly from 0.6 in each optimization process in Step 3.

Consider a case with $B = 1.0$. Optimization results starting with the frame with $(N_x^u, N_y^u) = (6, 3)$ are listed in the second column in Table 4.4. As is seen, the side constraint $X_1 - X_0 \geq D\ (= 1.0)$ is active and the nodes at $X = X_1$ should be removed. Similarly, the constraints $X_2 - X_1 \geq D$, $X_3 - X_2 \geq D$, $X_4 - X_3 \geq D$, $X_6 - X_5 \geq D$, $Y_2 - Y_1 \geq D$, and $Y_3 - Y_2 \geq D$ are satisfied with equality, and the nodes at $X = X_2, X_3, X_4, X_5, X_6$ and $Y = Y_2, Y_3$ are to be removed. Therefore, a frame with $(N_x^u, N_y^u) = (1, 1)$ has been found, and obviously no further optimization is needed. Note that the total cost C in Table 4.4 includes those of intermediate small values of A_i and B_k.

The optimal cross-sectional area was found for the pin-jointed truss with $(N_x^u, N_y^u) = (1, 1)$ for the purpose of verification. The values of A and C at

TABLE 4.4: Optimal cross-sectional area, nodal location, and total cost of plane rectangular frames.

(N_x^u, N_y^u)	(6,3)	(6,3)	(2,1)
B	1.0	0.1	0.1
A	1352.4	889.60	893.77
X_0	0.0	0.0	0.0
X_1	1.0	1498.1	1500.0
X_2	2.0	1499.1	3000.0
X_3	3.0	1501.0	—
X_4	4.0	1501.1	—
X_5	2999.0	2999.0	—
X_6	3000.0	3000.0	—
Y_0	0.0	0.0	0.0
Y_1	1498.0	1498.0	1500.0
Y_2	1499.0	1499.0	—
Y_3	1500.0	1500.0	—
C	3.7881	1.7346	1.7177

optimum are 1370.0 and 3.6925, respectively. The optimal pin-jointed truss has a little larger cross-sectional area and smaller total cost than the optimal frame, because there exist unnecessary members with small cross-sectional areas in the optimal frame.

Next, optimization is carried out for a smaller nodal cost $B = 0.1$. Optimization results are as listed in the third column of Table 4.4. In this case, the side constraints $X_2 - X_1 \geq D$, etc., are satisfied with equality, and the optimal topology of $(N_x^u, N_y^u) = (2,1)$ has been found as a result of the first optimization starting with the initial topology of $(N_x^u, N_y^u) = (6,3)$. Because the global optimal truss might have a smaller number of units, another optimization was carried out starting with $(N_x^u, N_y^u) = (2,1)$. The results are listed in the last column of Table 4.4, which verifies that the optimal solution has $(N_x^u, N_y^u) = (2,1)$.

For the pin-jointed truss with $(N_x^u, N_y^u) = (2,1)$, the optimal values of A, X_1, and C are 904.0, 1500.0, and 1.7327, respectively. Since a sufficiently small value has been given for the radius of gyration of area of the frame model, the differences in the optimal values between the truss and the frame are moderately small. Furthermore, we can see from the examples of $B = 1.0$ and 0.1 that a larger value of nodal cost leads to an optimal topology with a smaller number of nodes.

4.4 Generation of a link mechanism

4.4.1 Introduction

So far we have considered linear infinitesimal responses of trusses, where the strain-displacement relation is linear and the responses are obtained by solving the stiffness equation only once. Recently, extensive studies have been carried out for structural optimization under stability constraints considering geometrical nonlinearity (Ohsaki and Nakamura 1994; Ohsaki 2005a). The theoretical background and various methodologies of this field are summarized in the book by Ohsaki and Ikeda (2007). Ohsaki and Nishiwaki (2004, 2005) developed an optimization method for generating a multistable compliant bar-joint system, which produces the specified large deformation utilizing snapthrough behavior. This method was extended to utilize enumeration of statically determinate trusses (Ohsaki, Katoh, Kinoshita, Tanigawa, Avis, and Streinu 2009). Sekimoto and Noguchi (2001) presented an optimization approach to trace the desired nodal path for finite-element models. Saxena (2005) developed a path generation method of a compliant mechanism using a genetic algorithm. Jutte and Kota (2008) generated an arbitrary load-displacement path using spring elements. Eriksson (2008) presented an approach considering dynamic effect.

In this section, the optimization method is summarized for generating a link mechanism that has the specified path of the output node due to forced displacement at the input node (Ohsaki and Nishiwaki 2007a, 2007b). Many approaches have been presented for the design of link mechanisms, mainly based on trial-and-error modification of the design variables, including nodal locations (geometry) and member locations (topology). Kawamoto (2005) presented an approach using the truss model (Avilés, Ajuria, Vallejo, and Hernández 1997), where the nodal locations are modified to realize the desired mechanism. Kawamoto, Bendsøe, and Sigmund (2004b) utilized graph enumeration (Tuttle, Peterson, and Titus 1989), where the numbers of nodes and members should be assigned *a priori* (Kawamoto, Bendsøe, and Sigmund 2004a). Kim, Jang, Park, Hyun, and Nam (2005) used rigid bodies connected by springs, and the optimal topology with fixed geometry is obtained by removing the unnecessary springs.

4.4.2 Mechanical model of a link mechanism

Consider a link mechanism, as shown in Fig. 4.9. A forced displacement U_a is given at the input node 'a', and the final state is defined such that the displacement U_b of the output node 'b' reaches the specified value $\overline{U}_b$:

$$U_b = \overline{U}_b \qquad (4.9)$$

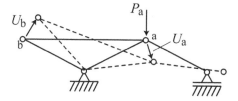

FIGURE 4.9: A link mechanism; solid line: undeformed shape, dotted line: deformed shape.

The solid and dotted lines in Fig. 4.9 are the configurations before and after deformation, respectively. Because the mechanism undergoes a large deformation, it is modeled as a bar-joint structure considering large nodal displacements.

Consider first a pin-jointed truss allowing elastic axial deformation of members. Let $\mathbf{P} = (P_1, \ldots, P_n)^\top$ and $\mathbf{N} = (N_1, \ldots, N_m)^\top$ denote the vectors of external loads and axial forces, where n is the number of degrees of freedom and m is the number of members. The equilibrium equation is written as

$$\mathbf{DN} = \mathbf{P} \qquad (4.10)$$

where the $n \times m$ matrix $\mathbf{D}$ is called the equilibrium matrix. The truss is characterized by the rank r of $\mathbf{D}$ as follows (Pellegrino and Calladine 1986):

- If $r = m$, then the truss is statically determinate, and the vector of axial forces $\mathbf{N}$ for a given $\mathbf{P}$ is found by solving the equilibrium equation (4.10).

- If $r < m$, then the truss is statically indeterminate, and there exist $q = m - r$ independent self-equilibrium modes of axial forces; hence, q is called statical indeterminacy.

- If $r < n$, then the truss is kinematically indeterminate (unstable), and there exist $h = n - r$ independent mechanisms; hence, h is called kinematical indeterminacy.

Therefore, the link mechanism that has one independent mode of deformation is classified as a truss with $h = 1$. For example, the mechanism in Fig. 4.9 has four members and five degrees of freedom. Hence, the kinematical indeterminacy is $5 - 4 = 1$, if the equilibrium matrix is full rank.

It should be noted that the mechanisms are classified into *infinitesimal mechanism* and *finite mechanism* (Calladine and Pellegrino 1991; Salerno 1992; Vassart, Laporte, and Motro 2000; Garcea, Formica, and Casciaro 2005). The bar-joint structure in Fig. 4.10(a) has an infinitesimal mechanism, because it is stable in the range of finite deformation that is illustrated by the dotted lines. In contrast, the structure in Fig. 4.10(b) has a finite

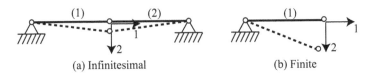

(a) Infinitesimal   (b) Finite

FIGURE 4.10: Classification of mechanisms; (a) infinitesimal, (b) finite.

mechanism, where the member rotates freely around the support without external load. The equilibrium equations for the mechanisms in Figs. 4.10(a) and (b) are written, respectively, as follows, using the axial forces N_1 and N_2 as well as the nodal loads P_1 and P_2:

$$\begin{pmatrix} 1 & -1 \\ 0 & 0 \end{pmatrix} \begin{pmatrix} N_1 \\ N_2 \end{pmatrix} = \begin{pmatrix} P_1 \\ P_2 \end{pmatrix}, \quad \begin{pmatrix} 1 & 0 \\ 0 & 0 \end{pmatrix} \begin{pmatrix} N_1 \\ N_2 \end{pmatrix} = \begin{pmatrix} P_1 \\ P_2 \end{pmatrix} \tag{4.11}$$

As is seen, the rank of the equilibrium matrix is 1, and the kinematical indeterminacy is 1 for both cases. Therefore, only the infinitesimal mechanism can be detected using the rank of the equilibrium matrix (Salerno 1992).

Our purpose here is to find the link mechanism that has the specified path of the output node under forced displacement at the input node. The approaches to this problem are classified into two categories based on multibody dynamics and finite element analysis, respectively.

In the first approach, the members are assumed to be rigid, and the path of the output node due to the forced displacement at the input node is traced using the Lagrange equations for multibody dynamics so that the length of each member is constant during deformation. The optimal shapes that minimize the error of the path of the output node from the target path can be found by considering the initial nodal locations as design variables.

However, in this approach, the numbers of members and nodes should be appropriately selected and fixed during optimization so that the kinematical indeterminacy is equal to 1 (Kawamoto, Bendsøe, and Sigmund 2004a). Furthermore, many geometrical constraints are needed to keep the member length constant during deformation. Therefore, this approach is not suitable for optimization that demands many analysis steps; hence, the second approach is used here (Avilés, Ajuria, Vallejo, and Hernández 1997); i.e., a conventional finite element analysis of truss elements is utilized.

Let L_i and A_i denote the length and cross-sectional area of member i, respectively. Define $\mathbf{C}$ as an $m \times m$ diagonal matrix of which the (i,i)-component is equal to $A_i E / L_i$, where E is the elastic modulus. The $n \times n$ stiffness matrix $\mathbf{K}$ is decomposed as

$$\mathbf{K} = \mathbf{D}\mathbf{C}\mathbf{D}^\top \tag{4.12}$$

The rank of $\mathbf{K}$ is equal to the rank r of $\mathbf{D}$, because $\mathbf{C}$ is full-rank.

Since elastic deformation is allowed for the truss at the intermediate state of optimization, the load-displacement relation is traced by a conventional displacement control or arc-length method of large-deformation analysis of trusses (Crisfield 1991). A mechanism is generated through optimization under constraint such that the maximum load vanishes throughout the deformation process.

For a kinematically indeterminate link mechanism with $h = 1$, there exists a mode (mechanism) $\boldsymbol{\Phi}$ that satisfies $\mathbf{K}\boldsymbol{\Phi} = \mathbf{0}$ corresponding to the vanishing eigenvalue of the stiffness matrix $\mathbf{K}$. Let $\mathbf{U}$ denote the nodal displacement vector. An unstable structure with $h = 1$ can be stabilized by constraining a component of $\mathbf{U}$ corresponding to a nonzero component of $\boldsymbol{\Phi}$. Suppose the truss is subjected to a set of proportional loads defined by $\mathbf{P} = \Lambda\mathbf{P}_0$ with the load factor Λ and the constant vector $\mathbf{P}_0$ of load pattern. Then, the path of the equilibrium states in the $\Lambda - \mathbf{U}$ space, called the *equilibrium path*, can be traced using a displacement component as a control parameter. This way, the *natural coordinate* can be successfully used for large-deformation analysis (Thompson and Sung 1986; Lio, Cossalter, and Lot 2000). Note that the rank of the equilibrium matrix should be checked at the deformed configuration to ensure that the truss behaves as a mechanism throughout the path.

Example 4.4

Consider a two-bar link mechanism, as shown in Fig. 4.11, which is assumed to represent a snapshot of a large-deformation process. The member numbers and the numbers of displacement components are defined in Fig. 4.11. The matrices $\mathbf{D}$, $\mathbf{C}$, and $\mathbf{K}$ at this deformed state are written as

$$
\mathbf{D} = \frac{1}{\sqrt{2}} \begin{pmatrix} -1 & 1 \\ 1 & 1 \\ 0 & -1 \end{pmatrix}, \quad \mathbf{C} = \frac{AE}{L} \begin{pmatrix} 1 & 0 \\ 0 & 1 \end{pmatrix}, \quad \mathbf{K} = \frac{AE}{2L} \begin{pmatrix} 2 & 0 & -1 \\ 0 & 2 & -1 \\ -1 & -1 & 1 \end{pmatrix} \quad (4.13)
$$

where L is the length of the members. As is seen, the ranks of $\mathbf{D}$, $\mathbf{C}$, and $\mathbf{K}$ are 2. Suppose a proportional load is given as $\mathbf{P} = (0, 0, \Lambda)^\top$ with the load factor Λ. The increments from the state in Fig. 4.11 are indicated by Δ.

The incremental equation to be solved in the process of path-tracing analysis is given as follows, if U_3 is chosen as the path parameter:

$$
\frac{AE}{L} \begin{pmatrix} 1 & 0 & 0 \\ 0 & 1 & 0 \\ -1/2 & -1/2 & 1 \end{pmatrix} \begin{pmatrix} \Delta U_1 \\ \Delta U_2 \\ \Delta\Lambda \end{pmatrix} = \frac{AE}{2L} \begin{pmatrix} 1 \\ 1 \\ -1 \end{pmatrix} \Delta U_3 \quad (4.14)
$$

Because the matrix in the left-hand side has full-rank, the increments ΔU_1, ΔU_2, and $\Delta\Lambda$ can be successfully found for the specified increment ΔU_3 as follows:

$$
\Delta U_1 = \frac{1}{2}\Delta U_3, \quad \Delta U_2 = \frac{1}{2}\Delta U_3, \quad \Delta\Lambda = 0 \quad (4.15)
$$

which verifies instability of the mechanism.

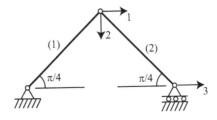

FIGURE 4.11: A two-bar mechanism under deformation.

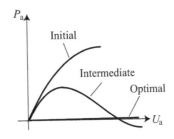

FIGURE 4.12: Relation between input force P_a and input displacement U_a of initial, intermediate, and optimal solutions.

4.4.3   Problem formulation

We generate link mechanisms by removing unnecessary members as a result of optimization from a highly connected ground structure consisting of elastic truss elements. The truss is stable in the early stage of optimization, if the maximum value $P_a^{\max}$ of the input load P_a applied in the specified direction at node 'a' is not zero before reaching the final state. Therefore, in order to naturally generate an unstable mechanism, a constraint is given such that $P_a^{\max}$ vanishes.

Fig. 4.12 shows the load-displacement relation at the input node from the initial undeformed state to the final state for the initial, intermediate, and optimal solutions, which illustrates that the maximum load is reduced to 0 through optimization. Since the maximum load is always nonnegative, as seen in Fig. 4.12, the constraint on the maximum load is given with inequality as

$$P_a^{\max} \le 0 \tag{4.16}$$

which is satisfied with equality at the optimal solution with a small tolerance ε as $P_a^{\max} \le \varepsilon$.

The design variables are the vectors of cross-sectional areas $\mathbf{A}$ and the nodal coordinates $\mathbf{X}$. In order to obtain a mechanism with a small number of members, the total structural volume $V(\mathbf{A}, \mathbf{X})$ is minimized as the objective function. If we assign the constraint (4.16) only, then all the cross-sectional areas vanish as a result of minimizing the total structural volume. Therefore, the stiffness constraints are given in terms of the nodal displacements as

$$U_{\mathrm{b}x}^{\mathrm{f0}} \leq \overline{U}_{\mathrm{b}x}^{\mathrm{f0}}, \quad U_{\mathrm{b}y}^{\mathrm{f0}} \leq \overline{U}_{\mathrm{b}y}^{\mathrm{f0}} \tag{4.17}$$

where $U_{\mathrm{b}x}^{\mathrm{f0}}$ and $U_{\mathrm{b}y}^{\mathrm{f0}}$ are the displacements of the output node 'b' in the x- and y-directions against the unit loads in the x- and y-directions at node 'b', respectively, after fixing the input degree of freedom, and $\overline{U}_{\mathrm{b}x}^{\mathrm{f0}}$ and $\overline{U}_{\mathrm{b}y}^{\mathrm{f0}}$ are their specified upper bounds. Note that the tangent stiffness at the deformed configuration is used for evaluating $U_{\mathrm{b}x}^{\mathrm{f0}}$ and $U_{\mathrm{b}y}^{\mathrm{f0}}$ (Ohsaki and Nishiwaki 2005).

The optimization problem for finding the link mechanism is formulated as

$$\text{Minimize} \quad V(\mathbf{A}, \mathbf{X}) \tag{4.18a}$$

$$\text{subject to} \quad P_{\mathrm{a}}^{\max}(\mathbf{A}, \mathbf{X}) \leq 0 \tag{4.18b}$$

$$U_{\mathrm{b}x}^{\mathrm{f0}}(\mathbf{A}, \mathbf{X}) \leq \overline{U}_{\mathrm{b}x}^{\mathrm{f0}} \tag{4.18c}$$

$$U_{\mathrm{b}y}^{\mathrm{f0}}(\mathbf{A}, \mathbf{X}) \leq \overline{U}_{\mathrm{b}y}^{\mathrm{f0}} \tag{4.18d}$$

$$\mathbf{A}^{\mathrm{L}} \leq \mathbf{A} \tag{4.18e}$$

$$\mathbf{X}^{\mathrm{L}} \leq \mathbf{X} \leq \mathbf{X}^{\mathrm{U}} \tag{4.18f}$$

where $\mathbf{X}^{\mathrm{U}}$ and $\mathbf{X}^{\mathrm{L}}$ are the upper and lower bounds for $\mathbf{X}$, respectively, and $\mathbf{A}^{\mathrm{L}} = (A_1^{\mathrm{L}}, \ldots, A_m^{\mathrm{L}})^{\top}$ defines the small lower bound for $\mathbf{A}$. Note that the member with $A_i = A_i^{\mathrm{L}}$ at the optimal solution is to be removed.

The optimization algorithm is given as follows:

Step 1: Assign initial values for $\mathbf{A}$ and $\mathbf{X}$.

Step 2: Trace the equilibrium path using the input displacement at node 'a' as the control parameter until reaching the final state satisfying condition (4.9).

Step 3: Evaluate the objective and constraint functions and their sensitivity coefficients.

Step 4: Update the variables according to the optimization algorithm.

Step 5: Go to Step 2 if the convergence criteria are not satisfied.

In Step 2, the optimization process is terminated if the output displacement is not in the specified direction, because, in this case, the desired mechanism is not likely to be obtained. Furthermore, the optimization problem is highly nonlinear, and many local optimal solutions are expected to exist. Therefore, the initial values of $\mathbf{A}$ and $\mathbf{X}$ are randomly generated, and path tracing is continued only for the case where the output displacement U_{b} is in the specified direction.

The optimal solution of the first-stage problem (4.18) is unstable and satisfies the requirements for a link mechanism. However, the displacement $\tilde{U}_{\mathrm{b}}$

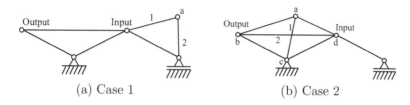

FIGURE 4.13:   Examples of unnecessary members.

of the output node in the direction perpendicular to the specified direction may be large, because no constraint has been given on the direction of the output displacement. Therefore, in the second-stage problem, the maximum absolute value $\tilde{U}_b^{\max}$ of $\tilde{U}_b$ is minimized by considering $\mathbf{X}$ as design variables, while $\mathbf{A}$ is fixed:

$$\text{Minimize} \quad \tilde{U}_b^{\max}(\mathbf{X}) \tag{4.19a}$$

$$\text{subject to} \ \ \mathbf{X}^L \le \mathbf{X} \le \mathbf{X}^U \tag{4.19b}$$

Hence, the link mechanisms with the desired property can be obtained as the solution of the two-stage problem.

Note again that the unnecessary members are removed in the first-stage problem for minimizing the total structural volume. However, because the problem is highly nonlinear, strict convergence to the global optimal solution may not be expected. Even when the global optimum is not found, a feasible solution is obtained in the first stage, and we can proceed to the second stage. Since the topology is fixed at the second stage, unnecessary members cannot be removed at this stage. However, an unnecessary member that can be removed without any effect on the load-displacement relation may be detected manually utilizing the following criteria:

1. If no stress exists in a member against the input force after constraining the output displacement, then the member is considered to be unnecessary. For example, if only two members, which are not colinear, are connected to a node except the input node, output node, and supports, these members are unnecessary, e.g., node 'a' in Fig. 4.13(a).

2. If the global property does not change after removing a member in a statically indeterminate substructure, then the member is unnecessary. For example, the quadrilateral substructure consisting of nodes 'a', 'b', 'c', and 'd' is statically indeterminate, and members 1 or 2 can be removed without loss of stability of the substructure.

Note that some members can also be added to replace the existing members to simplify the topology. The performance of the structure after removing the members is to be confirmed by path-tracing analysis.

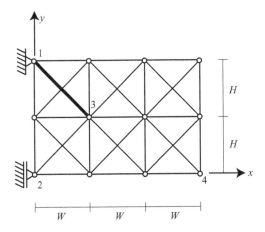

FIGURE 4.14: A 3 × 2 plane grid truss.

4.4.4 Numerical examples

Link mechanisms are found from the ground structure shown in Fig. 4.14, where $W = H = 200$ mm. The truss is pin-supported at node 1 and fixed in the x-direction at node 2. The ground structure has a 3 × 2 grid, and the intersecting diagonals are not connected with each other.

We obtain the mechanism so that node 4 moves 200 mm in the y-direction as the result of anticlockwise rotation of node 3 around support 1. For this purpose, a large value is given for the lower bound A_i^L of the cross-sectional area of the thick member in Fig. 4.14 connecting support 1 and node 3. Because the optimal solution is unstable, the value of A_i does not have any effect on the performance of the link mechanism obtained after optimization of cross-sectional areas and nodal locations. The anticlockwise rotation is realized by the forced y-directional displacement at node 3. Hence, the final state is defined as the deformation where the output displacement U_b in the y-direction of node 4 reaches 200 mm as the result of y-directional input displacement U_a at node 3. The upper-bound displacement against unit loads (1.0 N) in the first stage is given as $\overline{U}_{bx}^{f0} = \overline{U}_{by}^{f0} = 50$ mm. The displacement in the x-direction of node 4 is minimized in the second stage optimization problem (4.19).

The cross-sectional areas of members, except the thick member connecting support 1 and node 3, are considered as independent variables, for which the lower bounds are equal to 0.01 mm^2. The x-coordinate of node 3, the (x, y)-coordinates of node 4, and the coordinates in the constrained directions of the supports are fixed, and the remaining coordinates are the independent design variables. In order to prevent distortion of the topology, the bounds in the side constraints (4.18f) for the nodal coordinates are given as the square

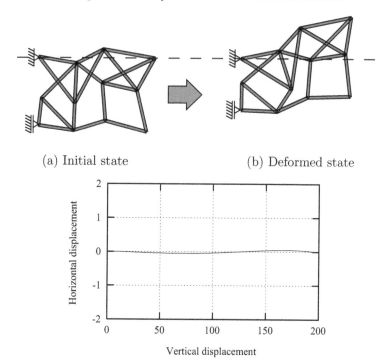

(a) Initial state   (b) Deformed state

(c) Relation between vertical and horizontal displacements (mm) of node 4

FIGURE 4.15: Optimal link mechanism before removing unnecessary members.

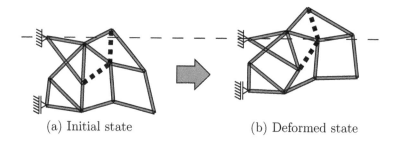

(a) Initial state   (b) Deformed state

FIGURE 4.16: Optimal link mechanism after removing unnecessary members.

region bounded by ±60 mm from the original locations defined in Fig. 4.14.

The engineering strain of a member is computed exactly from the deformed and undeformed lengths of the member. The elastic modulus E is 2.0 N/mm^2. Because the final mechanism is unstable, the value of E does not have any effect on the optimal solution. The y-directional displacement of node 3 is chosen as the parameter for path-tracing analysis.

The performance of the mechanism is confirmed by ADAMS 2005 (MSC Software 2005), which is a general purpose software package for multibody dynamics. The sequential quadratic programming program in IDESIGN Ver. 3.5 (Arora and Tseng 1987) is used for optimization, where the tolerance for the constraints is $\varepsilon = 1.0 \times 10^{-4}$. The sensitivity coefficients are evaluated by a finite difference approach.

In order to obtain various mechanisms, initial solutions are generated randomly. Since the ratios rather than the absolute values are important for the cross-sectional areas, the initial values are defined so as to distribute uniformly between 0 and an appropriate positive value A_0 as $A_i = A_0 R_i$, where $0 \leq R_i < 1$ is a uniform random number, and $A_0 = 10$ mm^2 in the following examples. The initial value for the nodal coordinate x_i (mm) of the ith node is given as

$$x_i = x_i^0 + 120.0(R_i - 0.5) \qquad (4.20)$$

where x_i^0 is the coordinate in Fig. 4.14. The y-coordinates are defined in a similar manner.

A mechanism obtained by using the proposed two-stage optimization algorithm is shown in Fig. 4.15(a). Fig. 4.15(b) shows the deformed final shape, where the dashed horizontal line is added to clearly indicate that the y-coordinate is fixed at support 1. Fig. 4.15(c) shows the relation between y-directional and x-directional displacements of output node 4. Note that the horizontal and vertical axes of the figure correspond to vertical (y-directional) and horizontal (x-directional) displacements, respectively, and their scales are different; therefore, node 4 moves almost straight in the y-direction.

The shape after removing the unnecessary members is shown in Fig. 4.16. The relation between y-directional and x-directional displacements of node 4 is the same as Fig. 4.15(c). The dotted members have been added to obtain a simple configuration. Note that the members that can be added have been assumed to be selected from the ground structure. However, a simpler configuration can be obtained if members can be added between any pair of nodes. The numbers of members and degrees of freedom are 18 and 19, respectively, after removing the unnecessary members. Therefore, the kinematical indeterminacy h is equal to 1, if the equilibrium matrix is full-rank. Several mechanisms can be obtained by carrying out optimization from different initial solutions (Ohsaki and Nishiwaki 2007a).

Chapter 5

Optimization of Building Frames

Optimum design of building frames was extensively studied in the 1960s, because analytical derivation of optimal solutions is possible for simple regular frames. Since the 1980s, with rapid development of computer technology, many methodologies and computer programs have been presented for optimizing real-world structures under practical design constraints. In this chapter, we present various optimization results for building frames. A historical review is presented in Sec. 5.1. The uniqueness of optimal solutions is investigated and approximate solutions are searched for in Sec. 5.2. Multidisciplinary optimization and parametric optimization approaches are presented in Sec. 5.3. A heuristic approach to multiobjective optimization is presented in Sec. 5.4, and finally, multiobjective seismic design optimization is discussed in Sec. 5.5.

5.1 Overview of optimization of building frames

5.1.1 Introduction

Optimum plastic design of building frames is a traditional problem investigated in the 1960s, because the optimal solutions are found analytically or by solving a linear programming problem, as we have seen in Sec. 1.4 (Hemp 1973). Various methods were developed in the 1970s mainly using optimality criteria approaches; see Sec. 1.7 for elastic design problems considering stress and displacement constraints.

However, practical design of building frames using sophisticated computational optimization techniques, including nonlinear programming (NLP) and heuristics, is a rather new field of research in spite of its importance in design practice, because various constraints and costs in view of structural and nonstructural requirements should be considered. Arora, Haug, and Rim (1975) applied a state-space optimal control approach to optimization of frames. Bhatti and Pister (1981) may be cited as one of the first papers on practically applicable optimization methods of building frames. Balling, Pister, and Ciampi (1983) presented an NLP approach under dynamic constraints, where geometrical nonlinearity is considered. Sui and Wang (1997) used geometric programming for frame optimization. Lagaros, Papadrakakis, and

Kokossalakis (2002) presented evolutionary approaches for frame optimization. Jármai, Farkas, and Kurobane (2006) used *particle swarm optimization* (PSO) (Kennedy 1997) for optimization of a plane steel frame, where the practical objective function, including the costs of connection and fabrication, is minimized under code-based design constraints. PSO has also been applied to the design of trusses (Li, Huang, Liu, and Wu 2007). Saka and Erdal (2009) applied a *harmony search* algorithm for frame optimization based on load and resistance factor design (LRFD) specifications.

Optimum design of real-world tall buildings has been studied since the 1980s. Grierson and Lee (1984) presented a two-level approach, where the problem with continuous variables is first solved to select the initial cross-sections for the second problem with discrete variables. Grierson and Chan (1993) developed an optimality criteria approach to the optimization of steel frames under lateral drift constraints, where the nodal displacements are explicitly written using the virtual force method, and the cross-sectional properties of each member are approximated as linear functions of the reciprocal of the cross-sectional area. Chan, Grierson, and Sherbourne (1995) optimized a 50-story steel frame using an optimality criteria approach to select the cross-sectional properties of members from a list of commercially available sections. Haque (1996) applied the *complex method* (Box 1965) for frame optimization with discrete variables. Kocer and Arora (1999) optimized a transmission tower using a simulated annealing and a genetic algorithm, where the material grades as well as the cross-sectional properties are considered as discrete design variables. Liu, Burns, and Wen (2006) presented a multiobjective programming approach for minimization of the number of different types of cross-sections. Bojczuk and Mróz (1998a) optimized the support locations of a continuous beam.

Recently, some attempts have been made to optimize the cross-sectional shapes of beams. Pan, Ohsaki, and Tagawa (2007) and Ohsaki, Tagawa, and Pan (2009) optimized the flange shape of a beam considering plastic energy dissipation under monotonic and cyclic loads. Erdal and Saka (2009) optimized the shape of the opening in the web of an I-beam using a heuristic approach called harmony search. Lagaros, Psarras, Papadrakakis, and Kokossalakis (2008) investigated the effect of the web opening on the optimum design of 3D frames.

5.1.2   Problem formulation

Consider a plane frame, e.g., as shown in Fig. 5.1, consisting of rigidly jointed beam-column (frame) elements. The assumption of a rigid floor is usually used; i.e., the horizontal displacements of the nodes in the same floor have the same value. Note that *rigid floor* means that the floor is rigid in the in-plane direction, and does not mean that it is rigid against out-of-plane bending/shear deformation.

Let n and n^{F} denote the number of degrees of freedom and the num-

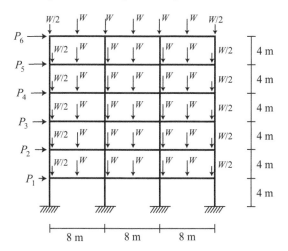

FIGURE 5.1: A six-story three-span frame.

FIGURE 5.2: Displacement numbers of a member of a plane frame.

ber of stories, respectively. The nodal load vector is denoted by $\mathbf{P}^{\mathrm{N}} = (P_1^{\mathrm{N}}, \ldots, P_n^{\mathrm{N}})^{\top}$, which includes the vertical static loads and the horizontal loads $P_1, \ldots, P_{n^{\mathrm{F}}}$ representing the seismic loads and the wind loads applied at the floors. The vertical loads include the self-weight of the beams and columns, the weights of the nonstructural components, and the live loads. Note that the horizontal load P_{i-1} is applied at the ith floor, as shown in Fig. 5.1, and the roof is denoted as the $(n^{\mathrm{F}} + 1)$st floor, at which the load $P_{n^{\mathrm{F}}}$ is applied. The $n \times n$ stiffness matrix is denoted by $\mathbf{K}$. Then the displacement vector $\mathbf{U}$ is found from the stiffness (equilibrium) equation

$$\mathbf{KU} = \mathbf{P}^{\mathrm{N}} \tag{5.1}$$

The displacement vector of member i is denoted by $\mathbf{u}^i = (u_1^i, \ldots, u_6^i)^{\top}$, where the local displacement numbers are indicated in Fig. 5.2. The vector of stresses at the member ends 'a', 'b', 'c', and 'd' in Fig. 5.2 is denoted by $\boldsymbol{\sigma}^i = (\sigma_{\mathrm{a}}^i, \sigma_{\mathrm{b}}^i, \sigma_{\mathrm{c}}^i, \sigma_{\mathrm{d}}^i)^{\top}$, which is written in terms of $\mathbf{u}^i$ as

$$\boldsymbol{\sigma}^i = \mathbf{d}^i \mathbf{u}^i \tag{5.2}$$

where $\mathbf{d}^i$ is a 4×6 constant matrix.

In most design codes, such as the Japanese Building Standard Law, the bound for the stress for a plane frame is given as follows as a summation of stress ratios:

$$\frac{|\sigma^A|}{\bar{\sigma}^A} + \frac{|\sigma^B|}{\bar{\sigma}^B} \leq 1 \qquad (5.3)$$

where σ^A is the stress due to axial force, σ^B is the stress due to bending moment, and $\bar{\sigma}^A$ and $\bar{\sigma}^B$ are their upper bounds. Note that different formulas and bounds are assigned in design codes for tensile and compressive states distinguished by the sign of σ^A. For a three-dimensional frame, the stresses due to bending moments around the two principal axes, denoted by subscripts $(\cdot)_y$ and $(\cdot)_z$, are incorporated as

$$\frac{|\sigma^A|}{\bar{\sigma}^A} + \frac{|\sigma^B_y|}{\bar{\sigma}^B_y} + \frac{|\sigma^B_z|}{\bar{\sigma}^B_z} \leq 1 \qquad (5.4)$$

One of the important aspects of the design of building frames is that the cross-sectional properties, including the area and second moment of inertia, cannot be defined as independent continuous variables; they should be selected from the catalog of available standard sections. Therefore, the optimization problem turns out to be a combinatorial problem. An example of the list of available sections is shown in Appendix A.8.

However, approximate optimal solutions are usually found by considering cross-sectional parameters as continuous variables. Let $\mathbf{A} = (A_1, \ldots, A_m)^\top$ denote the vector of cross-sectional areas of the members, where m is the number of members, including the beams and columns. In order to find an optimal solution using a formulation of the nonlinear programming (NLP) problem, considering $\mathbf{A}$ as continuous design variables, it is necessary that the cross-sectional properties, e.g., second moment of inertia and the section modulus of the ith member, are given as functions of A_i as $I_i(A_i)$ and $Z_i(A_i)$, respectively. Approximation is also needed for the plastic modulus Z_i^P if inelastic responses are to be considered. These values are usually approximated as $I_i(A_i) = aA_i^b$ for which the parameters a and b are to be identified from the list of sections (see Appendix A.8 for details). Note that these parameters should be found so that they accurately approximate the series of sections that is to be used in the respective design problem. Therefore, it is not important here to show the values of a and b used in the literature.

Alternatively, Z_i and I_i may be linearly interpolated with respect to A_i if the list of sections is given so that Z_i and I_i are increasing functions of A_i. Although, in this case, the sensitivity coefficients of Z_i and I_i with respect to A_i are discontinuous, optimal solutions are found successfully, as demonstrated in Example 5.1, if an NLP algorithm with line search is used.

In the design process of building frames, the upper bounds are usually given for the response displacements and stresses against static loads based on the design code. Therefore, the design requirements are generally formulated as follows as inequality constraints:

$$H_j(\mathbf{A}) \leq 0, \quad (j = 1, \ldots, n^I) \qquad (5.5)$$

where H_j is assumed to be an implicit function of $\mathbf{A}$ after eliminating the state variables, and n^{I} is the number of constraints. Hence, the optimization problem for minimizing the objective function, e.g., the total structural volume $V(\mathbf{A})$, is formulated as a standard NLP problem:

$$\text{Minimize} \quad V(\mathbf{A}) \tag{5.6a}$$

$$\text{subject to} \quad H_j(\mathbf{A}) \le 0, \quad (j = 1, \dots, n^{\mathrm{I}}) \tag{5.6b}$$

$$A_i^{\mathrm{L}} \le A_i \le A_i^{\mathrm{U}}, \quad (i = 1, \dots, m) \tag{5.6c}$$

where A_i^{L} and A_i^{U} are the lower and upper bounds for A_i, respectively.

The definition of the objective function based on the cost of the structure is the most ambiguous aspect of structural optimization in the field of civil engineering. Various definitions have been proposed, including structural and nonstructural costs (see Sec. 5.1.5 for details). The most realistic definition of the cost may be the life-cycle cost; however, the possible hazards and maintenance costs during the life cycle of the structure are highly unpredictable. Hence, it is very difficult to present a realistic and robust definition of the life-cycle cost. Therefore, the initial cost that is represented by the volume of the structural material is used for the objective function of a simple optimization problem of a building frame.

Another difficulty in the optimization of building frames is that various short-term loads, including seismic, wind, and snow loads, represented by equivalent static loads, should be considered in addition to long-term loads, including self-weight and live load. An optimum design should be found so that the structure remains in the elastic range for the long-term load and has enough safety with inelastic deformation against short-term loads (Grierson and Schmit 1982; Grierson and Chiu 1984). This concept was recently generalized as *performance-based design*, for which is the design process can be naturally formulated as an optimization problem (Foley 2002).

In view of practical application, some additional requirements should be considered for the construction process (Liu, Burns, and Wen 2006). For instance, construction cost depends on the complexity of the structure such as the number of different sections, number of parts to be assembled, and shape of the parts at the connections. Furthermore, requirements on width-thickness ratios of the flanges and webs of beams are given by the building codes. The strong column-weak beam criterion should usually be satisfied; i.e., the frame should have enough capacity of plastic energy dissipation at the plastic hinges at the beam ends before exhibiting total collapse due to the failure of columns.

Example 5.1

As a small example, optimal solutions are found for the six-story three-span plane steel frame shown in Fig. 5.1 (Ohsaki 2003a). The conventional assumption of a rigid floor is used. The vertical load $W = 120$ kN represents the

TABLE 5.1: List of available sections for beams and columns; cross-sectional area A ($\times 10^2$ mm^2), section modulus Z ($\times 10^3$ mm^4), and second moment of inertia I ($\times 10^6$ mm^6).

	Beam			Column		
	A	Z	I		A	Z	I
G1	61.4	610.0	106.0	C1	104.0	986.0	147.0
G2	83.4	1040.0	232.0	C2	162.0	1760.0	309.0
G3	116.0	1550.0	387.0	C3	186.0	2330.0	467.0
G4	142.0	2180.0	655.0	C4	277.0	3880.0	873.0
G5	184.0	3040.0	896.0	C5	309.0	4840.0	1210.0
G6	214.0	3560.0	1050.0	C6	599.0	8780.0	2190.0

self-weight and live load, and the elastic modulus is 200 kN/mm^2. The horizontal loads (kN) representing the seismic loads are given as $(P_1, \ldots, P_6) = (60.0, 110.0, 155.0, 195.0, 230.0, 260.0)$. The sections of beams and columns are selected from the list in Table 5.1, where A, Z, and I are the cross-sectional area, section modulus, and second moment of inertia, respectively. The total structural volume V is to be minimized. In the following, the units of length and force are mm and kN, respectively, which are omitted for brevity.

The columns and beams are modeled using Euler-Bernoulli beam elements, and each beam is divided into two elements in order to incorporate the stress constraints at the center of the beam. The upper bound 0.2 is given for the absolute value of the stresses at the edges of the two ends, denoted by 'a', 'b', 'c', and 'd' in Fig. 5.2, of the elements, including the beams and columns. In order to incorporate a realistic situation and to reduce the number of design variables, the beams and columns are classified into six groups, respectively, where the members in each group have the same section. The columns are classified into exterior and interior columns, and those in three pairs of stories (1,2), (3,4), and (5,6) have the same sections. The beams are classified into six groups similarly.

We first interpolate Z and I as piecewise linear functions of A using the values in Table 5.1. Note that the available sections have been chosen such that Z and I are increasing functions of A for both the columns and beams. The optimization software package IDESIGN Ver. 3.5 (Arora and Tseng 1987), which utilizes sequential quadratic programming, is used. The bounds for the cross-sectional areas are defined by the smallest and largest values in the lists for beams and columns, respectively. The optimal solution denoted by $\mathbf{A}^{\mathrm{R}}$ with continuous variables is listed in the second column of Table 5.2, where, e.g., 'Ext. beam 2, 3' indicates the beams in the exterior spans in the second and third floors, and 'Int. col. 1, 2' indicates the interior columns in the first and second stories, while the roof is denoted as the seventh floor. The total structural volume is also listed in Table 5.2, and the 'Max. stress' means the maximum absolute value of the stresses among all members. Note that

TABLE 5.2: Optimal solutions of the six-story three-span frame.

	Cont. var. $\mathbf{A}^{R}$	Greedy from nearest to $\mathbf{A}^{R}$	Enum. near $\mathbf{A}^{R}$	Greedy	Stingy
Ext. beam 2, 3	214.0	G6	G6	G6	G6
Ext. beam 4, 5	188.8	G6	G5	G5	G6
Ext. beam 6, 7	121.1	G3	G3	G4	G4
Int. beam 2, 3	206.2	G6	G6	G6	G6
Int. beam 4, 5	186.7	G5	G6	G6	G6
Int. beam 6, 7	122.2	G5	G4	G3	G4
Ext. col. 1, 2	289.1	C4	C4	C4	C5
Ext. col. 3, 4	186.2	C3	C3	C3	C3
Ext. col. 5, 6	116.2	C1	C3	C2	C2
Int. col. 1, 2	404.3	C6	C6	C6	C6
Int. col. 3, 4	282.8	C4	C4	C4	C5
Int. col. 5, 6	175.5	C3	C3	C2	C3
Total volume $(\times 10^{6})$	4.827	5.278	5.259	5.538	5.182
Max. stress	0.2	0.1971	0.1986	0.1949	0.1987

the stress constraint is satisfied with equality in at least one of the members in each group; i.e., the optimal solution is fully stressed.

An approximate optimal solution from the list of available sections can be found by selecting the nearest section from $\mathbf{A}^{R}$. If the stress constraint is not satisfied, then the section of the group with maximum violation is consecutively increased until the stress constraints are satisfied in all members. The solution obtained by using this simple greedy method is listed in the third column of Table 5.2; see Appendix A.3 for details of the greedy method. The maximum stress is 0.1971; i.e., the solution is not fully stressed, because the sections can be selected only from the discrete candidates. The total structural volume is 5.278×10^{6}, which is about 9.3% larger than $\mathbf{A}^{R}$.

If the available sections are restricted to the two nearest sections from $\mathbf{A}^{R}$ for each group, then we have $2^{12} = 4048$ candidate solutions. The optimal solution obtained by enumeration of these candidate solutions is listed in the fourth column of Table 5.2. Note that the total structural volume is almost the same as that found by the simple greedy method, although structural analysis has been carried out 4048 times.

We next use the greedy method starting with the smallest section for each group. The approximate solution obtained this way is shown in the fifth column of Table 5.2, which is worse than the previous two discrete solutions. In contrast, if we use the stingy method starting with the largest section for each group, the best approximate solution is obtained as shown in the sixth column of Table 5.2. Although the performances of the methods depend on the definition and parameter values of the problem, good approximate discrete solutions can be obtained using heuristic approaches.

5.1.3   Continuum approach

There are numerous papers on topology optimization of plates (sheets) discretized to finite elements subjected to in-plane loads. As a result of optimization, material layouts like frame structures, e.g., bridges and braced frames, are often obtained. Therefore, some attempts have been made to generate optimal layout of beams, columns, and braces of building frames utilizing continuum shape and topology optimization. Yang, Xie, Steven, and Querin (1999b) used the bi-directional evolutionary structural optimization (BESO) method (Yang, Xie, Steven, and Querin 1999a) for frequency optimization of frames, which was extended from the original evolutionary structural optimization (ESO) method allowing addition of the elements. Liu and Qiao (2009) optimized the layout of frames and bridge-type structures with different stiffnesses in tension and compression that are approximated by a smooth function. Mijar, Swan, Arora, and Kosaka (1998) presented a continuum optimization approach for generating braced frames utilizing the Voigt-Reuss mixing rule of constitutive relation. Rahmatalla and Swan (2003) developed a continuum-based conceptual design approach for bridges and transmission towers.

However, we do not discuss continuum approaches in this book, because there are many books on continuum shape/topology optimization (Haug and Cea 1981; Bendsøe and Sigmund 2003).

5.1.4   Semi-rigid connections and braces

Semi-rigid connections are widely used in American and European countries to concentrate the resistance to horizontal loads at the limited number of spans consisting of moment-resisting (rigidly jointed) frames. This way, the total cost of a frame is reduced, because the manufacturing and construction of rigid joints consisting of welding or high-tension bolts are more expensive than those of semi-rigid connections. In contrast, semi-rigid connections are not very popular in Asian communities, especially Japan, where columns with a box section are usually used with diaphragms welded to the flanges of the beams. However, rigorously speaking, most of the connection types are semi-rigid, and there is no strictly pin-jointed or rigidly jointed connection.

Various results of optimization of frames with semi-rigid connections are summarized by Xu (2002). Machaly (1986) optimized the cross-sectional dimensions of the members of an elastic frame with semi-rigid connections. Kameshki and Saka (2001b) and Hayalioglu and Degertekin (2005) used genetic algorithms for optimizing frames with semi-rigid connections, where the types of connections are fixed. In contrast, Kameshki and Saka (2003) also optimized the types of the semi-rigid connections.

The Frye–Morris polynomial model (Frye and Morris 1975) is often used

FIGURE 5.3: A beam-column element with semi-rigid connections.

for the moment-rotation relation of a semi-rigid connection. Machaly (1986) showed that the use of semi-rigid connections instead of rigid connections leads to a reduction of the total weight of a small-scale frame. Xu and Grierson (1993) used a continuous model for the stiffness of the connections. Doğan and Saka (2009) used a particle swarm optimization method for optimizing a frame under an AISC-LRFD specification (AISC 1999).

Hagishita and Ohsaki (2008a) developed a heuristic approach based on scatter search for optimization of braced frames with semi-rigid connections considering inelastic static responses. Kameshki and Saka (2001a) optimized the locations of braces using a genetic algorithm. Kameshki and Saka (2001b) optimized the types of braces of frames with semi-rigid connections.

Consider a beam-column element connected to semi-rigid connections (nodes 1 and 2) with elastoplastic springs, as shown in Fig. 5.3. The elastic modulus, second moment of inertia, and length of the element are denoted by E, I, and L, respectively. Let R_i denote the rotational stiffness of of the spring at node i, and define parameters r_a, r_b, r_c, a, b, and c as

$$r_a = \frac{1}{R^*}\left(4 + \frac{12EI}{LR_j}\right) \tag{5.7a}$$

$$r_b = \frac{1}{R^*}\left(4 + \frac{12EI}{LR_i}\right) \tag{5.7b}$$

$$r_c = \frac{2}{R^*} \tag{5.7c}$$

$$a = r_a + 2r_b + r_c \tag{5.7d}$$

$$b = r_a + r_b \tag{5.7e}$$

$$c = r_b + r_c \tag{5.7f}$$

where

$$R^* = \left(1 + \frac{4EI}{LR_i}\right)\left(1 + \frac{4EI}{LR_j}\right) - \left(\frac{EI}{L}\right)^2\left(\frac{4}{R_iR_j}\right) \tag{5.8}$$

The local displacement numbers are defined in Fig. 5.3. Let A denote the cross-sectional area of the member. Then the 6×6 elastic stiffness matrix $\mathbf{K}_E$

of a member of a plane frame is defined as (Chen 1991)

$$
\mathbf{K_E} =
\begin{pmatrix}
\dfrac{EA}{L} & 0 & 0 & -\dfrac{EA}{L} & 0 & 0 \\[2mm]
 & a\dfrac{EI}{L^3} & b\dfrac{EI}{L^2} & 0 & -a\dfrac{EI}{L^3} & c\dfrac{EI}{L^2} \\[2mm]
 & & r_a\dfrac{EI}{L} & 0 & -b\dfrac{EI}{L^2} & r_c\dfrac{EI}{L} \\[2mm]
 & & & \dfrac{EA}{L} & 0 & 0 \\[2mm]
 & \text{sym.} & & & a\dfrac{EI}{L^3} & -c\dfrac{EI}{L^2} \\[2mm]
 & & & & & r_b\dfrac{EI}{L}
\end{pmatrix}
\tag{5.9}
$$

For performance evaluation of the frames with semi-rigid connections, it is very important to consider inelastic responses, because the large rotational deformation is concentrated at the connections. For this purpose, the *refined plastic hinge method* (Liew, White, Chen, and Toma 1993a, 1993b), which is categorized as *advanced analysis*, is often used. This method incorporates geometrical and material nonlinearities, as follows, which are equivalent to the *load and resistance factor design* (LRFD) specification (AISC 1999; Chen and Kim 1997).

The geometrically nonlinear deformation of a two-dimensional beam-column element is modeled using corotational formulation (Crisfield 1991) so that the local coordinates are rotated in accordance with the global deformation and one of the local axes is always in the direction of the member axis. Hence, the 6×6 stiffness matrix of a beam-column element in global coordinates is reduced to a 3×3 matrix with respect to local coordinates. Let δM_i and $\delta \theta_i$ denote the incremental bending moment and rotation, respectively, of node i of the element. The increments of axial force and elongation are denoted by δP and δe, respectively. Then, the incremental form of the equilibrium (stiffness) equation is written as

$$
\begin{pmatrix} \delta M_1 \\ \delta M_2 \\ \delta P \end{pmatrix}
= \frac{EI}{L}
\begin{pmatrix} S_1 & S_2 & 0 \\ S_2 & S_1 & 0 \\ 0 & 0 & A/I \end{pmatrix}
\begin{pmatrix} \delta \theta_1 \\ \delta \theta_2 \\ \delta e \end{pmatrix}
\tag{5.10}
$$

where S_1 and S_2 are the stability functions to incorporate the secondary force and moment due to $P - \delta$ and $P - \Delta$ effects (Chen and Kim 1997). In order to incorporate the gradual yielding due to nonuniform distribution of the bending moment, the perfectly rigid-plastic model can be modified to an elastoplastic model in a manner similar to the LRFD formula.

There have been many models proposed for the relation between the moment M and rotation angle θ of semi-rigid connections. For example, Kishi

TABLE 5.3: Three parameters for five types of semi-rigid connections.

Connection type	M_u (kNm)	R_{init} (kNm/rad)	Shape parameter s
Single web angle	18.4	164.1	1.8
Double web angle	36.8	3282.2	2.7
Top and seat angle	54.0	23,409.4	0.8
Top and seat angle with double web angles	93.9	26,786.8	1.2
T-stub	150.0	60,000.0	2.0

and Chen (1990) presented the following simple three-parameter model:

$$\frac{M}{M_u} = \frac{\widetilde{\theta}}{(1 + \widetilde{\theta}^s)^{\frac{1}{s}}}, \quad \widetilde{\theta} = \frac{\theta}{\theta_0}, \quad \theta_0 = \frac{M_u}{R_{init}} \tag{5.11}$$

where the parameters M_u, R_{init}, and s are the ultimate bending moment, initial rotational stiffness, and shape parameter, respectively.

Semi-rigid connections are classified into five types: (1) single web angle, (2) double web angle, (3) top and seat angle, (4) top and seat angle with double web angles, and (5) T-stub. Types (1)–(4) use angle members at different locations for connecting the beam and the column. Simple formulas have been developed for the three parameters of these types, as shown in Table 5.3.

Frye and Morris (1975) presented the following odd-power polynomial model for the connections (a) end plate without column stiffness, (b) top angle seat with double web angle, and (c) top angle seat without double web angle:

$$\theta = C_1(KM) + C_2(KM)^3 + C_3(KM)^5 \tag{5.12}$$

where C_i ($i = 1, 2, 3$) are the curve-fitting parameters for each type of connection, and K is a parameter defined in view of the geometry of the connection.

Hagishita and Ohsaki (2008a) optimized a frame with semi-rigid connections and braces, as shown in Fig. 5.4, using a heuristic approach called scatter search (Laguna and Marti 2003), where the units of load and length are kN and mm, respectively. The types of semi-rigid connections are selected from the five models in Table 5.3. The locations, sections, and types of the braces, including X-brace, K-brace, and V-brace, are also optimized. Note that any type of brace can exist at each span of the story. The frame is subjected to vertical loads (long-term loads) and two types of horizontal loads (seismic loads), where the horizontal seismic loads of Levels 1 and 2 are shown, respectively, without and with parentheses in Fig. 5.4. The vertically aligned columns have the same section, and the standard assumption of a rigid floor is used. The sections of beams, columns, and braces are selected from the list of the available standard sections. Constraints are given for the interstory

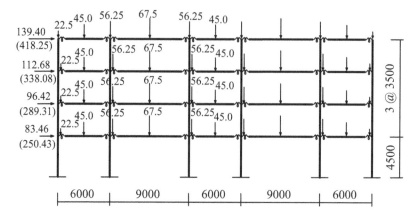

FIGURE 5.4: Geometry and loading conditions of a braced frame with semi-rigid connections.

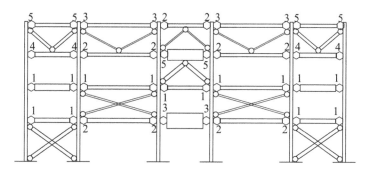

1. Single web angle   2. Double web angle   3. Top and seat angle

4. Top and seat angle with double web angle   5. T-stub

FIGURE 5.5: Optimal types of braces and semi-rigid connections.

drift angle under the seismic loads of Level 1 and the maximum load carrying capacity (ultimate plastic collapse load) under Level 2; see Hagishita and Ohsaki (2008a) for details. The cost of each semi-rigid connection is added to the total structural volume to formulate the objective function. An example of an optimal solution is shown in Fig. 5.5, where the width of each member is proportional to the height of the cross-section, and the types of semi-rigid connections are also indicated. As is seen, the beams in the short span have large sections. The braces are located so as to form a *hat truss* in the fourth story and to transmit the horizontal loads diagonally to the supports between the first and third stories.

5.1.5 Formulation of cost function

One of the most difficult aspects in the application of optimization approaches to practical design of building frames is that it is very difficult to formulate the objective (cost) function incorporating the various structural and nonstructural costs. As discussed in Sec. 1.1, we can obtain feasible designs through optimization even if the definition of the cost function is unrealistic. However, it is desired that good approximate optimal solutions are found with a realistic definition of the cost function.

Pavlovčič, Krajnc, and Beg (2004) investigated in detail the initial cost of a steel frame and classified it into the costs for steel material, welding manufacturing, welding material, assembly and trach welding, cutting, painting, surface preparation, flange aligning, joint, hole forming, bolting material, transportation, and erection. However, it is not possible to precisely define these costs as functions of the design variables.

One of the criticisms of the optimization of building frames is that it considers only the initial cost, or only the material cost, and does not include the cost throughout the service life of the building. Recently, some attempts have been made to optimize the frame considering the life-cycle cost. Sarma and Adeli (2002) presented a detailed classification of the cost and defined the life-cycle cost as the sum of the costs of initial construction, maintenance, inspection, repair, operation, failure, and dismantling.

Doi, Yoshimura, Nishiwaki, and Izui (2009) classified industrial products into short/long functional longevity and short/long usage period, where building frames are classified into the product with long usage period. They considered the design for '3R', i.e., *reduce*, *reuse*, and *recycling*, and formulated the total cost as the sum of manufacturing cost, recycling cost, and reuse cost. Wen and Kang (2001) optimized the frame considering the life-cycle cost under multiple hazards.

Liu, Wen, and Burns (2004) presented a multiobjective optimization approach considering the decision maker's preference on the acceptable risk for definition of the initial cost, and found Pareto solutions using a multiobjective genetic algorithm. Let T and $\mathbf{X}$ denote the service life of the structure and the vector of design variables, respectively. The annual occurrence rate of major seismic events is denoted by ν, which is modeled by a Poisson process. The annual monetary discount rate is denoted by λ; i.e., the value of money is reduced every year at the ratio of λ. The cost for the jth seismic damage state and its occurrence probability are denoted by C_j and P_j, respectively. Then the total cost $C(T, \mathbf{X})$ considering n^{d} damage states is formulated as (Liu, Wen, and Burns 2004)

$$C(T, \mathbf{X}) = \frac{\nu}{\lambda}(1 - e^{-\lambda T}) \sum_{j=1}^{n^{\mathrm{d}}} C_j P_j \qquad (5.13)$$

The cost function C_j is given as

$$C_j = C_j^{\text{damage}} + C_j^{\text{content}} + C_j^{\text{relocation}} + C_j^{\text{economic}} + C_j^{\text{injury}} + C_j^{\text{fatality}} \quad (5.14)$$

where C_j^{damage} is the direct structural/nonstructural damage and repair cost, C_j^{content} is the cost due to loss of contents, $C_j^{\text{relocation}}$ is the relocation cost, C_j^{economic} is the direct/indirect economic loss, C_j^{injury} is the human injury cost, and C_j^{fatality} is the human fatality cost.

Kocer and Arora (1999) defined the total initial cost C_I of a transmission tower, as follows, as the sum of the costs of material, painting, galvanizing, and welding:

$$C_I = C_{\text{sg}} C_W W + C_P A_{\text{surface}} + C_G W + C_{\text{weld}} L_{\text{weld}} \quad (5.15)$$

where C_{sg} is the coefficient indicating the steel grade, C_W is the cost per unit weight of the steel, W is the weight of steel, C_P is the cost of painting per unit area, A_{surface} is the surface area to be painted, C_G is the unit cost of galvanizing, C_{weld} is the cost of welding per unit length, and L_{weld} is the length to be welded. The welding cost C_{weld} includes the costs for labor, electrode, power, and equipment.

In another formulation (Cheng 2002), the total cost C_T is first classified into the initial cost C_I and the damage cost C_D as

$$C_T = C_I + C_D \quad (5.16)$$

and C_D is further classified into

$$C_D = C_r + C_c + C_e + C_s + C_f \quad (5.17)$$

where C_r is the repair or replacement cost of the structure, C_c is the loss of contents, C_e is the economic impact of structural damage, C_s is the cost of injuries caused by structural damage, C_f is the cost of fatalities from structural damage or collapse.

5.2 Local and global searches of approximate optimal designs

5.2.1 Introduction

Approximate optimal solutions under stress constraints have been conventionally found using the stress-ratio approach of fully-stressed design (FSD); see Sec. 1.6 for the FSD of trusses. Recently, it has been pointed out that there may exist many FSDs of building frames with almost the same total structural volume (Mueller, Liu, and Burns 2002; Liu and Burns 2003). Therefore, obtaining only one solution will not be sufficient for practical purposes, where several solutions satisfying stress constraints should be compared in view of other performance measures, including requirements in construction and fabrication processes. Furthermore, the objective function need not be strictly minimized; i.e., it will be helpful for the designer if several approximate optimal solutions with different distributions of cross-sectional areas are obtained.

The nonuniqueness of the optimal solution is generally conceived as a negative aspect in structural optimization, because it deteriorates the convergence property of the optimization algorithm; however, the nonuniqueness of the solutions, which is classified as follows, can be extensively utilized to generate many different approximate optimal solutions:

(a) Local nonuniqueness due to independence of the displacements and the constraint functions on the design variables, which is enhanced by regularity of the frame.

(b) Global nonuniqueness due to nonconvexity of the objective and/or constraint functions.

Simões (1989) studied isolated global optimality of the truss optimization problem using second-order optimality conditions. Similar nonuniqueness can be observed for a plate (sheet) discretized to finite elements and subjected to in-plane loads (Kutylowski 2002). Jog and Haber (1996) derived the conditions for the stability (uniqueness) of the optimal solution using an incremental form of the variational problem. They suggested that the nonuniqueness of the solution to a compliance optimization problem can be detected by the singular values of the matrix defined by the derivatives of the optimality conditions and the equivalent nodal load vector with respect to the design variables. Barthold and Gerzen (2009) utilized singular value decomposition of the sensitivity matrix to reduce the number of variables and to utilize second-order sensitivity coefficients in the optimization process. Petersson (1999) investigated convergence properties of the optimization process with respect to mesh size and noted that the objective function may be insensitive to the thickness variation for special simple loading conditions, such as uniaxial tension/compression and simple shear, if the element with a bilinear displacement interpolation function is used.

Watada and Ohsaki (2009a) investigated the nonuniqueness of the solution as the branching of the solution path based on the so-called *continuation method* with respect to the problem parameter. The continuation method for the optimization problem is basically the same as the parametric programming approach (Gal 1979; Fiacco 1983) or the homotopy method (Shin, Haftka, Watson, and Plaut 1988; Watson and Haftka 1989) for tracing the optimal solutions corresponding to the various parameter values (Nakamura and Ohsaki 1988).

In this section, three formulations are presented for local search of optimal solutions of a regular plane frame under an stress constraints. The distance between the solutions is then defined, and the approximate optimal solutions are globally searched consecutively so as to maximize the distance from the already found approximate solutions under an upper-bound constraint on the total structural volume (Ohsaki 2006a, 2006b).

5.2.2 Optimization problem and optimality conditions

Consider a rigidly jointed regular plane frame, as shown in Fig. 5.1, in Sec. 5.1.2. Let A_i^* denote the cross-sectional area of the ith member. The second moment of inertia I_i^* and section modulus Z_i^* are defined as continuous functions of A_i^* (see Appendix A.8). Hence, A_i^* can be considered as the only independent design variable of member i.

The members are classified into groups based on the symmetry and regularity properties of the frame, and the number of variables is reduced by using the so-called design variable linking approach. Let A_i denote the cross-sectional area of the members in the ith group, and the vector consisting of A_i of m groups is denoted by $\mathbf{A} = (A_1, \ldots, A_m)^\top$.

Let $\mathbf{P}$ denote the static load vector that is supposed to be independent of $\mathbf{A}$. The number of degrees of freedom is denoted by n. The nodal displacement vector $\mathbf{U}$ against $\mathbf{P}$ is found by solving the stiffness equation

$$\mathbf{K}(\mathbf{A})\mathbf{U} = \mathbf{P} \qquad (5.18)$$

where $\mathbf{K}(\mathbf{A})$ is the $n \times n$ stiffness matrix, which is a function of $\mathbf{A}$. Alternatively, for a given $\mathbf{U}$, the equivalent nodal load vector $\mathbf{F}(\mathbf{U}, \mathbf{A})$ is defined by

$$\mathbf{F}(\mathbf{U}, \mathbf{A}) = \mathbf{K}(\mathbf{A})\mathbf{U} \qquad (5.19)$$

For a frame modeled using the beam element, constraints are given for the stresses σ_a^i, σ_b^i, σ_c^i, and σ_d^i at the member ends 'a', 'b', 'c', and 'd', as shown in Fig. 5.2, of the members $i = 1, \ldots, m^*$, where m^* is the number of members. The same upper bound σ^U is given, for simplicity, for the absolute values of the stresses of all members. The conventional assumption of a rigid floor is used. Hence, there is no axial force in a beam, and the absolute values of the stresses satisfy $|\sigma_a^i| = |\sigma_b^i|$ and $|\sigma_c^i| = |\sigma_d^i|$. Therefore, the number of points at which the stresses are constrained is two for beams and four for columns that undergo axial deformation in addition to bending deformation. Constraints are also given for the nodal displacements.

Let $H_j(\mathbf{U}(\mathbf{A}), \mathbf{A})$ denote the jth inequality constraint function representing the bound on stress or displacement. The total length of the members in the ith group is denoted by L_i. The optimization problem for minimizing the total structural volume $V(\mathbf{A})$ is formulated as

$$\text{Minimize} \quad V(\mathbf{A}) = \sum_{i=1}^{m} A_i L_i \tag{5.20a}$$

$$\text{subject to} \quad H_j(\mathbf{U}(\mathbf{A}), \mathbf{A}) \leq 0, \quad (j = 1, \ldots, n^{\mathrm{I}}) \tag{5.20b}$$

$$A_i^{\mathrm{L}} \leq A_i \leq A_i^{\mathrm{U}}, \quad (i = 1, \ldots, m) \tag{5.20c}$$

where A_i^{L} and A_i^{U} are the lower and upper bounds for A_i, respectively, and n^{I} is the number of inequality constraints. Problem (5.20) is a nonlinear programming (NLP) problem that can be solved using a gradient-based optimization algorithm; see Appendix A.2.2.

Since Problem (5.20) is generally nonconvex, convergence to the global optimal solution is not guaranteed. However, if solutions with the same objective value are always found starting with different initial solutions, it is highly possible the solutions are globally optimal. Even if the solution is not globally optimal, it satisfies the Karush-Kuhn-Tucker (KKT) conditions that are the necessary conditions for local optimality (see Appendix A.2.2.3 for details).

Let $\boldsymbol{\mu} = (\mu_1, \ldots, \mu_{n^{\mathrm{I}}})^{\top} \ (\geq \mathbf{0})$ denote the vector of Lagrange multipliers for the inequality constraints. The constraint function with respect to $\mathbf{A}$ only is denoted as $\widetilde{H}_j(\mathbf{A}) = H_j(\mathbf{U}(\mathbf{A}), \mathbf{A})$; i.e., the displacements $\mathbf{U}$ are considered as implicit functions of $\mathbf{A}$. If we consider the side constraints separately from the general inequality constraints, the Lagrangian $S(\mathbf{A}, \boldsymbol{\mu})$ of Problem (5.20) is defined as

$$S(\mathbf{A}, \boldsymbol{\mu}) = \sum_{i=1}^{m} A_i L_i + \sum_{j=1}^{n^{\mathrm{I}}} \mu_j \widetilde{H}_j(\mathbf{A}) \tag{5.21}$$

The derivative S_i of S with respect to A_i is given as

$$S_i = L_i + \sum_{j=1}^{n^{\mathrm{I}}} \mu_j \frac{\partial \widetilde{H}_j}{\partial A_i}, \quad (i = 1, \ldots, m) \tag{5.22}$$

where

$$\frac{\partial \widetilde{H}_j}{\partial A_i} = \sum_{k=1}^{n} \frac{\partial H_j}{\partial U_k} \frac{\partial U_k}{\partial A_i} + \frac{\partial H_j}{\partial A_i} \tag{5.23}$$

The KKT conditions are written as

$$\begin{cases} S_i = 0 & \text{for} \quad A_i^{\mathrm{L}} < A_i < A_i^{\mathrm{U}}, \\ S_i \geq 0 & \text{for} \quad A_i = A_i^{\mathrm{L}}, \\ S_i \leq 0 & \text{for} \quad A_i = A_i^{\mathrm{U}} \end{cases} \quad (i = 1, \ldots, m) \tag{5.24}$$

with inequality conditions and complementarity conditions

$$\mu_j \geq 0, \quad H_j \leq 0, \quad \mu_j H_j = 0, \quad (j = 1, \ldots, n^{\mathrm{I}}) \tag{5.25}$$

If $\mathbf{A}$ is locally optimal, there exist $\mathbf{U}$ and $\boldsymbol{\mu}$ that satisfy the KKT conditions (5.24) and (5.25).

Let $\mathcal{J}^{\mathrm{E}}$ denote the set of indices of member groups satisfying $A_i^{\mathrm{L}} < A_i < A_i^{\mathrm{U}}$. Eq. (5.24) for $i \in \mathcal{J}^{\mathrm{E}}$ can be written as

$$L_i + \sum_{j=1}^{n^{\mathrm{I}}} \mu_j \frac{\partial \widetilde{H}_j}{\partial A_i} = 0, \quad (i \in \mathcal{J}^{\mathrm{E}}) \tag{5.26}$$

The number of groups in $\mathcal{J}^{\mathrm{E}}$ is denoted by n^{E}; i.e., we have n^{E} equations (5.26). Let $\mathcal{C}^{\mathrm{E}}$ and s^{E} denote the set of indices of independent active constraints in (5.20b) and the number of constraints in $\mathcal{C}^{\mathrm{E}}$, respectively; i.e., we have s^{E} unknown nonnegative Lagrange multipliers μ_j, and the remaining multipliers are 0. If the jth and kth constraints are active and dependent, i.e.,

$$\frac{\partial \widetilde{H}_k}{\partial A_i} = c \frac{\partial \widetilde{H}_j}{\partial A_i}, \quad (i \in \mathcal{J}^{\mathrm{E}}) \tag{5.27}$$

with c being a constant, the kth constraint is eliminated and $(1 + c)\mu_j$ is conceived as an independent multiplier in (5.26).

Note that the Lagrange multipliers are available from the output of the NLP algorithms, such as sequential quadratic programming. However, even if μ_j is not available from the optimization results, μ_j can be computed from (5.26) after obtaining $\mathbf{A}$ and $\mathbf{U}$ of the optimal solution and carrying out design sensitivity analysis to compute $\partial \widetilde{H}_j / \partial A_i$, because the existence of $\mu_j \geq 0$ ($j \in \mathcal{C}^{\mathrm{E}}$) is assured from the local optimality of the solution. In the following, we consider only the nondegenerate case in which μ_j is uniquely determined and the following conditions are satisfied:

$$S_i > 0 \quad \text{for} \quad A_i = A_i^{\mathrm{L}} \tag{5.28a}$$

$$S_i < 0 \quad \text{for} \quad A_i = A_i^{\mathrm{U}} \tag{5.28b}$$

$$\mu_j > 0 \quad \text{for} \quad H_j = 0 \tag{5.28c}$$

5.2.3 Local search of approximate optimal solutions

5.2.3.1 Approximate condition for a sandwich section

Consider the frame consisting of members with sandwich sections or solid sections with constant heights; i.e., the second moment of inertia and section modulus are proportional to the cross-sectional area. Let $\mathbf{A}^{\mathrm{E}}$ denote the vector consisting of the cross-sectional areas of the members in groups in $\mathcal{J}^{\mathrm{E}}$, and denote by $\Delta \mathbf{A}^{\mathrm{E}}$ the increment of $\mathbf{A}^{\mathrm{E}}$. If there is a direction $\Delta \mathbf{A}^{\mathrm{E}}$ that does not have any effect on $\mathbf{U}$, then H_j, including displacement constraints and stress constraints, do not explicitly depend on $\mathbf{A}^{\mathrm{E}}$, and the optimal solution is locally nonunique. Since $\mathbf{P}$ is independent of $\mathbf{A}^{\mathrm{E}}$, the incremental form of

(5.18) for variation $\Delta \mathbf{A}^{\mathrm{E}}$ of $\mathbf{A}^{\mathrm{E}}$ and the corresponding variation $\Delta \mathbf{U}$ of $\mathbf{U}$ is written as

$$\mathbf{K}(\mathbf{A}^{\mathrm{E}})\Delta \mathbf{U} + \mathbf{K}^{\mathrm{A}}(\mathbf{U})\Delta \mathbf{A}^{\mathrm{E}} = \mathbf{0} \tag{5.29}$$

where the (i, j)-component $K_{ij}^{\mathrm{A}}(\mathbf{U})$ of the $n \times n^{\mathrm{E}}$ matrix $\mathbf{K}^{\mathrm{A}}(\mathbf{U}, \mathbf{A}^{\mathrm{E}})$, which is called the *stiffness matrix with respect to cross-sectional areas*, is given using (5.19) as

$$K_{ij}^{\mathrm{A}}(\mathbf{U}, \mathbf{A}^{\mathrm{E}}) = \frac{\partial F_i(\mathbf{U}, \mathbf{A}^{\mathrm{E}})}{\partial A_j^{\mathrm{E}}} \tag{5.30}$$

The nonuniqueness of the optimal solution can be detected, as follows, using (5.29) on the basis of the rank of $\mathbf{K}^{\mathrm{A}}$.

Since I_i^* is proportional to A_i^* for each member with a sandwich section, $\mathbf{K}^{\mathrm{A}}$ does not explicitly depend on $\mathbf{A}^{\mathrm{E}}$. Let an n-vector $\mathbf{f}_i$ denote the nodal load vector representing the pair of loads equivalent to the nodal forces of member i and applied to the nodes connected to member i. Then the jth column of $\mathbf{K}^{\mathrm{A}}$ is equal to the sum of $(1/A_i^*)\mathbf{f}_i$ of the members in the jth group. Let ω_i ($\omega_1 \geq \omega_2 \geq \cdots$) denote the ith singular value of $\mathbf{K}^{\mathrm{A}}$, and define the *diagonal* rectangular matrix $\boldsymbol{\Omega}$ so that its (i, i)-component is equal to ω_i and the remaining components are 0. Then singular value decomposition (SVD) of $\mathbf{K}^{\mathrm{A}}$ leads to

$$\mathbf{K}^{\mathrm{A}} = \mathbf{Q}\boldsymbol{\Omega}\mathbf{R}^{\top} \tag{5.31}$$

where the ith columns of the $n \times n$ matrix $\mathbf{Q}$ and the $n^{\mathrm{E}} \times n^{\mathrm{E}}$ matrix $\mathbf{R}$ are the left and right singular vectors $\mathbf{Q}_i$ and $\mathbf{R}_i$, respectively, corresponding to ω_i, and the following relation is satisfied (see Appendix A.1.3 for details of SVD):

$$\mathbf{K}^{\mathrm{A}}\mathbf{R}_i = \omega_i \mathbf{Q}_i, \quad (i = 1, \ldots, n^{\mathrm{E}}) \tag{5.32}$$

It is assumed here that n^{E} is less than n, which is usually satisfied by a rigidly jointed regular frame even if the design variables are not linked and the cross-sectional areas of all the members are considered to be independent design variables. If $\mathbf{K}^{\mathrm{A}}$ is full column rank, i.e., rank $\mathbf{K}^{\mathrm{A}} = n^{\mathrm{E}}$, it can be observed from (5.29) that there is no solution $\Delta \mathbf{A}^{\mathrm{E}}$ ($\neq \mathbf{0}$) satisfying $\Delta \mathbf{U} = \mathbf{0}$ except for the case where there exists a member that is not deformed; however, such member has a lower-bound cross-sectional area and is not included in the set $\mathcal{J}^{\mathrm{E}}$ at the optimal solution. For a nonoptimal and nonregular general frame, the rank of $\mathbf{K}^{\mathrm{A}}$ is generally equal to n^{E}, because variation of $\mathbf{A}^{\mathrm{E}}$ in any pattern will lead to variation of $\mathbf{U}$. However, it often happens for a regular frame that rank $\mathbf{K}^{\mathrm{A}}$ is less than n^{E}.

Suppose rank $\mathbf{K}^{\mathrm{A}} = n^{\mathrm{E}} - 1$; i.e., $\omega_{n^{\mathrm{E}}} = 0$. In this case, (5.32) leads to

$$\mathbf{K}^{\mathrm{A}}\mathbf{R}_{n^{\mathrm{E}}} = \mathbf{0} \tag{5.33}$$

Therefore, $\Delta \mathbf{U} = \mathbf{0}$ is satisfied from (5.29) for $\Delta \mathbf{A}^{\mathrm{E}} = \mathbf{R}_{n^{\mathrm{E}}}$; hence, $\mathbf{U}$ is constant if $\mathbf{A}^{\mathrm{E}}$ is modified in the direction of $\mathbf{R}_{n^{\mathrm{E}}}$. The constraints are satisfied by the modified design, and there exist $\mu_j > 0$ ($j \in \mathcal{J}$) satisfying the

KKT conditions, if H_j does not contain explicit term with respect to $\mathbf{A}^{\mathrm{E}}$ and the sensitivity coefficients of $\widetilde{H}_j$ are continuous with respect to $\mathbf{A}^{\mathrm{E}}$. These conditions are rigorously satisfied for the stress constraints of a frame with a sandwich section. Therefore, in this case, the optimal solution is nonunique with first-order approximation and $\Delta \mathbf{A}^{\mathrm{E}}$ satisfies the condition

$$\sum_{i=1}^{n^{\mathrm{E}}} \Delta A_i^{\mathrm{E}} L_i^{\mathrm{E}} = 0 \qquad (5.34)$$

where L_i^{E} is the sum of the length of members in the ith group in $\mathcal{J}^{\mathrm{E}}$.

This way, optimal solutions can be locally searched using SVD of $\mathbf{K}^{\mathrm{A}}$ for a frame consisting of members with sandwich sections. Even if $\omega_{n^{\mathrm{E}}}$ in (5.32) is not equal to 0 and has a very small positive value compared to the maximum singular value ω_1, approximate optimal solutions can be found by searching $\mathbf{A}^{\mathrm{E}}$ in the direction of $\mathbf{R}_{n^{\mathrm{E}}}$. For optimal frames with a general cross-section under constraints that explicitly depend on the cross-sectional areas, the independence of $\mathbf{U}$ on $\mathbf{A}^{\mathrm{E}}$ may be regarded as a first-order approximate condition for local nonuniqueness.

5.2.3.2   Approximate condition for a general cross-section

More rigorous conditions for nonuniqueness of optimal frames with general cross-sections can be derived by relaxing the requirement of constant displacements. Suppose the optimal solution is fully stressed; i.e., $s^{\mathrm{E}} \geq n^{\mathrm{E}}$. Let $\mathbf{G}$ denote the $n^{\mathrm{I}} \times m$ sensitivity matrix for which the (j, i)-component G_{ji} is equal to the sensitivity coefficient of the jth constraint function with respect to the ith design variable:

$$G_{ij} = \frac{\partial \widetilde{H}_j}{\partial A_i} \qquad (5.35)$$

Suppose the constraints are numbered such that the first s^{E} constraints are active; i.e., the jth component of the vector $\mathbf{H}^{\mathrm{E}}$ of active constraints is defined as $H_j^{\mathrm{E}} = H_j$ $(i = 1, \ldots, s^{\mathrm{E}})$. The reduced $s^{\mathrm{E}} \times n^{\mathrm{E}}$ sensitivity matrix corresponding to the independent set of active constraints and the variables in the set $\mathcal{J}^{\mathrm{E}}$ only is denoted by $\mathbf{G}^{\mathrm{E}}$. Then, the first-order approximation $\Delta \mathbf{H}^{\mathrm{E}}$ of the incremental vector of active constraints $\mathbf{H}^{\mathrm{E}}$ is given as

$$\Delta \mathbf{H}^{\mathrm{E}} = \mathbf{G}^{\mathrm{E}} \Delta \mathbf{A}^{\mathrm{E}} \qquad (5.36)$$

Therefore, the search direction of the approximate optimal solutions satisfying the constraints and optimality conditions (5.26) can be found by the SVD of the reduced sensitivity matrix $\mathbf{G}^{\mathrm{E}}$. Note that the solution is locally nonunique if the singular value $\omega_{n^{\mathrm{E}}}$ is 0. In this case, H_j^{E} are insensitive to the variation of $\mathbf{A}^{\mathrm{E}}$ in the direction of the n^{E}th right singular vector of $\mathbf{G}^{\mathrm{E}}$. The approximate Lagrange multipliers μ_j^{E} corresponding to H_j^{E} $(i = 1, \ldots, s^{\mathrm{E}})$ for the solution after increment can be found from (5.26) assuming the continuity of the optimal solution and its sensitivity coefficients. This way, the SVD of sensitivity matrix can be effectively used for local search of the optimal solution of a frame with a general cross-section.

5.2.3.3   General conditions for nonuniqueness

General conditions for local nonuniqueness without restriction on the cross-sectional types and the numbers of s^{E} and n^{E} can be obtained by differentiating all the state equations and KKT conditions in a similar manner as the parametric programming approach in Appendix A.5. Let t denote an auxiliary parameter that defines the problem without any physical meaning. The (i, j)-components of $\mathbf{K}$ and $\mathbf{K}^{\mathrm{A}}$ are denoted by K_{ij} and K_{ij}^{A}, respectively. The derivatives of μ_j^{E}, A_i^{E}, and $\mathbf{U}$ of the optimal solution with respect to the parameter t are found from the following equations, which are derived by differentiating (5.18), (5.20), and (5.26):

$$
\sum_{j=1}^{n} K_{ij} \frac{\mathrm{d}U_j}{\mathrm{d}t} + \sum_{j=1}^{n^{\mathrm{E}}} K_{ij}^{\mathrm{A}} \frac{\mathrm{d}A_j^{\mathrm{E}}}{\mathrm{d}t} = \frac{\mathrm{d}P_i}{\mathrm{d}t}, \quad (i = 1, \ldots, n) \tag{5.37a}
$$

$$
\sum_{i=1}^{n} \frac{\partial H_j}{\partial U_i} \frac{\mathrm{d}U_i}{\mathrm{d}t} + \sum_{k=1}^{n^{\mathrm{E}}} \frac{\partial H_j}{\partial A_k^{\mathrm{E}}} \frac{\mathrm{d}A_k^{\mathrm{E}}}{\mathrm{d}t} = 0, \quad (j = 1, \ldots, s^{\mathrm{E}}) \tag{5.37b}
$$

$$
\sum_{j=1}^{s^{\mathrm{E}}} \frac{\partial \widetilde{H}_j}{\partial A_i^{\mathrm{E}}} \frac{\mathrm{d}\mu_j^{\mathrm{E}}}{\mathrm{d}t} + \sum_{j=1}^{s^{\mathrm{E}}} \sum_{k=1}^{n^{\mathrm{E}}} \mu_j^{\mathrm{E}} \frac{\partial^2 \widetilde{H}_j}{\partial A_i^{\mathrm{E}} \partial A_k^{\mathrm{E}}} \frac{\mathrm{d}A_k^{\mathrm{E}}}{\mathrm{d}t} = 0, \quad (i = 1, \ldots, n^{\mathrm{E}}) \tag{5.37c}
$$

where the sensitivity coefficients of $\widetilde{H}_j^{\mathrm{E}}$ are derived using the relation (5.23).

Let $\mathbf{X}$ denote the vector consisting of the unknown variables $\mathbf{U}$, A_i^{E}, and μ_j^{E}. Then, from (5.37), the derivative of $\mathbf{X}$ with respect to t is obtained by solving the linear equations in the following form:

$$
\mathbf{Z} \frac{\mathrm{d}\mathbf{X}}{\mathrm{d}t} = \mathbf{z} \tag{5.38}
$$

where $\mathbf{Z}$ and $\mathbf{z}$ are the constant matrix and vector, respectively. If $\mathbf{Z}$ is singular, then there exists a vector $\mathrm{d}\mathbf{X}/\mathrm{d}t$ corresponding to $\mathbf{z} = \mathbf{0}$. Therefore, the direction $\mathrm{d}A_i^{\mathrm{E}}/\mathrm{d}t$ of the nonunique solutions can be found from the singular vector corresponding to the zero singular value of $\mathbf{Z}$. Note that similar conditions can be obtained from the second-order optimality conditions (Simões 1989).

The condition above is more rigorous than the previous approximate conditions for nonuniqueness, because it utilizes the exact differentiation of the governing equations and KKT conditions. However, in this approach, the Hessian of H_j^{E} is needed, which is very difficult to obtain in the usual process of optimization.

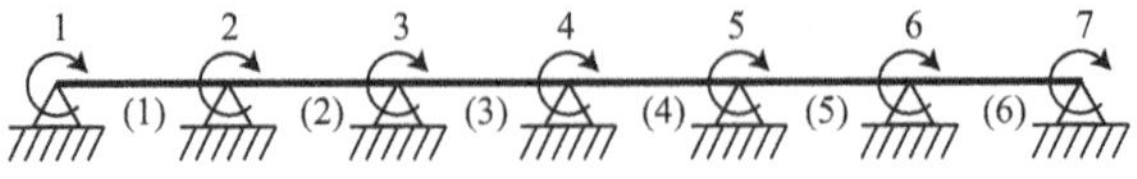

FIGURE 5.6: A six-span continuous beam.

5.2.4   Global search of approximate optimal solutions

Because Problem (5.20) is a nonconvex problem, there may exist many local optimal solutions that have slightly larger objective values than that of the global optimal solution. In view of practical application, the most preferred solution should be chosen from a set of approximate solutions in view of other performance measures, including constructability and stiffness, against other sets of design loads.

Let $\tilde{V}$ denote the optimal objective value of Problem (5.20), and assign the following requirement as an approximate optimal solution:

$$V(\mathbf{A}) \le \tilde{V} + \Delta V \tag{5.39}$$

where ΔV is assumed to be sufficiently small. Then the approximate solutions are successively found by solving the optimization problems. Let $\widehat{\mathbf{A}}^{(k)}$ denote the kth approximate optimal solution that has already been found. The distance $D^{(k)}$ between $\mathbf{A}$ and $\widehat{\mathbf{A}}^{(k)}$ is defined by the Euclidean norm as

$$D^{(k)} = \sqrt{\sum_{i=1}^{m}(A_i - \widehat{A}_i^{(k)})^2} \tag{5.40}$$

An auxiliary variable τ is introduced, and the following optimization problem is solved for maximizing the minimum distance from existing optimal solutions under constraints on the responses and the total structural volume:

Maximize τ

$$\text{subject to } \tau \le \sqrt{\sum_{i=1}^{m}(A_i - \widehat{A}_i^{(k)})^2}, \quad (k = 1, \ldots, q) \tag{5.41a}$$

$$V(\mathbf{A}) \le \tilde{V} + \Delta V \tag{5.41b}$$

$$H_j(\mathbf{U}(\mathbf{A}), \mathbf{A}) \le 0, \quad (j = 1, \ldots, n^{\mathrm{I}}) \tag{5.41c}$$

$$A_i^{\mathrm{L}} \le A_i \le A_i^{\mathrm{U}}, \quad (i = 1, \ldots, m) \tag{5.41d}$$

where q is the number of global and approximate solutions that have already been found. This way, approximate optimal solutions with various distributions of cross-sectional areas can be found consecutively by solving Problem (5.41).

Example 5.2
Nonuniqueness of the optimal solution is first investigated for a continuous beam consisting of m members (elements) and $m + 1$ nodes subjected to concentrated moment M_0 at each node. An example for $m = 6$ is shown in Fig. 5.6, where the numbers with and without parentheses are member numbers and node numbers, respectively. In the following examples, the units of force and length are kN and mm, respectively, which are omitted for brevity.

The length L of each element is 2000, the elastic modulus E is 200, and the upper bound $\bar{\sigma}$ for the absolute value of stress is 0.07. The lower bound A_i^L for the cross-sectional area is 100, whereas the upper bound is not given. The concentrated nodal moment is 10,000. Design variable linking is not used; i.e., the number of variables is m.

The beam has a sandwich section satisfying

$$I_i = h^2 A_i \tag{5.42}$$

where $h = 50$ is the distance between each flange and the center axis, which is the same for all members. Constraints are given for the stresses of the flanges at the two ends of each member.

Let θ_i denote the rotation of the ith node. To investigate the special case of nonunique optimal solution, we assign a periodic boundary condition such that $\theta_1 = \theta_{m+1}$. As a small example, consider the case $m = 2$. Using the periodic boundary condition, and assuming $A_i > A_i^L$ $(i = 1, 2)$, i.e., $\mathcal{J}^E = \{1, 2\}$, the matrices $\mathbf{K}$ and $\mathbf{K}^A$ are written as

$$\mathbf{K} = c(A_1 + A_2)\begin{pmatrix} 4 & 2 \\ 2 & 4 \end{pmatrix}, \quad \mathbf{K}^A = c\begin{pmatrix} 4\theta_1 + 2\theta_2 & 4\theta_1 + 2\theta_2 \\ 2\theta_1 + 4\theta_2 & 2\theta_1 + 4\theta_2 \end{pmatrix} \tag{5.43}$$

where $c_i = Eh/L$. Therefore, it is seen from the SVD of $\mathbf{K}^A$ that $\mathbf{U} = (\theta_1, \theta_2)^\top$ is insensitive to the design modification $\Delta \mathbf{A} = (1, -1)^\top$ in the direction of the right singular vector corresponding to zero singular value, and the absolute values of the stresses of the flanges of members 1 and 2 are equal to $\bar{\sigma}$, which satisfies

$$\bar{\sigma} = \frac{M_0}{h(A_1 + A_2)} \tag{5.44}$$

Therefore, the sensitivity matrix is obtained as

$$\mathbf{G} = -\frac{M_0}{h(A_1 + A_2)^2}\begin{pmatrix} 1 & 1 \\ 1 & 1 \end{pmatrix} \tag{5.45}$$

Hence, the singular vector corresponding to the zero singular value is obtained as $\Delta \mathbf{A} = (1, -1)^\top$ after normalization so that the maximum absolute value is 1, which is the same as the vector obtained by the SVD of $\mathbf{K}^A$.

Consider next the case $m = 6$ to investigate nonuniqueness through computational optimization. Optimization is carried out by IDESIGN Ver. 3.5 (Arora and Tseng 1987), where sequential quadratic programming is used. A

TABLE 5.4:   Optimization results of the continuous beam with $m = 6$.

	$\xi = 0$	$\xi = 0.5$	$\xi = 1.0$	$\xi = 2.0$
A_1	714.286	717.170	719.964	724.229
A_2	714.286	711.401	708.607	704.343
$V\ (\times 10^6)$	8.57143	8.57143	8.57143	8.57143

uniform random number $0 \le R_i < 1$ is generated to define the initial solution as

$$A_i = A_0(1 + \xi R_i), \quad (i = 1, \ldots, 6) \tag{5.46}$$

where ξ is a parameter, and $A_0 = 100$.

Table 5.4 shows the optimization results from various initial solutions generated by different values of ξ, where the solutions are periodic such that $A_1 = A_3 = A_5$ and $A_2 = A_4 = A_6$. Note that each optimal solution has been found with eight iterations of sequential quadratic programming and satisfies the stress constraint with equality at one of the flanges at an end of each member; i.e., the optimal solutions are fully stressed. Therefore, we can assume that the global optimal solution has been successfully found for all cases, although the optimization problem is nonconvex. The objective values are the same for all cases, although the cross-sectional areas are different. Hence, the optimal solution is nonunique. Note that the nodal rotations have the same value 4.6667×10^{-3} at all nodes of all solutions; i.e., the displacement vector is independent of the cross-sectional areas of the optimal solutions, and the optimal solutions can be locally searched using the SVD of $\mathbf{K}^A$.

The rank of $\mathbf{K}^A$ is 5, and the singular values are 14.000, 12.124, 12.124, 7.0000, 7.0000, 0.0. The singular vector $\mathbf{R}_6$ corresponding to the vanishing singular value is $(-1, 1, -1, 1, -1, 1)^\top$ after normalization. Therefore, the optimal solutions can be written with a parameter α as

$$\begin{aligned} A_1 = A_3 = A_5 = 714.286 + \alpha, \\ A_2 = A_4 = A_6 = 714.286 - \alpha \end{aligned} \tag{5.47}$$

and have the same objective value, which agrees with the result in Table 5.4.

If the requirement $\theta_1 = \theta_7$ of the periodic boundary condition is not given, the rank of $\mathbf{K}^A$ of the optimal solution is 6, and the singular values are 12.064, 11.151, 9.1058, 8.3791, 5.3656, 2.7534. Therefore, the rank deficiency of $\mathbf{K}^A$ is strongly related to the regularity of the frame.

5.2.5   Numerical example of a regular plane frame

5.2.5.1   Description of the frame model

Optimal solutions are locally and globally searched for a six-story six-span frame, as shown in Fig. 5.1, where $H = W = 4000$, $m^* = 84$, and $n = 97$. The

TABLE 5.5: Optimization results of the six-story six-span frame.

	$\xi = 0$	$\xi = 0.5$	$\xi = 1.0$	$\xi = 2.0$
A_1	22,047.3	22,041.2	22,040.5	22,036.4
A_2	14,026.0	13,992.8	14,008.7	13,983.8
A_3	8266.1	8265.3	8274.1	8271.7
A_4	4474.1	4486.2	4473.1	4484.5
$V~(\times 10^9)$	2.4824	2.4824	2.4824	2.4824

members are classified into 45 groups, i.e., $m = 45$, considering the symmetry condition. Note that the units of force and length are kN and mm also in this section. Only horizontal loads are applied, and $(P_1, P_2, P_3, P_4, P_5, P_6) = (50, 100, 150, 200, 250, 300)$ in Fig. 5.1. The elastic modulus E is 200 and the upper-bound stress $\overline{\sigma}$ is 0.07.

For the example of a sandwich section, $h = 250$ in (5.42) and $A_i^{\mathrm{L}} = 3000$ for all groups. For the frame with wide-flange beams and box columns, $A_i^{\mathrm{U}} = 5000$ for all groups, and the following relations are assumed for the second moment of inertia I_i and section modulus Z_i of members in the ith group so that only the cross-sectional areas are the design variables (Sawada and Matsuo 2003):

$$\begin{cases} \text{Columns}: & I_i = 1.076(A_i)^2, \quad Z_i = 0.804(A_i)^{1.5} \\ \text{Beams}: & I_i = 3.648(A_i)^2, \quad Z_i = 1.580(A_i)^{1.5} \end{cases} \tag{5.48}$$

The initial solutions are randomly generated in the same manner as (5.46) with $A_0 = A_i^{\mathrm{L}}$.

5.2.5.2 Local search

First we optimize the frame with a sandwich section under stress constraints only. The optimization results of Problem (5.20) are shown in Table 5.5, where the optimal objective value and the cross-sectional areas of the external columns of stories 1–4 denoted by $A_1, \ldots, A_4$ are listed. As is seen, the optimal objective values are 2.4824×10^9 for all cases, although the cross-sectional areas are different. Because the optimization from four different initial solutions converged to the same objective value, we can assume that the global optimal solutions have been successfully obtained, although uniqueness of the solution is not satisfied. The optimal solution for $\xi = 0$, denoted by Solution 1, is shown in Fig. 5.7, where the width of each member is proportional to its cross-sectional area. The number n^{E} of groups satisfying $A_i > A_i^{\mathrm{L}}$ is 36. The optimal solution is fully stressed even for the group with $A_i = A_i^{\mathrm{L}}$; i.e., the maximum absolute values of the stresses are equal to the upper bound $\overline{\sigma}$ in all groups. The number of active constraints at the optimal solution is 126, which is larger than n^{E} for all cases.

First, we carry out SVD for $\mathbf{K}^{\mathrm{A}}$ of Solution 1 in Fig. 5.7. The 20 lowest and the maximum singular values are listed in the second column of Table 5.6.

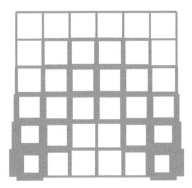

FIGURE 5.7:   Solution 1 of the six-story six-span frame.

Since there is a jump between ω_{25} and ω_{24}, the number of nearly zero singular values is 12.

Let σ_{ij} denote the stress at the jth point $(j = 1, \ldots, 4)$ in the ith member where the stress is constrained as shown in Fig. 5.2. The accuracy of an approximate solution is confirmed by the maximum stress ratio β defined by

$$\beta = \max_{i,j} \left\{ \frac{|\sigma_{ij}|}{\overline{\sigma}} \right\} \tag{5.49}$$

where $\beta = 1$ corresponds to a fully stressed design. The cross-sectional areas are parametrically varied in the direction of the right singular vector $\mathbf{R}_{36}$ corresponding to the smallest singular value. The unit increment $\Delta \mathbf{A}^{\mathrm{E}}$ is defined so that it is proportional to $\mathbf{R}_{36}$ and the maximum variation of cross-sectional area is equal to 1. Then the cross-sectional areas are varied from the optimal solution $\widehat{\mathbf{A}}^{\mathrm{E}}$ as $\widehat{\mathbf{A}}^{\mathrm{E}} + \eta \Delta \mathbf{A}^{\mathrm{E}}$ with cross-sectional parameter η. The ratio of V to $\widehat{V}$ of Solution 1 is denoted by γ. Variations of β and γ with respect to the cross-sectional parameter η are plotted in Fig. 5.8.

For example, if $\eta = 200$, i.e., if we allow maximum variation 200 of A_i, then $\beta = 1.0069$ and $\gamma = 1.0006$; hence, the ratios of increase of the objective function and the maximum stress ratio are sufficiently small compared with the variation of the cross-sectional areas. Note that the stress constraints are violated in some members, because the nodal displacements are not strictly constant for design variation in the direction of $\mathbf{R}_{36}$.

Optimal solutions are next found for the frame with wide-flange beams and box columns. In addition to the stress constraints, the upper bound 120 is given for the horizontal displacement at the roof level, which corresponds to an average drift angle of 1/120. The number n^{E} of groups satisfying $A_i > A_i^{\mathrm{L}}$ at the optimal solution is 36. The optimal solutions are fully stressed, and the displacement constraint is satisfied with equality; hence, the number of active constraints at the optimal solution is 127, which is larger than n^{E}. The

TABLE 5.6: Singular values of $\mathbf{K}^A$ for a sandwich section and $\mathbf{G}^E$ for wide-flange beams and box columns.

	$\mathbf{K}^A$	$\mathbf{G}^E$
ω_{36}	1.0960×10^{-2}	4.1060×10^{-6}
ω_{35}	1.8110×10^{-2}	8.6195×10^{-6}
ω_{34}	1.9496×10^{-2}	1.1614×10^{-5}
ω_{33}	2.4163×10^{-2}	1.3993×10^{-5}
ω_{32}	2.9663×10^{-2}	1.8806×10^{-5}
ω_{31}	3.8929×10^{-2}	2.3470×10^{-5}
ω_{30}	4.3074×10^{-2}	2.5060×10^{-5}
ω_{29}	5.1079×10^{-2}	3.3610×10^{-5}
ω_{28}	6.1228×10^{-2}	3.9091×10^{-5}
ω_{27}	7.8563×10^{-2}	4.1709×10^{-5}
ω_{26}	1.0079×10^{-1}	4.8968×10^{-5}
ω_{25}	7.0597×10^{-1}	5.1727×10^{-5}
ω_{24}	19.538	5.5647×10^{-5}
ω_{23}	27.132	6.2508×10^{-5}
ω_{22}	31.412	7.6444×10^{-5}
ω_{21}	33.566	7.8804×10^{-5}
ω_{20}	37.562	8.8555×10^{-5}
ω_{19}	44.657	9.8923×10^{-5}
ω_{18}	47.874	1.0774×10^{-4}
ω_{17}	54.960	1.1527×10^{-4}
$\cdots$	$\cdots$	$\cdots$
ω_1	124.01	4.8069×10^{-4}

results of the SVD of $\mathbf{G}^A$ are listed in the third column of Table 5.6. As is seen, no clear zero singular value is observed in this case. Variation of β with respect to the cross-sectional parameter η is as shown in Fig. 5.9 for the cross-sectional variation in the direction of the right singular vector $\mathbf{R}_{36}$ corresponding to the smallest singular value ω_{36}. Note that γ is not shown because it is very close to 1, e.g., 1.00008 at $\eta = 200$. Hence, approximate solutions with good accuracy can be obtained utilizing the SVD of $\mathbf{G}^E$ even for frames with general cross-sectional shapes.

5.2.5.3 Global search

Approximate optimal solutions are globally searched under an upper-bound constraint of V that is 2% larger than the optimal objective value $\widehat{V}$ of the global optimal solution in Fig. 5.7; i.e., $\Delta V = 0.02\widehat{V}$ in (5.41b).

Problem (5.41) has been solved to maximize the minimum distance from the existing approximate optimal solutions. The nine solutions indicated by Solutions 2–10, which are found consecutively, are shown in Figs. 5.10(a)–(i),

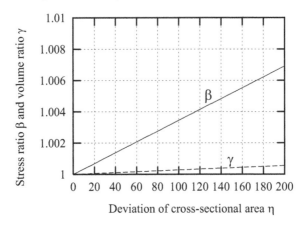

FIGURE 5.8:   Variation of the maximum stress ratio β (solid line) and the volume ratio γ (dashed line) with respect to cross-sectional parameter η using singular vector of $\mathbf{K}^{A}$ for a sandwich section.

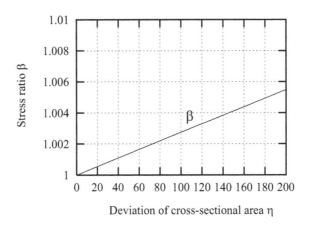

FIGURE 5.9:   Variation of the maximum stress ratio β with respect to cross-sectional parameter η using singular vector of $\mathbf{G}^{E}$ for wide-flange beams and box columns.

respectively. As is seen, approximate solutions with various distributions of cross-sectional areas have been successfully found. Note that the constraints on the total structural volume are satisfied with equality, i.e., $V = 1.02\widehat{V}$ for all solutions. The values of the minimum distance τ for Solutions 2–10 are plotted in Fig. 5.11, which confirms that τ decreases as more solutions are found.

This way, approximate solutions can be globally searched. However, it is possible to incorporate directly the preference of the designer under constraint

(a) Solution 2   (b) Solution 3   (c) Solution 4

(d) Solution 5   (e) Solution 6   (f) Solution 7

(g) Solution 8   (h) Solution 9   (i) Solution 10

FIGURE 5.10: Globally searched approximate optimal solutions.

(5.41b) on the structural volume with $\Delta V = 0.02\widehat{V}$. For example, if the variance of the cross-sectional areas is minimized, the cross-sectional areas shown in Fig. 5.12(a) are obtained. Fig. 5.12(b) shows the solution that minimizes the maximum cross-sectional area, which turns out to be 13878.8, which is about 60% of 22,047.3 of Solution 1.

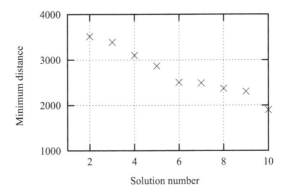

FIGURE 5.11:   Variation of minimum distance from the existing solutions.

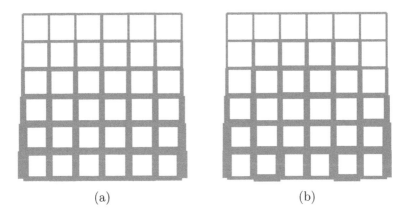

(a)   (b)

FIGURE 5.12:   Approximate optimal solutions obtained by direct consideration of the designer's preference on cross-sectional areas; (a) minimum variance, (b) minimum maximum value.

5.3   Parametric optimization of frames

5.3.1   Introduction

Most of the optimization methods developed for trusses and frames are verified for small-scale problems and may not always be applicable to large-scale problems due to difficulties in computational cost and convergence properties. To resolve these difficulties, some methods, e.g., multilevel decomposition, have been developed (Kirsch 1975; Friedman and Fuchs 1987). Two-level decomposition can be used in many fields of optimization, including resource allocation problems, min-max type multiobjective programming problems, and best approximation problems (Shimizu, Ishizuka, and Bard 1997).

Decomposition of an optimization problem was first studied for a linear programming problem (Dantzig and Wolfe 1960) and was applied to a plastic limit design problem (Woo and Schmit 1981; Kaneko and Ha 1983). Substructure approaches have also been developed (Kirsch 1972; Nguyen 1987; Svensson 1987); however, they utilize the specific properties of the structural models and the optimization problems; i.e., they are not developed in a general form.

A general framework for solving optimization problems of complex systems is called *multidisciplinary optimization* (MDO), which was developed in the 1990s in the fields of mechanical engineering and aeronautical engineering for application to optimization problems considering responses in several different disciplines (Sobieszczanski-Sobieski and Haftka 1996). Note that the concept of MDO is different from that of the conventional decomposition method for large-scale analysis, e.g., the substructuring method and the domain decomposition method (Smith, Bjørstad, and Gropp 1996). In MDO, the design process of a complex structural system is divided into subsystems based on the properties of responses, or *disciplines*, e.g., structural analysis, computational fluid dynamics, and electromagnetic analysis. For example, in automobile design, the simple approach of decomposition is based on *objects* such as powertrain, body, chassis, and electronics, whereas it can be partitioned in view of disciplines, e.g., durability, packaging, dynamics, safety, and noise-vibration-harshness (Kim, Michelena, Papalambos, and Jiang 2003). An interaction between aerodynamic analysis and structural analysis is illustrated in Fig. 5.13. The pressure load as an input to structural analysis is obtained as an output of aerodynamic analysis, whereas the initial condition of the aerodynamic analysis is determined using the deformation obtained by structural analysis.

The interactions among the different subsystems are modeled using parameters that define the design problem of each subsystem. An optimization problem that has parameters in addition to variables is called a *parametric programming* problem (Gal 1979; António 2002), which can be conceived as a general framework of an optimization method such as sequential unconstrained minimization techniques (SUMT) (Fiacco and Cormic 1968) and interior point methods (Kojima, Shindoh, and Hara 1997) (see Appendix A.5 for basic equations of parametric programming).

Parametric programming approaches have been widely applied to structural optimization, where the sensitivity of the optimal design with respect to the problem parameter, called *optimum design sensitivity* (Barthelemy and Sobieszczanski-Sobieski 1983; Schmit and Chang 1984; Sobieszczanski-Sobieski, James, and Dovi 1985; Vanderplaats and Yoshida 1985), is computed to investigate the properties of the optimal solutions and also to compute the sensitivity of the solution of the lower-level problem of a multilevel optimization problem with respect to the solution of the upper-level problem that is conceived as a parameter for the lower-level problem (Vanderplaats 1988;

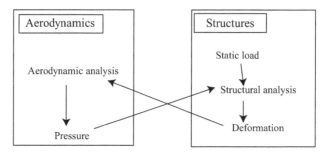

FIGURE 5.13:   Illustration of a multidisciplinary system with aerodynamics and structures.

Bloebaum, Hajela, and Sobieszczanski-Sobieski 1992). Two-level approaches have been extensively studied in the field of hybrid approaches to optimization with *anti-optimization* (Elishakoff, Haftka, and Fang 1994; Elishakoff and Ohsaki 2010), where optimal solutions are found considering the worst-case scenario of the uncertain parameters. A simple *cycle-based approach* can be used for this purpose (Gurav, Langhaar, Goosen, and van Keulen 2005).

There have been many papers on multilevel optimization of frames (Ding and Esping 1991; Friedman and Fuchs 1987), where the system variables, e.g., the nodal locations, are optimized in the upper level, and the component variables, e.g., cross-sectional properties, are optimized in the lower level. Salajegheh (1996) used a dual method with approximation for two-level optimization of frames. Nakamura and Ohsaki (1988) presented a parametric programming approach for generating a set of optimal trusses under multiple frequency constraints. In this respect, the parametric programming approach can be generalized in the framework of the continuation method (Mittelmann and Roose 1990) and the homotopy method (Shin, Haftka, Watson, and Plaut 1988; Watson and Haftka 1989). Ohsaki and Arora (1993) presented a general parametric programming approach to structural optimization.

In this section, we first describe the simple two-level decomposition approach to the design of frames and then present a general framework of non-hierarchical decomposition, which is applied to multilevel optimization of building frames (Ohsaki 1997a; Ohsaki, Nagano, and Wakamatsu 2000).

5.3.2   Two-level decomposition of frames

The simplest MDO approach is the two-level decomposition of the problem. Let $\mathbf{x}^i = (x_1^i, \ldots, x_{m^\times}^i)^\top$ denote the vector of global or system-level design variables of the ith member, representing, e.g., the cross-sectional area and second moment of inertia, where $m^\times$ is the number of global variables related to each member that is assumed to be the same for all members. The vector of all the global variables consisting of $\mathbf{x}^i$ ($i = 1, \ldots, m$) is denoted by $\mathbf{X} =$

$(X_1, \ldots, X_{m^g})^\top$, where m is the number of members, and $m^g = m \times m^x$ is the total number of global variables.

The vector of local or member-level design variables of the ith member is denoted by $\mathbf{y}^i = (y_1^i, \ldots, y_{m^y}^i)^\top$, which consists of the height, width, etc., of the cross-section, where m^y is the number of local variables for each member that is also assumed to be the same for all members. Suppose the global variables are defined by the local variables as

$$\mathbf{x}^i = \mathbf{f}(\mathbf{y}^i), \quad (i = 1, \ldots, m) \tag{5.50}$$

Note that the same vector $\mathbf{f} = (f_1, \ldots, f_{m^x})^\top$ is used for all members, for simplicity. If $n^x < n^y$, then the optimal value of $\mathbf{y}^i$ may be found through member-level optimization for a specified value of $\mathbf{x}^i$ as a result of global optimization. If $n^x = n^y$, then $\mathbf{y}^i$ is computed directly from $\mathbf{x}^i = \mathbf{f}(\mathbf{y}^i)$ without carrying out optimization. If $n^x > n^y$, then there might exist no feasible value of $\mathbf{y}^i$ satisfying $\mathbf{x}^i = \mathbf{f}(\mathbf{y}^i)$.

Let $\mathbf{U}$ denote the vector of nodal displacements that are conceived as the state variables. The vector of nodal displacements of the ith member is denoted by $\mathbf{u}^i$. The constraints on state variables, e.g., on nodal displacements in the global optimization problem, are generally given with inequalities $H_j(\mathbf{X}) \leq 0$ $(j = 1, \ldots, n^H)$, where n^H is the number of global constraints. Note that the state variables are implicitly included as a function of $\mathbf{X}$ in the definition of $H_j(\mathbf{X})$, and the derivatives of $H_j(\mathbf{X})$ are computed using the design sensitivity analysis presented in Chap. 2. The local constraints, e.g., the stress constraints, are given as $h_j^i(\mathbf{x}^i, \mathbf{u}^i(\mathbf{X}), \mathbf{y}^i) \leq 0$ $(j = 1, \ldots, n^h)$, where n^h is the number of local constraints that is the same for all members, and $\mathbf{u}^i$ is a function of $\mathbf{X}$. Only inequality constraints are considered, for simplicity.

Consider the following optimization problem for minimizing an objective function $F(\mathbf{X})$:

$$\text{Minimize} \quad F(\mathbf{X}) \tag{5.51a}$$

$$\text{subject to} \quad H_j(\mathbf{X}) \leq 0, \quad (j = 1, \ldots, n^H) \tag{5.51b}$$

$$h_j^i(\mathbf{x}^i, \mathbf{u}^i(\mathbf{X}), \mathbf{y}^i) \leq 0, \quad (j = 1, \ldots, n^h; \ i = 1, \ldots, m) \tag{5.51c}$$

$$X_i^L \leq X_i \leq X_i^U, \quad (i = 1, \ldots, m^g) \tag{5.51d}$$

$$\mathbf{x}^i = \mathbf{f}(\mathbf{y}^i), \quad (i = 1, \ldots, m) \tag{5.51e}$$

$$y_j^{iL} \leq y_j^i \leq y_j^{iU}, \quad (j = 1, \ldots, n^y; \ i = 1, \ldots, m) \tag{5.51f}$$

where the superscripts $(\cdot)^U$ and $(\cdot)^L$ denote the upper and lower bounds, respectively.

The optimal solution of Problem (5.51) is to be found by iteratively solving the upper-level (global or system level) and lower-level (local or member level) problems. There are many approaches to the definition of these problems (Sobieszczanski-Sobieski, James, and Dovi 1985). The most serious difficulty for solving the lower-level problem is that it often happens that the problem is infeasible due to unrealistic value of the parameters defined by the variables in the upper-level problem. One possible approach to avoid infeasibility is to maximize margin, or minimize violation, of the constraints instead of satisfying all the constraints. Another approach is to define the objective function $C^i(\mathbf{h}^i)$ of the lower-level problem of the ith member as the penalty for violating the local constraints; i.e., the objective function of the lower-level problem is not related to the objective function $F(\mathbf{X})$ of the upper-level problem. Hence, *cumulative penalty function* $C^i(\mathbf{h}^i)$, defined as follows, may be minimized in each member:

$$C^i(\mathbf{h}^i) = \sum_{j=1}^{n^{\mathrm{h}}} C_j^i(h_j^i) \tag{5.52}$$

where $C_j^i(h_j^i)$ is defined so that $C_j^i(h_j^i) > 0$ for $h_j^i > 0$, and $C_j^i(h_j^i) = 0$ for $h_j^i \leq 0$ (see Appendix A.2.2.4 for details of penalty function approaches). Note that $C^i(\mathbf{h}^i)$ should be a continuously differentiable and non-decreasing function of $\mathbf{h}^i$. Then the lower-level problem for the ith member is formulated as

$$\text{Minimize} \quad C^i(\mathbf{h}^i) \tag{5.53a}$$

$$\text{subject to} \quad \mathbf{x}^i = \mathbf{f}(\mathbf{y}^i) \tag{5.53b}$$

$$h_j^i(\mathbf{u}^i(\mathbf{X}), \mathbf{y}^i) \leq 0, \quad (j = 1, \ldots, n^{\mathrm{h}}) \tag{5.53c}$$

$$y_j^{i\mathrm{L}} \leq y_j^i \leq y_j^{i\mathrm{U}}, \quad (j = 1, \ldots, n^{\mathrm{y}}) \tag{5.53d}$$

where $\mathbf{y}^i$ is the variable vector, and $\mathbf{x}^i$ and $\mathbf{u}^i$ are conceived as parameter vectors that are specified as the result of upper-level optimization. Therefore, the optimal solution and the corresponding optimal objective value of the lower-level problem are found for each specified value of $\mathbf{x}^i$ and $\mathbf{u}^i$. Hence, the optimal solution and the optimal objective value may be conceived as functions of $\mathbf{x}^i$ and $\mathbf{u}^i$, consequently, as functions of $\mathbf{X}$, which are denoted by $\widetilde{\mathbf{y}}^i(\mathbf{X})$ and $\widetilde{C}^i(\mathbf{X})$, respectively. Then the sensitivity coefficients of $\widetilde{\mathbf{y}}^i(\mathbf{X})$ and $\widetilde{C}^i(\mathbf{X})$ with respect to $\mathbf{X}$, which are called *parametric sensitivity coefficients*, can be found using the techniques of parametric programming or optimum design sensitivity analysis, as described in Appendix A.5. The values of the solution at the kth iteration of the upper-level problem for optimizing $\mathbf{X}$ are denoted by the superscript $(\cdot)^{(k)}$. Then, $\widetilde{C}^i(\mathbf{X})$ is linearly approximated as

$$\widetilde{C}^i(\mathbf{X}) = \widetilde{C}^i(\mathbf{X}^{(k)}) + \left(\frac{\partial \widetilde{C}^i}{\partial \mathbf{X}} \right)^{\mathsf{T}} (\mathbf{X} - \mathbf{X}^{(k)}) \tag{5.54}$$

Similar approximation is given for $\widetilde{\mathbf{y}}^i(\mathbf{X})$. Hence, the upper-level optimization

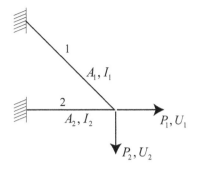

FIGURE 5.14: A two-bar frame.

problem is formulated as

$$\text{Minimize} \quad F(\mathbf{X}) \tag{5.55a}$$

$$\text{subject to} \quad H_j(\mathbf{X}) \le 0, \quad (j = 1, \dots, n^{\mathrm{H}}) \tag{5.55b}$$

$$\widetilde{C}^i(\mathbf{X}) \le 0, \quad (i = 1, \dots, m) \tag{5.55c}$$

$$X_i^{\mathrm{L}} \le X_i \le X_i^{\mathrm{U}}, \quad (i = 1, \dots, m^{\mathrm{g}}) \tag{5.55d}$$

$$y_j^{i\mathrm{L}} \le \widetilde{y}_j^i(\mathbf{X}) \le y_j^{i\mathrm{U}}, \quad (j = 1, \dots, n^{\mathrm{y}}; \ i = 1, \dots, m) \tag{5.55e}$$

In the optimization process, the initial value $\mathbf{X}^{(0)}$ is first given for $\mathbf{X}$, and the state variables $\mathbf{U}$ are computed. Then $\mathbf{x}^i$ and $\mathbf{u}^i$ are transmitted to the lower-level problems as parameters, and the optimal local variables as well as the Lagrange multipliers are found. Then the upper-level problem is solved, where the sensitivities of optimal variables and objective value are effectively incorporated to improve the convergence property.

Example 5.3
An example is given for the definition of variables and constraints for a rigidly jointed frame that consists of two members, as shown in Fig. 5.14. The relation among the variables, parameters, and constraints of the global system (frame) and the subsystems (members) are illustrated in Fig. 5.15. Suppose the two members have rectangular cross-sections, and let a_i and b_i $(i = 1, 2)$ denote the height and width, respectively, of the section of the ith member. The cross-sectional area and second moment of inertia of member i are denoted by A_i and I_i, respectively, which are functions of a_i and b_i.

The global variables are $\mathbf{x}^1 = (A_1, I_1)^{\top}$ and $\mathbf{x}^2 = (A_2, I_2)^{\top}$, which are combined as $\mathbf{X} = (A_1, I_1, A_2, I_2)^{\top}$. The global constraints are given for the nodal displacements U_1 and U_2, which are functions of $\mathbf{X}$; i.e.,

$$\begin{aligned} H_1(\mathbf{X}) &= U_1(\mathbf{X}) - U_1^{\mathrm{U}}, \quad H_2(\mathbf{X}) = U_1^{\mathrm{L}} - U_1(\mathbf{X}), \\ H_3(\mathbf{X}) &= U_2(\mathbf{X}) - U_2^{\mathrm{U}}, \quad H_4(\mathbf{X}) = U_2^{\mathrm{L}} - U_2(\mathbf{X}) \end{aligned} \tag{5.56}$$

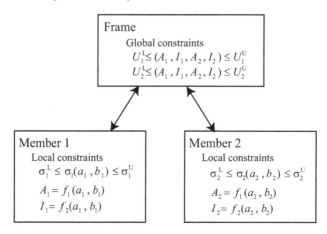

FIGURE 5.15: Illustration of two-level decomposition of the two-bar frame.

where upper and lower bounds are denoted by $(\cdot)^{\mathrm{U}}$ and $(\cdot)^{\mathrm{L}}$, respectively. The global variables are defined with respect to the local variables as

$$
\begin{aligned}
A_1 = f_1(a_1, b_1), \quad I_1 = f_2(a_1, b_1),\\
A_2 = f_1(a_2, b_2), \quad I_2 = f_2(a_2, b_2)
\end{aligned}
\tag{5.57}
$$

Local constraints are given for the maximum absolute value σ_i of the stress $(i = 1, 2)$, which is regarded as a function of a_i and b_i by fixing the displacements $\mathbf{u}^i$ as parameters. Let σ_i^{L} and σ_i^{U} denote the lower and upper bounds for σ_i, respectively. Then the local constraints are formulated as follows using the variable vectors $\mathbf{y}^1 = (a_1, b_1)^{\top}$ and $\mathbf{y}^2 = (a_2, b_2)^{\top}$:

$$
\begin{aligned}
h_1^1(\mathbf{u}^1, \mathbf{y}^1) = \sigma_1(\mathbf{u}^1, \mathbf{y}^1) - \sigma_1^{\mathrm{U}}, \quad h_2^1(\mathbf{u}^1, \mathbf{y}^1) = \sigma_1^{\mathrm{L}} - \sigma_1(\mathbf{u}^1, \mathbf{y}^1),\\
h_1^2(\mathbf{u}^2, \mathbf{y}^2) = \sigma_2(\mathbf{u}^2, \mathbf{y}^2) - \sigma_2^{\mathrm{U}}, \quad h_2^2(\mathbf{u}^2, \mathbf{y}^2) = \sigma_2^{\mathrm{L}} - \sigma_2(\mathbf{u}^2, \mathbf{y}^2)
\end{aligned}
\tag{5.58}
$$

5.3.3   General concept of decomposition to subsystems

The simplest approach to decomposition of large complex systems into subsystems is a *hierarchical decomposition*, as we have seen in the previous section. Kirsch (1975) defined coordinate variables between the first (lower) level and the second (upper) level problems, and demonstrated optimization of cross-sectional areas and nodal coordinates of a small truss, where the nodal coordinates are used as coordination variables. The general concept of coordination of subsystems was presented by Mesarović, Macko, and Takahara (1970).

If the subsystems are optimized individually, it is very difficult to satisfy feasibility in the system-level (global) constraints. Therefore, only the local constraints are to be satisfied when optimizing the subsystems. The feasibility of the local solution with fixed global variables is called *single discipline feasibility*, and a solution satisfying feasibility in every discipline (subsystem) is said to have *individual discipline feasibility* (Cramer, Dennis, Frank, Lewis, and Shubin 1994).

Kim, Michelena, Papalambos, and Jiang (2003) proposed *target cascading* for decomposing the optimization problem of a complex system. Consider a problem of minimizing the deviation of the responses from their target values. Let $\mathbf{t}_{ij}$ denote the target parameter vector transmitted from the jth subsystem in the ith level, while its response is denoted by $\mathbf{r}_{ij}$. Then consistency conditions are formulated with respect to the difference vector $\mathbf{c}_{ij} = \mathbf{t}_{ij} - \mathbf{r}_{ij}$, and the penalty for violation of consistency is added in the objective function for optimizing the subsystem with some appropriate equality and inequality constraints. Tosserams, Etman, Papalambos, and Rooda (2006) extended the target cascading approach to utilize the augmented Lagrangian method.

It should be noted that most of the decomposition approaches are heuristic and convergence of the iterative procedure is not proved. Even if the solution converges, the convergence property is not very good if the dependencies among the subsystems are highly nonlinear and global optimality of the global system is not guaranteed.

If the system is *separable*, then the variable vector is partitioned to the vectors $\mathbf{y}^i$ of local variables without any duplication, and the objective function $F(\mathbf{y}^1, \ldots, \mathbf{y}^m)$ is formulated as the sum of the objective functions $F_i(\mathbf{y}^i)$ of subsystems as

$$F(\mathbf{y}^1, \ldots, \mathbf{y}^m) = \sum_{i=1}^{m} F_i(\mathbf{y}^i) \tag{5.59}$$

and the constraints of the ith subsystem are defined using $\mathbf{y}_i$ only. In this case, the optimal solution of the global system is found successfully by optimizing each subsystem. If the objective function includes the term defined by the global variables, then the constraints with respect to the global variables are to be satisfied. Haftka and Watson (2005) defined quasiseparable subsystems and formulated the optimization problem as

$$\text{Minimize} \quad F(\mathbf{X}, \mathbf{y}^1, \ldots, \mathbf{y}^m) = F_0(\mathbf{X}) + \sum_{i=1}^{m} F_i(\mathbf{y}^i) \tag{5.60a}$$

$$\text{subject to} \quad H_j(\mathbf{X}) \leq 0, \quad (j = 1, \ldots, n^{\mathrm{H}}) \tag{5.60b}$$

$$h_j^i(\mathbf{y}^i) \leq 0, \quad (j = 1, \ldots, n^{\mathrm{h}}; \; i = 1, \ldots, m) \tag{5.60c}$$

In this case, if the lower-level problem is formulated to maximize the margin of the local constraints, then the lower-level problem is always feasible, and the optimal solution can be obtained by iteratively solving the upper- and lower-level problems.

However, for general complex systems, it is not always possible to decompose the system hierarchically to a tree-type structure of the subsystems, because the subsystems are highly likely to interact with each other. There-

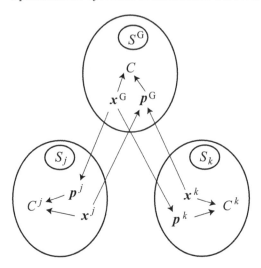

FIGURE 5.16: Relation between parameters of a global system and two subsystems.

fore, a non-hierarchical system is required to model general systems (Balling and Sobieszczanski-Sobieski 1996).

5.3.4 Parametric multidisciplinary optimization problem

In this section, we generalize the concept of two-level parametric optimization to problems consisting of hierarchical and non-hierarchical subsystems. Suppose a large structural system is divided into several subsystems, and optimization of the global system is carried out iteratively optimizing the subsystems. In order to improve the convergence property, we can utilize parametric programming approaches, as discussed in the previous section. The global system S is divided into n^S subsystems $S_1, \ldots, S_{n^S}$. Generally, each pair of subsystems has interacting parameters and variables, and the parameters for S_j will be modified after optimizing another subsystem S_k.

Consider first a hierarchical system, where n^S subsystems exist below the global system. A simple case of $n^S = 2$ is illustrated in Fig. 5.16, which is a generalization of the two-bar frame in Example 5.3. Let $\mathbf{x}^G$ and $\mathbf{p}^G$ denote the vectors of design variables and parameters of the global system, for which the optimization problem is formally written as

$$\text{Minimize}\quad C(\mathbf{x}^G, \mathbf{p}^G) \tag{5.61a}$$

$$\text{subject to}\quad \mathbf{x}^G \in \mathcal{X}^G(\mathbf{p}^G) \tag{5.61b}$$

where $\mathcal{X}^G$ is the feasible region for $\mathbf{x}^G$.

Let $\mathbf{x}^j = (x_1^j, \ldots, x_{m^j}^j)^\top$ denote the vector of m^j design variables of subsystem S_j. The objective function of S_j is given as $C_j(\mathbf{x}^j, \mathbf{p}^j)$, where $\mathbf{p}^j = (p_1^j, \ldots, p_{n^P}^j)^\top$ is the vector of n^P parameters. The problem to be solved in the jth subsystem is simply written as

$$\text{Minimize} \quad C_j(\mathbf{x}^j, \mathbf{p}^j) \tag{5.62a}$$

$$\text{subject to} \quad \mathbf{x}^j \in \mathcal{X}_j(\mathbf{p}^j) \tag{5.62b}$$

where $\mathcal{X}_j$ is the feasible region for $\mathbf{x}^j$. Note that $C_j(\mathbf{x}^j, \mathbf{p}^j)$ may be the same as $C(\mathbf{x}^G, \mathbf{p}^G)$, or may be the penalty function for the local constraints, as discussed in the previous section.

Suppose $\mathbf{p}^G$ and $\mathbf{p}^j$ $(j = 1, \ldots, n^S)$ are functions of $\mathbf{x}^j$ $(j = 1, \ldots, n^S)$ and $\mathbf{x}^G$, respectively. Then, the vector $\mathbf{p}^G = (p_1^G, \ldots, p_{n^P}^G)^\top$ is conceived as an implicit function of $\mathbf{x}^G$ denoted by $\widetilde{\mathbf{p}}^G(\mathbf{x}^G)$ through optimization of the subsystems, and the objective function with respect to the vector $\mathbf{x}^G = (x_1^G, \ldots, x_{m^G}^G)^\top$ only is defined as

$$\widetilde{C}(\mathbf{x}^G) = C(\mathbf{x}^G, \widetilde{\mathbf{p}}^G(\mathbf{x}^G)) \tag{5.63}$$

Accordingly, the sensitivity coefficient of $\widetilde{C}$ with respect to x_i^G is written as

$$\frac{\partial \widetilde{C}}{\partial x_i^G} = \frac{\partial C}{\partial x_i^G} + \sum_{k=1}^{n^P} \frac{\partial C}{\partial p_k^G} \frac{\partial \widetilde{p}_k^G}{\partial x_i^G}, \quad (i = 1, \ldots, m^G) \tag{5.64}$$

If p_k^G is an explicit function of $\mathbf{x}^j$ $(j = 1, \ldots, n^S)$, the sensitivity of p_k^G with respect to $\mathbf{x}^j$ is easily computed. Furthermore, sensitivity of the optimal value of $\mathbf{x}^j$ with respect to $\mathbf{p}^j$ is found using the parametric programming approach, and the dependence of $\mathbf{p}^j$ on $\mathbf{x}^G$ is also assumed to be known. Hence, the sensitivity coefficient $\partial \widetilde{p}_k^G / \partial x_i^G$ can be easily obtained. By utilizing the sensitivity coefficient $\partial \widetilde{C} / \partial x_i^G$ in (5.64), instead of using only $\partial C / \partial x_i^G$ in the right-hand side of (5.64), convergence of a gradient-based optimization process may be improved.

Next, we consider non-hierarchical systems. Suppose, for simple presentation of formulations, that we have two subsystems j and k. Fig. 5.17 illustrates the dependencies of the variables and parameters between two subsystems. The parameter vector $\mathbf{p}^j$ of S_j is a function of $\mathbf{x}^k$, and $\mathbf{p}^k$ is a function of $\mathbf{x}^j$. If we use the simple cyclic approach, $\mathbf{p}^k(\mathbf{x}^j)$ is first computed by assigning $\mathbf{x}^j$, and S_k is to be optimized to find the optimal value of $\mathbf{x}^k$. Because the optimal solution of S_k is found for each specified value of $\mathbf{p}^k$, it is a function of $\mathbf{p}^k(\mathbf{x}^j)$, and accordingly, is a function of $\mathbf{x}^j$ denoted as $\widetilde{\mathbf{x}}^k(\mathbf{x}^j)$. Then $\mathbf{p}^j$ is updated from $\mathbf{p}^j(\mathbf{x}^k)$, and the subsystem S_j is optimized. Hence, the parameter $\mathbf{p}^j$ may be conceived as an implicit function of $\mathbf{x}^j$ denoted by $\widetilde{\mathbf{p}}^j(\mathbf{x}^j)$, and the objective function of $\mathbf{x}^j$ only is defined as $\widetilde{C}^j(\mathbf{x}^j) = C^j(\mathbf{x}^j, \widetilde{\mathbf{p}}^j(\mathbf{x}^j))$.

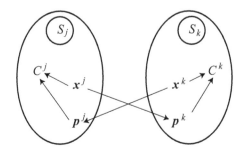

FIGURE 5.17: Relation between parameters of two non-hierarchical subsystems.

The sensitivity coefficients of $\widetilde{C}^j(\mathbf{x}^j)$ considering the variation of the parameters of the subsystems with respect to x_i^j is obtained for general non-hierarchical systems as

$$
\begin{aligned}
\frac{\partial \widetilde{C}^j}{\partial x_i^j} &= \frac{\partial C^j}{\partial x_i^j} + \sum_{r=1}^{n_j^{\mathrm{P}}} \frac{\partial C^k}{\partial p_r^j} \frac{\partial \widetilde{p}_r^j}{\partial x_i^j} \\
&= \frac{\partial C^j}{\partial x_i^j} + \sum_{r=1}^{n_j^{\mathrm{P}}} \left(\frac{\partial C^j}{\partial p_r^j} \sum_{s=1}^{m^k} \frac{\partial \widetilde{p}_r^j}{\partial x_s^k} \frac{\partial \widetilde{x}_s^k}{\partial x_i^j} \right), \quad (i = 1, \ldots, m^j)
\end{aligned}
\tag{5.65}
$$

The relation (5.65) is utilized to improve the convergence property of optimization of subsystems, which may be carried out sequentially or simultaneously. The second-order sensitivity coefficients can be incorporated, if necessary, as presented in Appendix A.5.

5.3.5   Optimization of plane frames

5.3.5.1   Problem formulation

In this section, a parametric programming approach is presented for optimization of plane frames, which are divided into subsystems based on analysis types, i.e., static analysis and dynamic analysis (Ohsaki 1997a). The static analysis is carried out using the frame model consisting of beams and columns, while the dynamic analysis is carried out using the shear model.

As Subsystem 1, we formulate the optimization problem of the shear model under constraint on the eigenvalue of vibration. The objective function $C^{\mathrm{D}}(\mathbf{D})$ is the total structural volume, which is a function of the vector of story stiffnesses $\mathbf{D} = (D_1, \ldots, D_{n^{\mathrm{F}}})^{\top}$, where n^{F} is the number of stories. The lower bounds of the lowest eigenvalue $\Omega_1(\mathbf{D})$ and D_i are denoted by Ω^{L} and D_i^{L}, respectively. Then the optimization problem of the shear model is formulated as

$$\text{Minimize} \quad C^{\mathrm{D}}(\mathbf{D}, \mathbf{p}^{\mathrm{D}}) \tag{5.66a}$$

$$\text{subject to} \quad \Omega_1(\mathbf{D}) \geq \Omega^{\mathrm{L}} \tag{5.66b}$$

$$D_i \geq D_i^{\mathrm{L}}, \quad (i = 1, \ldots, n^{\mathrm{F}}) \tag{5.66c}$$

where $\mathbf{p}^{\mathrm{D}}$ is the parameter vector defining the relation between the structural volume and story stiffnesses, which can be obtained from the stiffnesses of the members as the solution of the optimization problem of Subsystem 2 described below.

Stress constraints under the specified static loads are considered in the optimization problem of Subsystem 2 with the cross-sectional areas $\mathbf{A} = (A_1, \ldots, A_m)^\top$ as design variables. The length of the ith member is denoted by L_i. The constraints are given for the stresses σ_{ij} at the four points $(j = 1, \ldots, 4)$ of the ith member $(i = 1, \ldots, m)$, as described in Fig. 5.2, and their upper and lower bounds are denoted as σ_i^{U} and σ_i^{L}, respectively. Let $\Delta_i(\mathbf{A})$ and Q_i denote the interstory drift and shear force of the ith story under static horizontal loads. Then, the stiffness D_i of the ith story is defined as $D_i = Q_i/\Delta_i$ $(i = 1, \ldots, n^{\mathrm{F}})$, and the optimization problem for minimizing the total structural volume $C^{\mathrm{A}}(\mathbf{A})$ is formulated as

$$\text{Minimize} \quad C^{\mathrm{A}}(\mathbf{A}) = \sum_{i=1}^{m} A_i L_i \tag{5.67a}$$

$$\text{subject to} \quad \sigma_i^{\mathrm{L}} \leq \sigma_{ij}(\mathbf{A}) \leq \sigma_i^{\mathrm{U}}, \quad (i = 1, \ldots, m; \ j = 1, \ldots, 4) \tag{5.67b}$$

$$\frac{Q_i}{\Delta_i(\mathbf{A})} = p_i^{\mathrm{A}}, \quad (i = 1, \ldots, n^{\mathrm{F}}) \tag{5.67c}$$

$$A_i \geq A_i^{\mathrm{L}} \tag{5.67d}$$

where $p_i^{\mathrm{A}} = D_i$ is the specified value of the stiffness of the ith story, which is the optimal solution of Subsystem 1, and is conceived as the parameter of Subsystem 2.

It is easily seen from (A.122) in Appendix A.5 that the parametric sensitivity coefficients of the objective function of Problem (5.67) with respect to p_i^{A} $(= D_i)$ are equal to the Lagrange multipliers λ_i for the constraints (5.67c). The second-order sensitivity coefficients of the objective function, denoted by $W_{ij} = \partial^2 \tilde{C}^{\mathrm{A}}/\partial p_i^{\mathrm{A}} \partial p_j^{\mathrm{A}}$, are found from (A.124). Let the superscript $(\cdot)^{(k)}$ denote the value at the kth iteration. Since C^{D} is the total structural volume, which is an implicit function of $\mathbf{D}$, the parameter vector $\mathbf{p}^{\mathrm{D}}$ consists of λ_j and W_{ij}, and the objective function $C^{\mathrm{D}}(\mathbf{D}, \mathbf{p}^{\mathrm{D}})$ of Problem (5.66) is approximated as follows using $p_i^{\mathrm{A}} = D_i$:

$$\tilde{C}^{\mathrm{D}}(\mathbf{D}) \simeq C^{\mathrm{D}}(\mathbf{D}^{(k)}) + \sum_{i=1}^{n^{\mathrm{F}}} \lambda_i (D_i - D_i^{(k)})$$

$$+ \frac{1}{2} \sum_{i=1}^{n^{\mathrm{F}}} \sum_{j=1}^{n^{\mathrm{F}}} W_{ij} (D_i - D_i^{(k)})(D_j - D_j^{(k)}) \qquad (5.68)$$

$$+ \tau \sum_{i=1}^{n^{\mathrm{F}}} (D_i - D_i^{(k)})^4$$

where τ is the parameter for preventing divergence in the iterative process. A large value for τ leads to a large penalty for increments of D_i. This penalty is not needed if the matrix $\mathbf{W} = (W_{ij})$ is positive definite.

The optimization algorithm is summarized as follows:

Step 1: Assign the coefficients τ, λ_i, and W_{ij} for (5.68) and the initial value $D_i^{(0)}$ for D_i. Set the iteration counter $k = 0$.

Step 2: Solve Problem (5.66) to obtain the optimal value of $D_i^{(k)}$ by assigning the parameters $\mathbf{p}^{\mathrm{D}}$ for the definition of the objective function $C^{\mathrm{D}}(\mathbf{D}, \mathbf{p}^{\mathrm{D}})$.

Step 3: Incorporate the optimal value of $D_i^{(k)}$ to $p_i^{\mathrm{A}(k)}$, and solve Problem (5.67) to obtain the optimal value of $A_i^{(k)}$. Then compute the first- and second-order parametric sensitivity coefficients λ_i and W_{ij}, respectively, of the objective function with respect to $\mathbf{p}^{\mathrm{D}(k)} = \mathbf{D}^{(k)}$.

Step 4: Update the iteration counter as $k \leftarrow k + 1$ and go to Step 2 if not converged.

5.3.5.2   Numerical examples

Optimal solutions are found for a six-story three-span plane frame with base beams, as shown in Fig. 5.18, subjected to horizontal loads representing seismic loads. The elastic modulus is 205.8 kN/mm^2, the bounds for stress are $\sigma_i^{\mathrm{U}} = -\sigma_i^{\mathrm{L}} = 68.6$ N/mm^2, and $A_i^{\mathrm{L}} = 5.0 \times 10^{-3}$ m^2 for all members. The mass of each story is 4.0×10^4 kg. The horizontal loads (kN) for Problem (5.67) are given as $(P_1, \ldots, P_6) = (31.578, 45.469, 60.168, 76.442, 96.785, 159.96)$. Note that the self-weight is not considered.

The initial values are $D_i^{(0)} = 98.0$ kN/m and $\lambda_i = 1.0$ m^4/N for all stories, and $W_{ij} = 1.0$ m^5/N^2 for $i = j$ and 0 for $i \neq j$. It is confirmed that the optimal solution does not depend on the initial solution. The parameter τ is fixed at 1.0412×10^{-2} mm^5/N^2. Each member consists of a sandwich section, for simplicity, where the distance between the flanges is 0.6 m. The number of

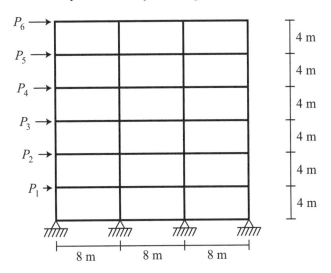

FIGURE 5.18: A six-story three-span frame.

TABLE 5.7: Optimal cross-sectional areas A_i ($\times 10^4 \text{mm}^2$).

Floor/Story	Ext. beam	Int. beam	Ext. column	Int. column
1	2.6761	0.5000	2.8547	3.1243
2	4.9633	1.4121	2.3583	3.2674
3	3.7934	3.0018	1.5722	3.5504
4	2.5645	4.0827	1.0598	3.1192
5	1.6073	3.9255	0.5733	2.4236
6	1.0577	2.6093	0.5000	1.2400
7	0.5000	0.7432		

design variables is 26 considering the symmetry conditions. The optimization software package IDESIGN (Arora and Tseng 1987) is used.

The lower bound for the fundamental eigenvalue is 40.0 rad^2/s^2. The optimal cross-sectional areas are obtained as shown in Table 5.7. Note that the beams in the first and seventh floor means the base beams and those in the roof level, respectively. The optimal story stiffnesses $D_1, \ldots, D_6$ ($\times 10^8$ N/m) are 3.4730, 3.2573, 2.9363, 2.4114, 1.7521, 1.0322, and the optimal objective value is 6.0994 m^3, which is close to 6.0956 m^3 obtained without using the parametric programming approach. The number of cycles in Steps 2–4 is 2, and the CPU time is 59.6% of that without the parametric programming approach.

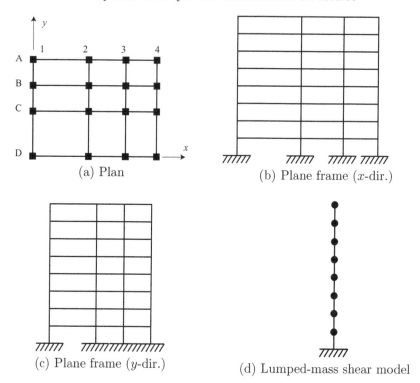

(a) Plan

(b) Plane frame (x-dir.)

(c) Plane frame (y-dir.)

(d) Lumped-mass shear model

FIGURE 5.19:   A 4×4 span eight-story three-dimensional frame.

5.3.6   Optimization of a three-dimensional frame

5.3.6.1   Definition of seismic load and story stiffness

Consider a three-dimensional regular frame, as shown in Fig. 5.19, which is divided into plane frames, as shown in Figs. 5.19(b) and (c), in the x- and y-directions, respectively. The optimal solution that minimizes the total structural volume is found under constraints on the responses against the seismic motions in the two horizontal directions. In the process of evaluating seismic responses, the plane frame is further simplified into a lumped-mass shear model, as shown in Fig. 5.19(d). The conventional assumption of a rigid floor is used.

Let $\mathbf{D}^* = (D_1^*, \dots, D_{n^F}^*)^\top$ denote the vector of story stiffnesses in x- or y-direction of the three-dimensional frame, which are the design variables of the shear model. The story mass includes the non-structural mass and the structural mass, which is a function of the cross-sectional areas of members. The story stiffness of the plane frame is to be found so that the mean-maximum interstory drift $\delta_i(\mathbf{D}^*)$ for the specified displacement response spectrum $S_{\mathrm{D}}(\Omega_r)$ evaluated by the square-root-of-sum-of-squares (SRSS) method is equal to the specified value $\bar{\delta}_i$:

$$\delta_i(\mathbf{D}^*) = \bar{\delta}_i, \quad (i = 1, \ldots, n^{\mathrm{F}}) \tag{5.69}$$

The details of the response spectrum approach are shown in Appendix A.7. Because the numbers of design variables D_i^* and the constraints (5.69) are the same for the shear model, the values of D_i^* are easily determined from (5.69) using any optimization technique, including the stress-ratio approach for fully-stressed design (see Sec. 1.6).

After obtaining D_i^*, the equivalent static story shear force Q_i^* and the horizontal load P_i^* to be applied to the plane frame in order to optimize the member cross-sectional areas are defined as follows in terms of the interstory drift and the story stiffness:

$$Q_i^*(\mathbf{D}^*) = D_i^*(\mathbf{D}^*)\bar{\delta}_i, \quad (i = 1, \ldots, n^{\mathrm{F}}) \tag{5.70a}$$

$$P_i^*(\mathbf{D}^*) = Q_{i+1}^*(\mathbf{D}^*) - Q_i^*(\mathbf{D}^*), \quad (i = 1, \ldots, n^{\mathrm{F}} - 1) \tag{5.70b}$$

$$P_{n^{\mathrm{F}}}^*(\mathbf{D}^*) = Q_{n^{\mathrm{F}}}^*(\mathbf{D}^*) \tag{5.70c}$$

In the following, the argument $\mathbf{D}^*$ is not written explicitly for simple presentation of the formulations.

5.3.6.2 Optimization for specified story stiffness

The cross-sectional areas of members are optimized under constraints on the story stiffness and the stresses against the specified static loads. Let a^* and a^j denote the total area of each floor and the area covered by the jth plane frame in the x- or y-direction. In the following, the value for the jth plane frame is indicated with the superscript j. Then the story stiffness $\overline{D}_i^j$ to be specified, horizontal load P_i^j, and story shear force Q_i^j of the ith story of the jth plane frame are given as

$$\overline{D}_i^j = \frac{a^j}{a^*}D_i^*, \quad P_i^j = \frac{a^j}{a^*}P_i^*, \quad Q_i^j = \frac{a^j}{a^*}Q_i^* \tag{5.71}$$

In this case, the plane frames in each direction deform without interaction between them.

Let A_i^j, $I_i^j(A_i^j)$, and $Z_i^j(A_i^j)$ denote the cross-sectional area, second moment of inertia, and section modulus of the ith member of the jth plane frame. Note that $I_i^j(A_i^j)$ and $Z_i^j(A_i^j)$ are functions of A_i^j. Two load vectors $\mathbf{P}^{\mathrm{S}j}$ and $\mathbf{P}^{\mathrm{H}j}$, corresponding to the vertical live load and self-weight, and the horizontal loads that represent the seismic loads, respectively, are considered.

The stress σ_i^j of the ith member of the jth plane frame is defined as the maximum absolute value of the stresses at the two edges of two ends against $\mathbf{P}^{\mathrm{S}j}$ and $\mathbf{P}^{\mathrm{H}j}$ that are applied simultaneously. The set of loads $\mathbf{P}^{\mathrm{S}j}$ and $-\mathbf{P}^{\mathrm{H}j}$ is also applied if the frame is not symmetric. The bending stresses due to the distributed vertical loads along the beams are also considered; however, axial force does not exist in the beams because of the assumption of a rigid floor.

The interstory drift of the ith story of the jth frame consisting of m^j members is denoted by $\delta_i^j(\mathbf{A}^j)$, which is a function of $\mathbf{A}^j = (A_1^j, \ldots, A_{m^j}^j)^\top$. Then the story stiffness $D_i^j(\mathbf{A}^j)$ of the jth frame is calculated from

$$D_i^j(\mathbf{A}^j) = \frac{Q_i^j}{\delta_i(\mathbf{A}^j)}, \quad (i = 1, \ldots, n^{\mathrm{F}}) \tag{5.72}$$

Let $\sigma_i^{\mathrm{U}j}$ denote the upper bound for the maximum stress $\sigma_i^j(\mathbf{A}^j)$ of the ith plane frame, which is assumed to be positive. The lower bound for A_i^j and the length of the ith member are denoted by $A_i^{\mathrm{L}j}$ and L_i^j, respectively. Then the problem for minimizing the total structural volume $C^j(\mathbf{A}^j)$ of the jth plane frame is formulated as

$$\text{Minimize} \quad C^j(\mathbf{A}^j) = \sum_{i=1}^{m^j} A_i^j L_i^j \tag{5.73a}$$

$$\text{subject to} \quad \sigma_i^j(\mathbf{A}^j) \leq \sigma_i^{\mathrm{U}j}, \quad (i = 1, \ldots, m^j) \tag{5.73b}$$

$$D_i^j(\mathbf{A}^j) = \overline{D}_i^j, \quad (i = 1, \ldots, n^{\mathrm{F}}) \tag{5.73c}$$

$$A_i^j \geq A_i^{\mathrm{L}j}, \quad (i = 1, \ldots, m^j) \tag{5.73d}$$

5.3.6.3 Successive optimization of plane frames

A three-dimensional frame is optimized by successively optimizing the plane frames in the two directions. A serious difficulty arises, however, from the fact that the columns belong to two plane frames in different directions; hence, the cross-sectional properties of the columns cannot be modified independently in the process of optimizing a plane frame.

Therefore, the cross-sectional areas of the columns are fixed in the lower-level problem for optimizing the beams of the plane frames. After optimization of all the plane frames is completed, the cross-sectional areas of the columns are modified using the parametric sensitivity coefficients of the optimal solutions of the plane frames with respect to the cross-sectional areas of the columns, which are regarded as the parameters for the optimization problem of a plane frame. Then the beams of the plane frames are optimized for the updated values of the cross-sectional areas of the columns.

Suppose the member numbers are assigned so that the members $1, \ldots, m_{\mathrm{b}}^j$ are beams and $m_{\mathrm{b}}^j + 1, \ldots, m^j$ are columns. The vector of cross-sectional areas of beams is given as $\mathbf{A}_{\mathrm{b}}^j = (A_1^j, \ldots, A_{m_{\mathrm{b}}^j}^j)^\top$. Let $\boldsymbol{\lambda}^j = (\lambda_1^j, \ldots, \lambda_{m^j}^j)^\top$, $\boldsymbol{\mu}^j = (\mu_1^j, \ldots, \mu_{n^{\mathrm{F}}}^j)^\top$, and $\boldsymbol{\eta}^j = (\eta_1^j, \ldots, \eta_{m_{\mathrm{b}}^j}^j)^\top$ denote the vectors of non-negative Lagrange multipliers for the constraints (5.73b), (5.73c), and (5.73d), respectively. Then, the Lagrangian for Problem (5.73a) with fixed cross-sectional areas of columns is defined as

$$
\psi^j(\mathbf{A}_b^j, \boldsymbol{\lambda}, \boldsymbol{\mu}, \boldsymbol{\eta}) = \sum_{i=1}^{m_b^j} A_i^j L_i^j + \sum_{i=1}^{m^j} \lambda_i^j(\sigma_i^j(\mathbf{A}_b^j) - \sigma_i^{\mathrm{U}j})
$$
$$
+ \sum_{i=1}^{n^{\mathrm{F}}} \mu_i^j(D_i^j(\mathbf{A}_b^j) - \overline{D}_i^j) + \sum_{i=1}^{m_b^j} \eta_i^j(A_i^{\mathrm{L}j} - A_i^j)
$$

(5.74)

The multipliers of the optimal solution are usually available if a gradient-based nonlinear programming approach is used. Even if the multipliers are not available, they are easily calculated from the optimality conditions after the solution has converged; see Sec. 5.2.

The member numbers are assigned also for the three-dimensional frame so that the members $1, \ldots, m_b$ are beams and $m_b + 1, \ldots, m$ are columns, where m is the total number of members. In the following, the values for the three-dimensional frame are indicated without superscript j.

From (A.122) in Appendix A.5, the sensitivity coefficients of the objective function C of the three-dimensional frame with respect to the cross-sectional areas A_k $(k = m_b+1, \ldots, m)$ of the columns, which are the same as the cross-sectional areas A_k^j $(k = m_b^j + 1, \ldots, m^j)$ of the columns of the jth frame, are calculated from

$$
\frac{\partial C}{\partial A_k} = L_k + \sum_{j=1}^{n^{\mathrm{S}}} \sum_{i=1}^{m_b^j} \lambda_i^j \frac{\partial \sigma_i^j}{\partial A_k} + \sum_{j=1}^{n^{\mathrm{S}}} \sum_{i=1}^{n^{\mathrm{F}}} \mu_i^j \frac{\partial D_i^j}{\partial A_k}, \quad (k = m_b + 1, \ldots, m) \quad (5.75)
$$

where n^{S} is the number of plane frames.

The optimization algorithm is summarized as follows:

Step 1: Optimize each plane frame considering the cross-sectional areas of beams and columns as independent design variables. Define A_i for each column, e.g., as the mean or maximum value of the optimal cross-sectional area of the corresponding column in the plane frames in two different directions. Skip this process if default initial values are given for the cross-sectional areas of columns.

Step 2: Find optimal cross-sectional areas of beams by optimizing each plane frame for the fixed cross-sectional areas of columns, and evaluate the Lagrangian multipliers for the constraints. Note that the optimization of plane frames can be carried out in parallel.

Step 3: Calculate the parametric sensitivity coefficients of the objective function with respect to A_i of the columns using (5.75), and modify A_i based on the steepest descent method. Reduce the move limit ΔA for preventing divergence to $\beta \Delta A$ $(0 < \beta < 1)$ at each iteration. If there is no feasible solution, modify A_i for a column as follows for each constraint that is violated:

$$\text{Interstory drift:} \quad A_i = \gamma_1 A_i \quad (\gamma_1 > 1) \quad\quad (5.76\text{a})$$

$$\text{Stress of beam:} \quad A_i = \gamma_2 A_i \quad (\gamma_2 < 1) \quad\quad (5.76\text{b})$$

$$\text{Stress of column:} \quad A_i = \gamma_3 A_i \quad (\gamma_3 > 1) \quad\quad (5.76\text{c})$$

where γ_1, γ_2, and γ_3 are the specified parameters. Note that if the stress constraint of a beam is violated, it can be satisfied by reducing the cross-sectional area of the columns connected to the beam so that the bending deformation of the beam is reduced.

Step 4: Go to Step 2 if the solution is not converged.

5.3.6.4   Numerical examples

An optimum design is found for a 27-story three-dimensional frame with a plan view, as shown in Fig. 5.19(a). The span lengths (m) in the x- and y-directions are $(W_{12}, W_{23}, W_{34}) = (12, 10, 10)$ and $(W_{AB}, W_{BC}, W_{CD}) = (8, 8, 10)$, respectively, where, e.g., W_{12} is the distance between lines 1 and 2. The elastic modulus is 205.8 kN/mm^2, the mass at each floor is 500 kg/m^2, and the mass density of the steel beams and columns is 7.86×10^3 kg/m^3. The beams and columns consist of wide-flange and box sections, and the values of I_i^j and Z_i^j are defined as functions of A_i^j as

$$\text{Column}: \quad I_i^j = 1.2(A_i^j)^2, \quad Z_i^j = 1.0(A_i^j)^{1.5} \quad\quad (5.77\text{a})$$

$$\text{Beam}: \quad I_i^j = 8.0(A_j^j)^2, \quad Z_i^j = 2.0(A_j^j)^{1.5} \quad\quad (5.77\text{b})$$

The lower bound for the cross-sectional area is 0.05 m^2.

The response spectrum by Newmark and Hall (1982) is used for representing seismic motions; see its definition and parameter values in Appendix A.7. The six lowest modes are considered for evaluating the seismic responses of the shear model. Stiffness-proportional damping is used with the damping ratio 0.02 for the lowest mode.

The columns of the three-dimensional frame and the beams of each plane frame are classified, respectively, into 15 and 5 groups with the same cross-sectional areas; see Ohsaki, Nagano, and Wakamatsu (2000) for details. The method of modified feasible directions in the library DOT 5.0 (VR&D 1999) is used for optimization. The upper bounds for the maximum interstory drift and stresses are 0.02 m and 323.4 N/mm^2, respectively. The parameters for modification of the cross-sectional areas of the columns are $\gamma_1 = 1.15$, $\gamma_2 = 1.05$, $\gamma_3 = 1.25$, and $\beta = 0.5$.

The cross-sectional area of each column at the first stage is defined to be equal to the larger value in the optimal plane frames in two directions. The optimal solution at the sixth step for $\Delta A = 0.02$ m^2 is as shown in Fig. 5.20,

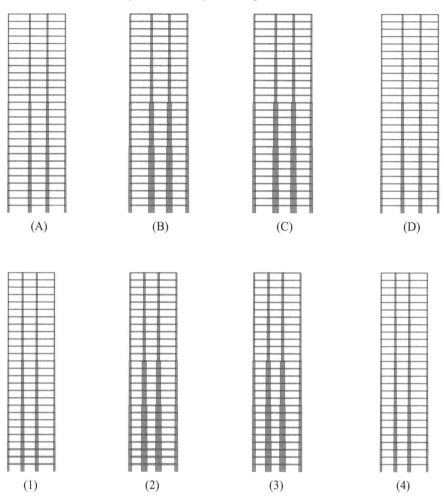

FIGURE 5.20: Optimization result of the 27-story three-dimensional frame.

where (A)–(D) and (1)–(4) in the figure are defined in Fig. 5.18, and the width of each member is proportional to the cross-sectional area. It may be observed from Fig. 5.20 that the columns in the center span have larger cross-sectional areas than the exterior columns.

5.4   Local search for multiobjective optimization of frames

5.4.1   Introduction

As we have discussed in the previous sections, the cross-sectional properties of the frames are usually selected from a list or catalog of the available standard sections. Therefore, the optimization problems are formulated as a combinatorial problem with continuous objective function, constraint functions, and state variables (see Sec. 1.3.2 for general formulation and classification of structural optimization problems with discrete variables).

It is very easy to solve a combinatorial optimization problem by enumerating all the possible combinations of the variables if the number of variables is small. However, the computational cost increases as an exponential function of problem size represented by numbers of variables and constraints; therefore, it is not possible to solve a practical problem using the enumeration approach within a practically admissible computational time.

An important aspect of the practical design process is that it may be sufficient to obtain an approximate optimal design rather than obtaining the global optimal solution. Another aspect of practical design optimization is that multiple performance measures are to be considered, and the problem turns out to be a multiobjective programming (MOP) problem.

As defined in Appendix A.4.1, the approaches to MOP are classified into those with and without *a priori* information of preference. In the latter approach, which is also called *a method with posteriori articulation of preferences* (Marler and Arora 2004), a set of Pareto optimal solutions is first generated, and the most preferred solution is selected from the set by the decision maker. If we regard the solutions that are dominated by some other feasible solutions are not acceptable as candidate designs, the number of acceptable solutions is simply reduced by generating the set of Pareto optimal solutions. Therefore, in this approach, it is important that the Pareto optimal solutions with enough number and diversity be generated, or preferably all the Pareto solutions are enumerated.

Heuristics have been extensively applied to obtain Pareto solutions of combinatorial MOP problems in many fields of engineering, including structural design (Jones, Mirrazavi, and Tamiz 2002). However, an efficient optimization approach is desired for a structural optimization problem, because computationally expensive structural analysis should be carried out for evaluation of the objective and/or constraint functions. Therefore, in this section, we concentrate on simulated annealing and tabu search, which utilize local search and are categorized as single-point-search heuristics. Application of population-based approaches, including the genetic algorithm (GA) and particle swarm optimization (PSO) to MOP problems are discussed in various books and papers (Goldberg 1989; Coello Coello, Pulido, and Lechuga 2004; Coello Coello and Pulido 2005). Ray and Liew (2002) utilized Pareto dominance to maintain diversity of the solutions, and used *multilevel sieve* to handle constraints for PSO. As an extension of GA, an immune algorithm (IA) (Fukuda, Mori, and Tsukiyama 1999) can also be applied to MOP. Luh and Chueh (2004) presented an IA for constrained MOP of trusses.

In this section, we first present several approaches of single-point-search heuristics to MOP problems. Then, a simple local search method is presented for combinatorial multiobjective structural optimization (Ohsaki 2008) (see Appendix A.3 for details of single-point-search heuristic approaches to single-objective problems and Appendix A.4 for various methodologies of MOP problems).

5.4.2 Heuristic approaches to combinatorial multiobjective programming

5.4.2.1 Problem formulation and basic algorithm

Suppose the following list $\mathcal{X}_i$ is given as a set of r_i available values for the variable x_i:

$$\mathcal{X}_i = \{x_1^i, \ldots, x_{r_i}^i\} \tag{5.78}$$

Let $\mathbf{J} = (J_1, \ldots, J_m)^\top$ denote the vector of integer variables, where m is the number of variables. The jth value x_j^i in $\mathcal{X}_i$ is assigned to x_i if $J_i = j$.

Consider, for simplicity, an unconstrained problem, and let $F_i(\mathbf{J})$ denote the ith objective function to be minimized. Then an MOP problem with n^{F} objective functions is formulated as

$$\text{minimize} \quad F_1(\mathbf{J}), \ldots, F_{n^{\mathrm{F}}}(\mathbf{J}) \tag{5.79a}$$

$$\text{subject to} \quad J_i \in \{1, \ldots, r_i\}, \quad (i = 1, \ldots, m) \tag{5.79b}$$

Problem (5.79) is classified as a combinatorial MOP problem because it has integer variables and multiple objective functions.

In a single-point-search heuristic approach, only one solution, called the *seed solution*, is obtained at each step of iteration. Let $\mathbf{J}^{(k)}$ denote the seed solution at the kth step. A solution $\mathbf{J}^*$ in the neighborhood of $\mathbf{J}^{(k)}$ is accepted or rejected as the seed solution of the next step in accordance with each specific algorithm; see Appendix A.3 for details.

5.4.2.2 Simulated annealing

Simulated annealing (SA) is very effective in problems with many local optimal solutions, because it allows a move to a non-improving solution. Although SA was originally developed for single-objective problems, it has been recently extended to MOP problems.

Let $\mathbf{J}^{(k)}$ denote the solution at the kth step of iteration. For a neighborhood solution $\mathbf{J}^*$, define ΔF as

$$\Delta F = \frac{\alpha}{n^{\mathrm{F}}} \left(\sum_{i=1}^{n^{\mathrm{F}}} \frac{F_i(\mathbf{J}^*) - F_i(\mathbf{J}^{(k)})}{\delta_i} \right) \tag{5.80}$$

where α is the scaling parameter, and δ_i is a possible range of F_i for conversion of the ith objective value to a nondimensional value. A single Pareto optimal solution can be found if ΔF is used for the criterion for acceptance of the neighborhood solution $\mathbf{J}^*$ in the same manner as the SA for a single-objective problem.

Ho, Yang, Wong, and Ni (2003) used the fitness function, which is similar to the Pareto ranking approach of a multiobjective GA (Goldberg 1989). A sharing function approach that is also used for multiobjective GA (Goldberg and Richerdson 1987) is incorporated to generate the Pareto solutions with enough diversity. The main drawback of sharing function for MOP is that a dominated solution tends to have large fitness because many Pareto solutions are stored in the list and a new Pareto solution will have a large penalty with the sharing function. Ho, Yang, Wong, and Ni (2003) proposed an efficient fitness function to avoid this drawback. Viana and de Sousa (2000) applied a multiobjective SA to project scheduling using a weighted Chebychev metric.

Jilla and Miller (2001) compared several methods of multiobjective SA and presented the following algorithm:

Algorithm 5.1
Step 1: Randomly generate the initial solution $\mathbf{J}^{(0)}$, and set the temperature parameter as $T = T_0$. Initialize the Pareto candidate set as empty, assign δ_i for defining the range of F_i, and set the iteration counter $k = 0$.

Step 2: Randomly generate the neighborhood solution $\mathbf{J}^*$ of $\mathbf{J}^{(k)}$.

Step 3: If there is no solution in the Pareto candidate set that dominates $\mathbf{J}^*$, then accept $\mathbf{J}^*$ as $\mathbf{J}^{(k+1)} = \mathbf{J}^*$ and store it in the Pareto candidate set; otherwise, accept $\mathbf{J}^*$ with the probability P defined as

$$P = \exp\left(-\frac{\Delta F}{T_k} \right) \tag{5.81}$$

Add $\mathbf{J}^{(k+1)}$ to the Pareto candidate list if $\mathbf{J}^*$ has been accepted; otherwise, let $\mathbf{J}^{(k+1)} = \mathbf{J}^{(k)}$.

Step 4: Decrease T_k to T_{k+1} by the specified rule called the *cooling schedule*. In the most popular approach of the cooling schedule, T_k is updated to $T_{k+1} = \eta T_k$ with a specified constant η. Several steps of neighborhood search can be done at the same temperature.

Step 5: If the termination condition is not satisfied, set $k \leftarrow k+1$ and go to Step 2; otherwise, remove the dominated solutions in the Pareto candidate list, output the remaining Pareto optimal solutions, and terminate the process.

This way, a set of Pareto solutions can be generated by conducting the algorithm only once.

Czyzak and Jaszkiewicz (1998) presented a method called Pareto simulated annealing that maintains multiple solutions at each step like GA. The weights λ_j for the objective functions are updated through iteration to keep diversity of the solutions. The probability of acceptance of a neighborhood solution may be defined using the Chebychev norm:

$$P = \min \left\{ \exp \left[\max_{j=1,\ldots,n^F} \left(\frac{\lambda_j(F_j(\mathbf{J}^{(k)}) - F_j(\mathbf{J}^*))}{T^{(k)}} \right) \right] \right\} \tag{5.82}$$

or the sum of the increments of the objective functions

$$P = \min \left\{ \exp \left[\sum_{j=1}^{n^F} \left(\frac{\lambda_j(F_j(\mathbf{J}^{(k)}) - F_j(\mathbf{J}^*))}{T^{(k)}} \right) \right] \right\} \tag{5.83}$$

The solution is updated to $\mathbf{J}^{(k+1)}$ that is closest to $\mathbf{J}^{(k)}$ and is not dominated by $\mathbf{J}^{(k)}$. Then the weights are modified as

$$\lambda_j^{(k+1)} = \begin{cases} \kappa\lambda_j^{(k)} & \text{if } F_j(\mathbf{J}^{(k+1)}) \geq F_j(\mathbf{J}^{(k)}) \\ \lambda_j^{(k)}/\kappa & \text{if } F_j(\mathbf{J}^{(k+1)}) < F_j(\mathbf{J}^{(k)}) \end{cases} \tag{5.84}$$

where $\alpha \, (> 1)$ is a specified parameter.

Whidborne, Gu, and Postlethwaite (1997) presented two approaches of SA to multiobjective control design. Their objective is not optimization but satisfaction of inequality constraints given as $g_j(\mathbf{J}) \leq \varepsilon_j \ (j = 1, \ldots, n^I)$, where ε_j is the specified upper bound and n^I is the number of constraints. Their first method transforms the MOP problem to a problem with the single objective function $F(\mathbf{J})$ using a goal programming based on a Chebychev norm as

$$F(\mathbf{J}) = \max_{j=1,\ldots,n^I} \left(\frac{g_j - \varepsilon_j}{\lambda_j}, 0 \right) \tag{5.85}$$

where λ_j is the weight parameter for the jth constraint. Their second algorithm utilizes Pareto optimality for defining the acceptance probability of the solution. A function $c_j(\mathbf{J})$ is first defined as

$$c_j(\mathbf{J}) = \begin{cases} \varepsilon_j & \text{if } g_j(\mathbf{J}) \leq \varepsilon_j \\ g_j(\mathbf{J}) & \text{if } g_j(\mathbf{J}) > \varepsilon_j \end{cases} \tag{5.86}$$

For two solutions $\mathbf{J}_1$ and $\mathbf{J}_2$, the following three properties are defined:

P1: $\mathbf{J}_1$ is said to be superior to $\mathbf{J}_2$ if and only if $c_j(\mathbf{J}_1) \leq c_j(\mathbf{J}_2)$ for all $j \in \{1, \dots, n\}$.

P2: $\mathbf{J}_1$ is said to be inferior to $\mathbf{J}_2$ if and only if $c_j(\mathbf{J}_1) \geq c_j(\mathbf{J}_2)$ for all $j \in \{1, \dots, n\}$ and $c_j(\mathbf{J}_1) > c_j(\mathbf{J}_2)$ for at least one j.

P3: $\mathbf{J}_1$ is noninferior to $\mathbf{J}_2$ if and only if $\mathbf{J}_1$ is not either inferior or superior to $\mathbf{J}_2$.

If a neighborhood solution $\mathbf{J}^*$ is superior to $\mathbf{J}^{(k)}$, then accept $\mathbf{J}^*$ unconditionally as $\mathbf{J}^{(k+1)} = \mathbf{J}^*$. Otherwise, define a function φ_j as

$$\varphi_j = \frac{1}{\lambda_j}(c_j(\mathbf{J}^*) - c_j(\mathbf{J}^{(k)})) \tag{5.87}$$

and compute ΔF as the sum of φ_j for j such that $c_j(\mathbf{J}^*) > c_j(\mathbf{J}^{(k)})$. The probability of acceptance is given by ΔF using the standard process of SA.

5.4.2.3   Tabu search

Tabu search (TS), which is originally developed for single objective problems (Glover 1989), has also been shown to be effective for MOP problems (Hansen 1997). In TS, the best solution in the neighborhood is selected as the next candidate, and a tabu list is used to prevent a local cyclic search among a small number of different solutions. Armentano and Arroyo (2004) presented an algorithm for MOP that has multiple solutions at each step. Baykasoglu (2006) presented a multiobjective TS and applied it to simple structural optimization problems.

The single-point-search algorithm for MOP by Baykasoglu, Owen, and Gindy (1999b) is summarized as follows:

Algorithm 5.2

Step 1: Randomly generate the initial solution $\mathbf{J}^{(0)}$, which is chosen as the *seed solution*. Initialize the Pareto list $\mathcal{P}$, tabu list $\mathcal{T}$, and Pareto candidate list $\mathcal{C}$ as $\mathcal{P} = \mathcal{T} = \mathcal{C} = \{\mathbf{J}^{(0)}\}$. Set the iteration counter as $k = 0$.

Step 2: Generate the set of q neighborhood solutions $\mathcal{N} = \{\mathbf{J}_j^{\mathrm{N}} \mid j = 1, \dots, q\}$ of the current seed solution $\mathbf{J}^{(k)}$. Let $\mathcal{S}$ denote the set of candidate solutions in $\mathcal{N}$ that are not dominated by any solution in $\mathcal{N}, \mathcal{P}$, and $\mathcal{C}$, and are not included in $\mathcal{T}$.

Step 3: Randomly select a solution $\mathbf{J}^*$ from $\mathcal{S}$. If $\mathcal{S}$ is empty, select the oldest solution in $\mathcal{C}$ as $\mathbf{J}^*$.

Step 4: Remove the solutions in $\mathcal{P}$ and $\mathcal{C}$ that are dominated by a solution in $\mathcal{S}$.

Step 5: Add $\mathbf{J}^*$ to $\mathcal{P}$ and $\mathcal{T}$, and add other candidate solutions to $\mathcal{C}$. Set $\mathbf{J}^{(k+1)} = \mathbf{J}^*$.

Step 6: If $\mathcal{C}$ is empty and there exists no new candidate solution, or if the number of iterative steps exceeds the specified upper bound, terminate the process; otherwise, let $k \leftarrow k + 1$ and go to Step 2.

The algorithm by Hansen (1997) maintains multiple solutions in a similar manner as the GA, and each solution has its own tabu list. Let δ_i denote the range of the ith objective function F_i, and define $c_i = 1/\delta_i$. The range parameter π_i is defined as

$$\pi_i = \frac{c_i}{\bar{c}}, \quad \bar{c} = \sum_{j=1}^{n^{\mathrm{F}}} c_i \qquad (5.88)$$

The distance $s_j(\mathbf{J}^*)$ from the solution $\mathbf{J}^*$ to the set S is defined using the Chebychev distance as

$$s_j(\mathbf{J}^*) = \min_{\mathbf{J}_i \in S}\{\max_k[\lambda_k(F_k(\mathbf{J}^*) - F_k(\mathbf{J}_i))]\} \qquad (5.89)$$

where λ_k is a weight coefficient. The algorithm is summarized as follows:

Algorithm 5.3
Step 1: Generate an initial set of solutions S, empty the tabu list, set the iteration counter $k = 0$, and initialize the weight coefficients as $\lambda_i = 1$ $(i = 1, \ldots, n^{\mathrm{F}})$.

Step 2: For each $\mathbf{J}$ in S, determine an appropriate search direction by updating λ_j in view of the relation between $\mathbf{J}$ and the Pareto optimal solutions in S, so that the Pareto solutions will be equally spaced in the objective function space. Then select the best seed solution $\mathbf{J}^*$, which is in the neighborhood of $\mathbf{J}$ and is not included in the tabu list.

Step 3: Add $\mathbf{J}^*$ to the tabu list and the Pareto set, and remove the solutions in the Pareto set that are dominated by $\mathbf{J}^*$. Replace a randomly selected solution in S with a randomly generated solution through the procedure called *drift* to maintain the diversity of Pareto solutions.

Step 4: Go to Step 2 if the termination condition is not satisfied; otherwise, output a set of Pareto optimal solutions and terminate the process.

Baykasoglu, Owen, and Gindy (1999a) presented a TS based on goal programming. It maintains only one solution at each step, and reaches a Pareto optimal solution; i.e., it cannot generate a Pareto optimal set with a single run. The algorithm is summarized as follows:

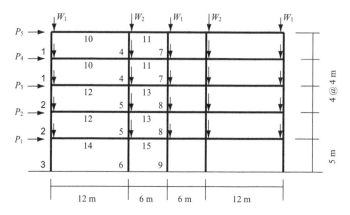

FIGURE 5.21: A five-story four-span frame.

Algorithm 5.4

Step 1: Generate an initial solution $\mathbf{J}^{(0)}$, and regard $\mathbf{J}^{(0)}$ as the best solution. Arrange the objective functions in nonincreasing order of priority; i.e., $F_1(\mathbf{J})$ has the largest priority. Set the iteration counter $k = 0$.

Step 2: Generate a set of neighborhood solutions of $\mathbf{J}^{(k)}$.

Step 3: Evaluate the deviations of the objective functions from their goals in the objective function space, and determine the rank of the solutions based on the priorities of the objective functions.

Step 4: Move to the best neighborhood solution $\mathbf{J}^*$ that is not included in the tabu list.

Step 5: Update the solution as $\mathbf{J}^{(k+1)} = \mathbf{J}^*$, and replace the best solution by $\mathbf{J}^*$ if it is better than the previous best solution.

Step 6: Construct the tabu list containing the recent best solutions.

Step 7: Set $k \leftarrow k + 1$ and go to Step 2 if the termination condition is not satisfied.

5.4.3   Local search for multiobjective structural optimization

Consider a plane frame, as shown in Fig. 5.21, where the members are classified into m groups. Let $\mathbf{J} = (J_1, \ldots, J_m)^\top$ denote the vector of integer variables that represent the cross-sectional parameters. The jth value in the specified list of available sections is assigned to the ith group if $J_i = j$ (see Sec. 5.4.2.1 and Appendix 5.4.2 for details of the problem formulation).

If we use SA or TS for a multiobjective structural optimization problem, a set of Pareto optimal solutions can be found by carrying out structural analysis many times, say, several thousand times. However, for practical applications, the number of analyses should be reduced to several hundreds at the most. Furthermore, because we have many design variables, the performance of the algorithms applied to mathematical toy problems does not suggest any performance measure for a structural optimization problem. In this regard, the use of a tabu list in TS is not very effective, and we may be able to generate some approximate Pareto solutions using a simple local search (LS), as described below.

In order to show the effectiveness of the local search in a simple manner, we consider a problem without constraints. The algorithm is an extension of Algorithm 5.2 in Sec. 5.4.2.3 for multiobjective TS by Baykasoglu, Owen, and Gindy (1999b). The objective functions F_i are normalized by the range parameter δ_i as

$$F_i^* = \frac{F_i}{\delta_i}, \quad (i = 1, \ldots, n^{\mathrm{F}}) \tag{5.90}$$

Let $\mathbf{F}^*$ and $\mathbf{F}^{*(i)}$ denote the objective values of the current seed solution and the ith solution in the set $\mathcal{S}$ of the candidate solutions, respectively.

To obtain the Pareto optimal set with good accuracy and diversity, the following four strategies are used for the selection of the seed solution in Step 3 of Algorithm 5.2:

Strategy 1: Randomly select a seed solution from $\mathcal{S}$. Note that this strategy corresponds to Step 3 of the original algorithm.

Strategy 2: Select the solution that minimizes $\sum_{j=1}^{n^{\mathrm{F}}}(F_j^{*(i)} - F_j^*)$.

Strategy 3: Select the solution that minimizes $F_j^{*(i)} - F_j^*$ for a specific objective function F_j.

Strategy 4: Define the density φ_i of the solutions near $\mathbf{J}_i \in \mathcal{S}$ in the set of Pareto solutions as

$$\varphi_i = \sum_{\mathbf{J}_j \in \mathcal{P}} s(d(\mathbf{J}_i, \mathbf{J}_j)) \tag{5.91}$$

where $d(\mathbf{J}_i, \mathbf{J}_j)$ is the distance between $\mathbf{J}_i$ and $\mathbf{J}_j$ in the space of objective functions or design variables. In the following examples, the Euclidean distance in the objective function space is used. The sharing function $s(d)$, which is often used in a multiobjective GA, is defined as

$$s(d) = \max\left\{0, 1 - \frac{d}{\gamma}\right\} \tag{5.92}$$

where γ is the parameter called *sharing radius* or *niche size*. Then select the solution in $\mathcal{S}$ that has the smallest value of φ_i.

5.4.4   Properties of Pareto optimal solutions

Let $\mathcal{P}$ denote the correct Pareto optimal set obtained by enumeration of all solutions. The size of approximate Pareto optimal set $\mathcal{A}$ obtained using a heuristic approach is denoted by n^{P}. Various measures have been presented for evaluation of the properties of the approximate Pareto optimal set (Coello Coello, Pulido, and Lechuga 2004). We use the following measures:

- **Error ratio:**
  Define e_i so that $e_i = 0$ if the solution i exists in the set $\mathcal{P}$ and $e_i = 1$ if not. Then the error ratio ER is defined as

$$\mathrm{ER} = \frac{1}{n^{\mathrm{A}}} \sum_{i=1}^{n^{\mathrm{A}}} e_i \tag{5.93}$$

  The approximate set is included in the correct set if $\mathrm{ER} = 0$.

- **Generational distance:**
  Let d_i denote the Euclidean distance from the ith approximate solution to its nearest solution in $\mathcal{P}$:

$$d_i = \min_j \sqrt{\sum_{k=1}^{n^{\mathrm{F}}} (F_k^{*(i)} - \widetilde{F}_k^{*(j)})^2} \tag{5.94}$$

  where $F_k^{*(i)}$ and $\widetilde{F}_k^{*(j)}$ are the values of F_k^* of the ith solution in $\mathcal{A}$ and the jth solution in $\mathcal{P}$, respectively. Then, define the generational distance GD as

$$\mathrm{GD} = \frac{1}{n^{\mathrm{A}}} \sqrt{\sum_{i=1}^{n^{\mathrm{A}}} d_i^2} \tag{5.95}$$

  The approximate set is included in the correct set if $\mathrm{GD} = 0$.

- **Spacing:**
  The minimum Manhattan distance g_i between the ith approximate solution and a solution in $\mathcal{P}$ is defined as

$$g_i = \min_j \sum_{k=1}^{n^{\mathrm{F}}} |F_k^{*(i)} - \widetilde{F}_k^{*(j)}| \tag{5.96}$$

  Let $\bar{g}$ denote the mean value of g_i, and define the spacing SP as

$$\mathrm{SP} = \sqrt{\frac{1}{n-1} \sum_{i=1}^{n^{\mathrm{A}}} (\bar{g} - g_i)^2} \tag{5.97}$$

  If the approximate solutions are uniformly distributed in the objective function space, then $\mathrm{SP} = 0$.

- **Span:** Span SN is defined as the maximum distance between two approximate solutions:

$$\text{SN} = \sum_{k=1}^{n^{\text{A}}} \max_{i,j}\{|F_k^{*(i)} - F_k^{*(j)}|\} \tag{5.98}$$

If SN has a large value, then the solutions are widely distributed.

5.4.5 Numerical examples

5.4.5.1 Mathematical problem

Consider first the following problem with two objective functions with respect to four variables (Coello Coello, Pulido, and Lechuga 2004; Coello Coello and Pulido 2005):

$$\text{Minimize} \quad F_1(\mathbf{x}) = 200(2x_1 + \sqrt{2x_2} + \sqrt{x_3} + x_4) \tag{5.99a}$$

$$F_2(\mathbf{x}) = \frac{1}{100}\left(\frac{2 + 2\sqrt{2}}{x_2} - \frac{2\sqrt{2}}{x_3} + \frac{2}{x_4}\right) \tag{5.99b}$$

$$\text{subject to} \quad 1 \le x_1 \le 3 \tag{5.99c}$$

$$\sqrt{2} \le x_2 \le 3 \tag{5.99d}$$

$$\sqrt{2} \le x_3 \le 3 \tag{5.99e}$$

$$1 \le x_4 \le 3 \tag{5.99f}$$

The real variable x_i is converted to an integer variable J_i by the following relation:

$$x_i(J_i) = x_i^{\text{L}} + \Delta x_i \times J_i \tag{5.100a}$$

$$\Delta x_i = \frac{x_i^{\text{U}} - x_i^{\text{L}}}{t_i - 1} \tag{5.100b}$$

where x_i^{U} and x_i^{L} are the upper and lower bounds of x_i, and the number t_i of the integer values is $2^{14} = 16,384$ for x_1 and x_4, and $2^{13} = 8192$ for x_2 and x_3. In each of the following figures, the exact Pareto optimal set, called the *Pareto front*, obtained by enumeration is plotted in a solid line.

The neighborhood solutions are generated using the random variable of normal distribution with 0 mean. The standard deviation is 800 for J_1 and J_4, and 400 for J_2 and J_3. Note that all the variables are modified simultaneously in the process of generating neighborhood solutions. If $J_i < 1$ or $J_i > t_i$ is satisfied as the result of updating J_i using the random number, J_i is replaced by 1 or t_i, respectively. The range parameters of the objective values in (5.90) are $\delta_1 = 600$ and $\delta_2 = 0.03$.

Fig. 5.22 shows the objective values of the 100 neighborhood solutions generated from a seed solution, indicated by the filled square. As is seen, the

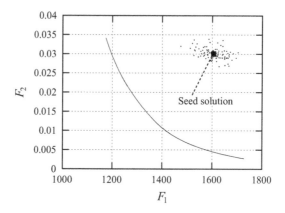

FIGURE 5.22: Distribution of the neighborhood solutions of the mathematical problem.

TABLE 5.8: Comparison of performances of the mathematical problem by TS and LS with various numbers of analyses.

	ER	GD ($\times 10^{-3}$)	SP ($\times 10^{-2}$)	SN
LS (8000)				
Minimum	0.139	0.011	0.111	1.270
Maximum	0.574	0.360	0.506	2.071
Mean	0.399	0.070	0.275	1.703
Standard deviation	0.102	0.072	0.104	0.214
TS (8000)				
Minimum	0.155	0.009	0.126	1.295
Maximum	0.574	0.754	0.643	2.267
Mean	0.407	0.100	0.259	1.693
Standard deviation	0.107	0.178	0.106	0.260
LS (400)				
Minimum	0.296	0.528	0.536	0.145
Maximum	1.000	43.61	3.644	1.427
Mean	0.737	5.255	1.764	0.944
Standard deviation	0.194	9.453	0.665	0.287

neighborhood solutions are uniformly distributed around the seed solution. Therefore, the Pareto front can be easily reached using Strategy 1 with random selection of the seed solution.

The optimization results with various numbers of analyses are listed in Table 5.8, where the minimum, maximum, and mean as well as the standard deviation are computed with 30 trials from different initial random seeds. The number of analyses for a single trial of each case is indicated in parentheses, e.g., LS (8000) means a local search with 8000 analyses. The number q of the

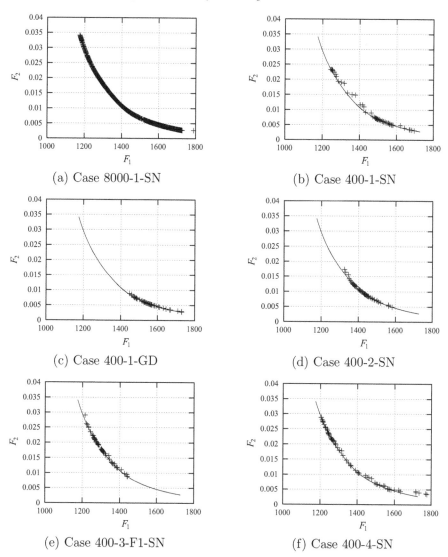

FIGURE 5.23: Pareto solutions of the mathematical example.

neighborhood solutions is 10, which leads to 800 and 40 cycles for the cases of 8000 and 400 analyses, respectively, of updating a seed solution. Strategy 1 is used for selection of the seed solutions.

The results of TS with the tabu list of length 1000 are also listed for comparison purposes. Since the average number of solutions that were rejected with the use of the tabu list with 8000 analyses is only 28, the use of the tabu list does not have any strong effect on the results. Furthermore, about 40% of the solutions by TS and LS are included in the correct Pareto set. Therefore, in the following examples, we use LS without the tabu list.

The results are presented by using an abbreviation based on the number of analyses, the strategy number for selection of the seed solution, and the performance measure used for selection of the results from the 30 trials; e.g., Case 400-1-SN indicates that the number of analyses is 400, and the Pareto set with maximum SN is selected from the 30 results by Strategy 1. For Strategy 3, the objective function to be minimized is indicated as Case 400-3-F1-SN, which shows that the solution with minimum F_1 is to be selected.

The '+' marks in Fig. 5.23(a) are the approximate Pareto solutions of Case 8000-1-SN. Note that the Pareto solutions with good accuracy and diversity can be obtained by 8000 analyses in all 30 trials irrespective of the initial random seed. The performance measures for various cases are listed in Table 5.8. The result of 400 analyses with Strategy 1 with maximum SN is shown in Fig. 5.23(b). As is seen from Table 5.8, the four measures of 400 analyses are worse than those of 8000 analyses; however, it is observed from Fig. 5.23(b) that many diverse approximate Pareto solutions with good accuracy have been found within 400 function evaluations. The Pareto solutions for minimum GD are shown in Fig. 5.23(c), which shows that diversity of solutions is sacrificed if accuracy is increased.

The result of maximum SN by Strategy 2 is shown in Fig. 5.23(d), which also has a small range of solutions. It may be observed from Figs. 5.23(b)–(d) that the solutions with small F_1 may not be found if the two objective functions are equally evaluated. Therefore, we next minimize F_1 using Strategy 3. If we maximize SN, the Pareto set obtained is shown in Fig. 5.23(e), which shows that the number of solutions in the small range of F_2 has been sacrificed as a result of minimizing F_1. Finally, if we use Strategy 4 with the sharing function, the results of maximum SN are as shown in Fig. 5.23(f), where the radius δ is 0.1 for the normalized objective functions F_1^* and F_2^*. In this example, the use of the sharing function is very effective in view of both accuracy and diversity.

5.4.5.2   Optimization of a plane frame

Consider a five-story four-span plane frame, as shown in Fig. 5.21, subjected to static loads, where the horizontal loads (kN) are $(P_1, P_2, P_3, P_4, P_5) = (7.5, 8.4, 10.1, 13.1, 31.5)$, and the vertical loads are $W_1 = 245$ kN and $W_2 = 343$ kN. The elastic modulus is 205.8 kN/mm^2.

In the following example, the total structural volume and the compliance defined in Sec. 1.8 are chosen as the two objective functions $F_1(\mathbf{J})$ and $F_2(\mathbf{J})$, respectively. The members are classified into 15 groups considering symmetry conditions as indicated in Fig. 5.21. The variables J_i $(i = 1, \ldots, 15)$ are selected from the predefined lists of available sections. The lists for the columns and beams are shown in Table A.2 and Table A.3 in Appendix A.8, respectively. Note that the number t_i of the available sections is 8 for columns and

TABLE 5.9: Comparison of performances of frame optimization with various numbers of analyses.

	GD ($\times 10^{-3}$)	SP ($\times 10^{-2}$)	SN
Strategy 1, 5000			
Minimum	1.349	0.968	1.190
Maximum	2.784	3.610	2.299
Mean	1.794	1.576	1.815
Standard deviation	0.314	0.516	0.207
Strategy 1, 200			
Minimum	9.215	1.112	0.584
Maximum	25.52	3.674	1.463
Mean	14.31	2.198	1.033
Standard deviation	0.351	0.671	0.208
Strategy 2, 200			
Minimum	7.765	0.923	0.555
Maximum	33.17	4.931	1.314
Mean	12.46	2.448	0.988
Standard deviation	4.342	0.897	0.180
Strategy 3, minimum F_1, 200			
Minimum	7.733	0.922	0.721
Maximum	30.16	4.865	1.467
Mean	15.82	2.501	1.072
Standard deviation	5.495	0.823	0.191
Strategy 3, minimum F_2, 200			
Minimum	7.341	1.188	0.633
Maximum	36.88	6.248	1.275
Mean	14.32	2.509	0.955
Standard deviation	5.320	0.998	0.141
Strategy 4, 200			
Minimum	7.140	1.186	0.859
Maximum	27.81	6.658	1.650
Mean	13.10	2.261	1.207
Standard deviation	4.074	1.008	0.222

9 for beams.

The range parameters of the objective values in (5.90) are $\delta_1 = 4.0$ and $\delta_2 = 30.0$. A uniform random number $0 \leq r < 1$ is generated, and the section is increased or decreased by 1 for $r \geq 0.5$ or $r < 0.5$, respectively. If the value of J_i generated by the random number turns out to be less than 1 or greater than t_i, then J_i is replaced by 1 or t_i, respectively. All variables are modified simultaneously when generating a neighborhood solution.

Thirty sets of Pareto solutions are generated from different initial random seeds. The solid line in each of the following figures shows the Pareto front obtained by carrying out a local search 90 times with 5000 analyses, which is conceived as a good approximate Pareto front. The optimization results

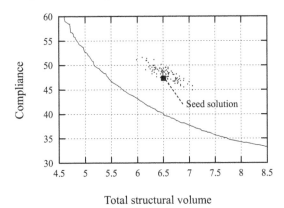

FIGURE 5.24: Distribution of the neighborhood solution of the plane frame.

with various numbers of analyses are listed in Table 5.9, where the minimum, maximum, mean, and standard deviation are computed from 30 trials from different initial random numbers. Note that ER is not computed, because the accurate Pareto set is not available. The number q of neighborhood solutions is 10 for 5000 analyses and 4 for 200 analyses. It is confirmed from Table 5.9 that a smaller number of analyses leads to larger GD and SP as well as smaller SN.

The optimization results are presented by using an abbreviation, e.g., Case 200-2-SN and Case 400-3-F2-GD, in the same manner as the examples of the mathematical problem. The '+' marks in Fig. 5.24 show the 100 neighborhood solutions in the objective function space of a seed solution indicated by the filled square. In this example, there is no neighborhood solution that improves both of the two objective functions. Therefore, it seems that approaching the Pareto front is more difficult for this frame optimization problem than for the mathematical problem.

The result of Case 5000-1-SN is shown in Fig. 5.25(a), which shows that the Pareto set of good accuracy and diversity has been found. In contrast, no solution was found in the correct set of Pareto solutions for Case 200-1-GD, as shown in Fig. 5.25(b). If we decrease q to 2 to increase the number of selection steps of the seed solution for Case 200-1-GD, it turns out that there is no candidate solution before reaching the final 200th step. The result of Case 200-2-SN is shown in Fig. 5.25(c). In this case, the use of Strategy 2 does not have a significant effect, as expected from the distribution of the neighborhood solutions in Fig. 5.24. Fig. 5.25(d) shows the Pareto set for Case 200-3-F1-SN, which verifies that the solutions with small F_1 are successfully found. In contrast, if we minimize F_2, then the solutions with small F_2 are found as shown in Fig. 5.25(e). Finally, the result of Case 200-4-SN using the sharing

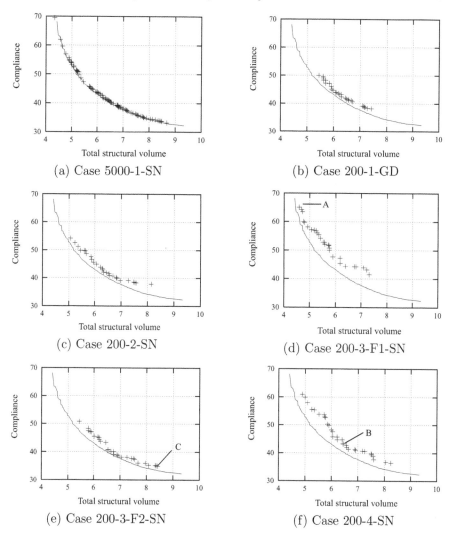

(a) Case 5000-1-SN

(b) Case 200-1-GD

(c) Case 200-2-SN

(d) Case 200-3-F1-SN

(e) Case 200-3-F2-SN

(f) Case 200-4-SN

FIGURE 5.25: Pareto solutions of a plane frame.

function with $\gamma = 0.1$ is shown in Fig. 5.25(f). As is seen, Pareto solutions with good diversity can be found using the sharing function, although accuracy is sacrificed. The average number of approximate Pareto solutions for the 30 trials is 26.8. Therefore, a good number of solutions are found within 200 analyses.

The frames corresponding to the solutions A, B, and C indicated in the objective function space in Figs. 5.25(d)–(f) are shown in Fig. 5.26, where the width of each member is proportional to its cross-sectional area. It is seen from these figures that the cross-sectional areas are almost uniform if

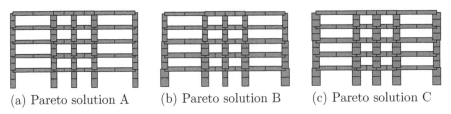

(a) Pareto solution A   (b) Pareto solution B   (c) Pareto solution C

FIGURE 5.26: Cross-sectional areas of Pareto optimal solutions.

the total structural volume is small, and an increase of the total structural volume in the Pareto set leads to an increase of cross-sectional areas from the lower stories. This way, the designer can select the most preferred design from the set of Pareto solutions in view of other structural and nonstructural performance measures.

5.5   Multiobjective seismic design of building frames

5.5.1   Introduction

Building frames have been conventionally designed in the framework of specification-based design, where the bounds on geometries and material properties are given in design codes for the members and connections. However, since the 1990s, performance-based design (PBD) has been widely accepted as a structural design philosophy in which the design criteria are expressed in terms of achieving explicit performance requirements when the structure is subjected to various specified levels of seismic motions (Mahin, Malley, and Hamburger 2002). The structural performance of an ordinary building frame is usually defined as: (1) resist an occasional strong earthquake without structural damage, (2) allow repairable structural damage against a rare major earthquake, and (3) resist a maximum credible earthquake without total collapse.

The performance measures may include the response stresses, maximum load-carrying capacity, interstory drifts, responses and energy dissipation at the limit state or the target (demand) damage state, etc. In the simple procedure of allowable stress design, the stresses under design loads should be within the specified bound, and the ductility ratio obtained by pushover analysis under monotonic loading is often used as the performance measure. However, for PBD, it is not reasonable to define performance by the ductility ratio of the story, because yielding may not lead to a total collapse, and a frame can resist seismic excitation by energy dissipation under cyclic plastic deformation (Leelataviwat, Goel, and Stojadinović 2002). Therefore, performances against seismic motions are defined in view of displacements or global deformation under specified seismic motions (Whittaker, Constantinou, and Tsopelas 1998). Xue and Chen (2003) presented a displacement-based PBD using a capacity-spectrum method. Hasan, Xu, and Grierson (2002) presented a method of pushover analysis for measuring the plasticity factors at the demand deformation level. However, time-history analysis is so far believed to be the most accurate methodology for evaluating structural performance in the inelastic range against seismic motions.

Optimization methods can be effectively used for PBD, because structural performances can be naturally incorporated as objective functions or constraints into the optimization problem (Foley 2002). Bhatti and Pister (1981) may be cited as the first paper that considered responses for two levels of seismic motions in the optimization problem. Ganzerli, Pantelides, and Reaveley (2000) utilized the convex model to account for uncertainty of the input motions. Since structural responses should be evaluated many times in the optimization process, the seismic responses may be approximately evaluated by a response spectrum approach. Mohammadi, El Naggar, and Moghaddam (2004) presented an equivalent linearization approach with an inelastic response spectrum for PBD. The capacity spectrum approach is also commonly used, because the fundamental eigenmode dominates in the response of a building frame (Freeman 2004). However, rapid development in computer technology has enabled us to use time-history analysis for a mathematical programming approach to structural optimization (Balling, Pister, and Ciampi 1983).

As mentioned earlier, several performances should be simultaneously considered in PBD. Therefore, the optimization problem should be formulated as a multiobjective programming (MOP) problem. Liu, Wen, and Burns (2005) presented a multiobjective optimization method for PBD of frames. Xu, Gong, and Grierson (2006) presented a seismic multiobjective optimization method considering uniform interstory drift under equivalent static loads as one of the objective functions.

In MOP for PBD, some of the performance measures are constrained strictly as *hard constraints* based on a building code, i.e., those performance measures should not exceed the specified upper bounds regardless of the structural cost. On the other hand, some responses and performance measures called *soft targets* are to be minimized or maximized only when other targets are not too much sacrificed. Note that a soft target is equivalent to the *flexible constraint* for the constraint satisfaction problem within the framework of decision making in the fuzzy set theory (Dubois and Fortemps 1999).

In this section, an optimization method based on time-history analysis and MOP is presented for PBD of steel frames (Pan, Ohsaki, and Kinoshita 2007).

5.5.2   Formulation of the multiobjective programming problem

Optimal designs are found for plane frames subjected to recorded ground motions. The following performance measures may be considered:

1. Total structural volume that is regarded as representing the structural material cost; a more rigorous definition of the cost, which was described in Sec. 5.1.5, may be used, if possible.

2. Maximum interstory drift angles under seismic motions scaled to Levels 1, 2, and 3, corresponding to occasional, rare, and maximum credible seismic motions.

3. Maximum floor accelerations under seismic motions scaled to Levels 1, 2, and 3.

Since the upper bounds for the maximum interstory drifts are given strictly by the design code, they are regarded as *hard constraints* and are assigned as inequality constraints in the optimization problem. In contrast, the total structural volume and the maximum floor accelerations should preferably be minimized, and their bounds are not given explicitly. Hence, they are regarded as soft targets that are included as objective functions of an MOP problem.

The beams and columns are classified into groups, and the members in each group have the same cross-sectional properties. Let A_i denote the cross-sectional area of the members in the ith group. From the available standard sections of the square tube columns and wide flange beams, the second moment of inertia I_i and the plastic modulus Z_i^{P} for the members in the ith group are defined as functions of A_i, as presented in Appendix A.8.

Let L_i denote the sum of the lengths of members in the ith group. The total structural volume $V(\mathbf{A})$ is given as a function of the design variable vector $\mathbf{A} = (A_1, \ldots, A_m)^\top$ as

$$V(\mathbf{A}) = \sum_{i=1}^{m} A_i L_i \qquad (5.101)$$

where m is the number of groups. The maximum interstory drift angle of the jth story and the maximum acceleration of the jth floor against the Level k input motion are denoted, respectively, by $d_j^{(k)}(\mathbf{A})$ and $a_j^{(k)}(\mathbf{A})$, which are functions of $\mathbf{A}$. Note that $d_j^{(k)}(\mathbf{A})$ and $a_j^{(k)}(\mathbf{A})$ are defined as the largest values, respectively, among the earthquake motions of the specified level if multiple motions are considered. The maximum floor acceleration $a_{\max}^{(k)}(\mathbf{A})$ among all floors for the Level k input is given as

$$a_{\max}^{(k)}(\mathbf{A}) = \max_{j=1,\ldots,n^{\mathrm{F}}} a_j^{(k)}(\mathbf{A}) \qquad (5.102)$$

where n^{F} is the number of stories (floors). The upper bound for $d_j^{(k)}(\mathbf{A})$ is denoted by $d^{(k)\mathrm{U}}$. The multiobjective optimization problem considering input motions of three levels is formulated as

$$\text{Minimize} \quad V(\mathbf{A}), \ a_{\max}^{(1)}(\mathbf{A}), \ a_{\max}^{(2)}(\mathbf{A}), \ \text{and} \ a_{\max}^{(3)}(\mathbf{A}) \tag{5.103a}$$

$$\text{subject to} \quad d_j^{(k)}(\mathbf{A}) \le d^{(k)\mathrm{U}}, \quad (j = 1, \dots, n^{\mathrm{F}}; \ k = 1, 2, 3) \tag{5.103b}$$

$$A_i^{\mathrm{L}} \le A_i \le A_i^{\mathrm{U}}, \quad (i = 1, \dots, m) \tag{5.103c}$$

where A_i^{L} and A_i^{U} are the lower and upper bounds for A_i, respectively.

5.5.3 Optimization method

Responses to the recorded motions are computed by time-history analysis considering geometrical and material nonlinearities. Note that the sensitivity coefficients of maximum responses are discontinuous with respect to the design variables, because the occurrence time of the maximum response varies discontinuously due to design modification. It is possible to give many constraints at discretized time steps to avoid difficulties due to discontinuity (Hsieh and Arora 1984; Paeng and Arora 1989; Reemtsen and Rückmann 1998). For example, the constraint on $d_j^{(k)}(\mathbf{A})$ can be replaced as

$$d_j^{(k)}(\mathbf{A}, t_i) \le d^{(k)\mathrm{U}}, \quad (i = 1, \dots, n^{\mathrm{t}}) \tag{5.104}$$

where t_i is the ith specified time step, and n^{t} is the number of time steps at which the constraints are given. However, the computational cost for this formulation is very large, because we should have many constraints for an accurate estimation of the maximum acceleration. Therefore, the formulation of Problem (5.103) is used, and a finite difference approach is utilized for the sensitivity analysis of maximum responses in the numerical examples. Note that nonlinear programming (NLP) with line search can converge to a local optimum even though the sensitivity coefficients are discontinuous. There have been many studies on sensitivity analysis of elastoplastic responses (Ohsaki and Arora 1994), including dynamic responses (Kleiber 1997). However, these methods are computationally expensive for optimization of complex structures.

Furthermore, the optimization problem considered here is highly nonlinear, and there may exist many local optimal solutions. However, an approximate solution can be successfully obtained using an NLP algorithm starting with several initial solutions and selecting the best solutions among the converged solutions (Kim, Haftka, Mason, Watson, and Grossman 2002).

A constraint approach is used for solving MOP problem; see Appendix A.4 for details. If minimization of $V(\mathbf{A})$ is more important than that of $a_{\max}^{(k)}(\mathbf{A})$, the following single objective problem (SOP) is first solved:

$$\text{Minimize} \quad V(\mathbf{A}) \tag{5.105a}$$

$$\text{subject to} \quad d_j^{(k)}(\mathbf{A}) \le d^{(k)\mathrm{U}}, \quad (j = 1, \dots, n^{\mathrm{F}}; \ k = 1, 2, 3) \tag{5.105b}$$

$$A_i^{\mathrm{L}} \le A_i \le A_i^{\mathrm{U}}, \quad (i = 1, \dots, m) \tag{5.105c}$$

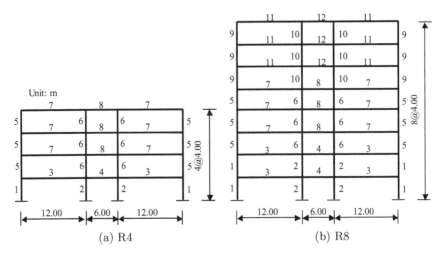

FIGURE 5.27:   Three-span plane frame models R4 and R8.

Let $\widetilde{V}$ denote the optimal objective value of Problem (5.105). A small relaxation parameter for $V(\mathbf{A})$ is denoted by ΔV. We may increase $V(\mathbf{A})$ to $\widetilde{V} + \Delta V$ if one of $a_{\max}^{(r)}(\mathbf{A})$ $(r \in \{1, 2, 3\})$ can be drastically reduced. Suppose minimizing $a_{\max}^{(s)}(\mathbf{A})$ is more important than minimizing $a_{\max}^{(r)}(\mathbf{A})$ $(r \neq s)$. Then, the following SOP is next solved:

$$\text{Minimize} \quad a_{\max}^{(s)}(\mathbf{A}) \tag{5.106a}$$

$$\text{subject to} \quad d_j^{(k)}(\mathbf{A}) \leq d^{(k)\mathrm{U}}, \quad (j = 1, \ldots, n^{\mathrm{F}}; \ k = 1, 2, 3) \tag{5.106b}$$

$$V(\mathbf{A}) \leq \widetilde{V} + \Delta V \tag{5.106c}$$

$$A_i^{\mathrm{L}} \leq A_i \leq A_i^{\mathrm{U}}, \quad (i = 1, \ldots, m) \tag{5.106d}$$

Problem (5.106) can be solved consecutively; i.e., we next minimize $a_{\max}^{(t)}(\mathbf{A})$ $(t \neq s)$ under constraint on the relaxed upper bound of $a_{\max}^{(s)}(\mathbf{A})$.

5.5.4   Numerical examples

5.5.4.1   Descriptions of plane frame models

Optimal designs are found for two three-span plane steel frames, as shown in Figs. 5.27(a) and (b), designated R4 and R8, respectively. Weights of 60 kN/m are distributed on the floors. Considering the symmetry and practical requirements in the construction process, the columns and beams are classified into 8 and 12 groups for R4 and R8, respectively, as shown in Fig. 5.27, where the members in each group have the same cross-sectional properties. The cross-sections of columns and beams are taken to be square tubes and wide flanges, respectively. The second moment of inertia I_i and the plastic modulus

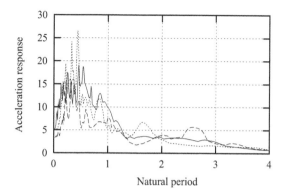

FIGURE 5.28: Pseudoacceleration response spectra for 5% damping of the adopted ground motions scaled to Level 2; solid line: El Centro NS, dashed line: Hachinohe NS, dotted line: Taft EW.

Z_i^{p} are assumed to be functions of A_i, as shown in Appendix A.8. The upper and lower bounds for A_i are 0.1600 m^2 and 0.0060 m^2, respectively, for all the groups of columns and beams.

Three recorded near-fault ground motions are adopted, namely, the NS component of 1940 El Centro, the NS component of 1968 Hachinohe, and the EW component of 1952 Taft. Three performance levels, designated as Levels 1, 2, and 3, commonly adopted in Japanese design practice are considered. The peak ground velocities (PGVs) are specified as 0.25 m/s, 0.50 m/s, and 0.75 m/s, respectively, for Levels 1, 2, and 3. Hence, $d_j^{(k)}(\mathbf{A})$ and $a_j^{(k)}(\mathbf{A})$ for each seismic level are the maximum values among the three recorded motions. The pseudoacceleration response spectra of the adopted ground motions scaled to Level 2 are plotted in Fig. 5.28 for 5% damping. The upper bounds $d^{(k)\mathrm{U}}$ of the maximum interstory drift angles are 0.005, 0.01, and 0.02 for Levels 1, 2, and 3, respectively, for all stories.

A structural analysis program called CLAP (Ogawa and Tada 1994) is used for nonlinear dynamic response analysis. This program adopts the concentrated plastic hinges assigned at member ends, where the interaction between axial force and bending moment is considered. Gravity loads are applied prior to the dynamic analysis, and its influence on plastification and the $P-\Delta$ effect is included.

The modified method of feasible directions in library DOT Ver. 5.0 (VR&D 1999) is used for optimization, and the forward finite difference approach is used for design sensitivity analysis. Because the problem considered here has many local optimal solutions, optimization is carried out starting with 15 different initial solutions, where uniform random numbers are generated to assign the initial values of the design variables within the feasible regions.

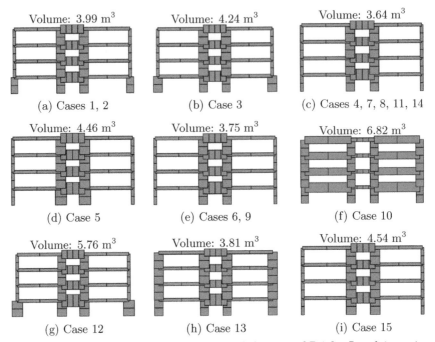

FIGURE 5.29: Single-objective optimal designs of R4 for Level 1 motions starting with 15 different initial solutions.

5.5.4.2 Single-objective optimization

We first solve the single-objective optimization problem for minimizing the total structural volume under constraints on interstory drift angles against Level 1 input only in order to verify the convergence property of the optimization algorithm. The optimal designs of R4 are plotted in Figs. 5.29(a)–(i) for the 15 different initial solutions indicated by Cases 1–15. The width of each member in the figures is proportional to its cross-sectional area. As is seen, nine different local optimal solutions have been obtained. The optimal objective values are also shown in Figs. 5.29(a)–(i). Cases 4, 7, 8, 11, and 14 converge to the identical optimal design that is also the best solution. Hence, this solution is taken as the approximate global optimal solution. In each of the following optimization problems, the best solution is first selected from the local solutions with 15 randomly selected initial solutions.

Optimal solutions for ground motions of Levels 1, 2, and 3 separately are as shown in Figs. 5.30(a), (b), and (c), respectively. The total structural volume of R4 is 3.64 m^3 for Level 1, which is 1.28 and 1.62 times as large as those for Levels 2 and 3, respectively. Similarly, the total structural volume of R8 is 9.01 m^3 for Level 1, which is 1.55 and 1.87 times as large as those for Levels 2 and 3, respectively. Note that both of the PGV and allowable drift angle for Level 2 are twice as large as those for Level 1; however, the total

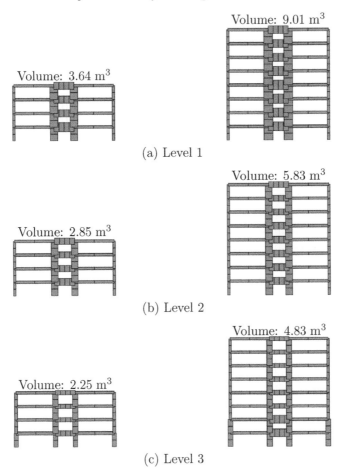

Volume: 9.01 m³

Volume: 3.64 m³

(a) Level 1

Volume: 5.83 m³

Volume: 2.85 m³

(b) Level 2

Volume: 4.83 m³

Volume: 2.25 m³

(c) Level 3

FIGURE 5.30: Single-objective optimal designs of R4 and R8 for Levels 1, 2, and 3.

structural volume of Level 2 is smaller than that of Level 1 owing to plastic energy dissipation.

The interior columns and beams have larger cross-sections than the exterior members for both structures for all input levels. This is mainly because increasing the cross-sectional areas of the beams and columns at shorter spans is more effective for achieving large lateral stiffness than increasing those of longer spans. However, such an effect is less significant for Levels 2 and 3 than for Level 1. The cross-sections of lower exterior columns have large values for Level 3, because the lowest exterior columns have the largest axial force and bending moment due to lateral dynamic loads as well as the $P - \Delta$ effect.

The members in the lower stories have larger cross-sections than those in

TABLE 5.10:   Maximum acceleration $a^{\max}$ (m/s^2) and first
natural period T (s) of the multiobjective optimal designs of R4
and R8 with relaxed upper-bound total structural volume.

	R4		R4	
	$a^{\max}$	T	$a^{\max}$	T
Single objective	10.68	0.56	7.05	0.98
Relax 10%	9.96	0.53	5.02	1.26
Relax 20%	9.50	0.48	4.80	1.27

the higher stories; i.e., no additional constraint is needed so that the column and beam have smaller cross-sections, respectively, than those directly below them, because the upper-bound drift angles $d^{(k)\mathrm{U}}$ are the same for all stories, whereas the shear and axial forces in the lower stories are larger than those in the higher stories. We can see from Fig. 5.30 that the columns generally have larger cross-sections than the beams, which conforms with the design philosophy of *strong column-weak beam*.

Optimization is next carried out considering Levels 1, 2, and 3 inputs simultaneously. The optimal objective values for R4 and R8 are 3.65 m^3 and 9.04 m^3, respectively. The distribution of cross-sectional areas is similar to Fig. 5.30(a) considering Level 1 only, because constraints on the interstory drift angle (0.005) of Level 1 are significantly stricter than those specified in Levels 2 and 3.

5.5.4.3 Multiobjective optimization

The constraint approach to MOP is applied considering Level 1 input only. We investigate the results of relaxation of the total structural volume by 10% and 20%; i.e., ΔV in Problem (5.106) is $0.1\tilde{V}$ and $0.2\tilde{V}$, respectively. The maximum floor accelerations $a^{\max}$ (m/s^2) are listed in Table 5.10. As is seen, the maximum floor accelerations are reduced by 6.7% and 11%, while sacrificing 10% and 20% of the structural volume, respectively, for R4, whereas they are reduced by 30% and 32% for R8.

The values of T (s) are 0.56 and 0.98 for the original designs of R4 and R8, respectively. They are shifted to 0.53 and 1.26 by relaxing 10% of the structural volume for R4 and R8, respectively, and to 0.48 and 1.27 by relaxing 20% of the structural volume. The maximum floor acceleration is reduced mainly due to the shift of the natural period. As is seen from the response spectra in Fig. 5.28, the response acceleration is a roughly decreasing function of T in the region $T > 1.0$; hence, a large reduction in acceleration is observed for R8 as a result of the increase of T from 0.98 to 1.26.

Chapter 6

Optimization of Spatial Trusses and Frames

In this chapter, we present various optimization results of spatial frames, namely, latticed domes and long-span arches. Following the historical review in Sec. 6.1, we carry out sensitivity analysis and optimization of arch-type trusses and a double-layer cylindrical grid in Sec. 6.2, as illustrative examples. In Sec. 6.3, single-point-search heuristic approaches, e.g., greedy method, simulated annealing, and tabu search, are applied to the optimal design of a spatial frame, and their performances are compared. In Sec. 6.4, an approach is presented for incorporating the designer's preference of shape of an arch-type frame that is described using a Bézier curve. Multiobjective shape optimization of a single-layer latticed shell is presented in Sec. 6.5. In Sec. 6.6, a method based on the genetic algorithm is presented for configuration optimization of arch-type trusses incorporating explicit geometrical constraints. In Sec. 6.7, a parametric programming approach is presented for estimating the effect of spatial variation of seismic motions on optimal solutions. Finally, a substructure approach is presented in Sec. 6.8 to optimize a roof truss without carrying out analysis of the whole structure at each optimization step.

6.1 Introduction

Spatial trusses and frames, including arches, reticulated shells, and cable-supported frames, are designed and constructed for covering large spaces for stadiums, arenas, etc. For regular building frames, the locations of members are determined by architects mainly in view of planning of the floors, and the cross-sectional properties of the members are determined by engineers. In contrast, mechanical properties such as displacements, stresses, and buckling loads against specified design loads play key roles for determination of the shapes and topologies of spatial trusses and frames. Although the preferred shape is drawn by an architect, the beauty of a spatial structure is closely related to the efficiency in the load-carrying capacity of the structure. Therefore, structural optimization plays an important role in the design of spatial structures.

Approaches to designing mechanically efficient spatial structures are classified into *form finding* and *structural optimization*. The former is aimed at designing the shape of flexible structures, including membrane roof structures and cable networks, where computational approaches are developed for obtaining, e.g., the catenary and minimal surface that are optimal for simple loading conditions (Otto 1967; Krishna 1979). Computational approaches were also developed for finding an ideal shape of equilibrium under complex and realistic boundary conditions (Haber and Abel 1982). In contrast, structural optimization was mainly developed in the fields of mechanical engineering and aeronautical engineering (Shield 1960; Haug and Cea 1981; Bennett and Botkin 1986; Guillet, Noël, and Léon 1996). In this book, we focus on structural optimization methods for application to structures in civil and architectural engineering. It should be noted here that form finding and structural optimization were recently combined to develop a field called *structural morphology*. Readers can consult many books and papers published by researchers in the society of spatial structures, namely, the International Association of Shell and Spatial Structures, e.g., Pugnale and Sassone (2007), Kimura and Ohmori (2008).

Because spatial trusses and frames have many nodes and members, their optimization methods have been developed in accordance with the development of computer technologies. Since the 1970s, many papers have been published on the optimization of spatial trusses; e.g., Pedersen (1973), Saka and Ulker (1991), Krishnamoorthy, Venkatesh, and Sudarshan (2002), Saka (2007). Recently, some *evolutionary* strategies have been applied to find optimal cross-sectional areas, topology, and geometry of spatial frames (Ebenau, Rottschäfer, and Thierauf 2005; Rajasekaran, Mohan, and Khamis 2004). Kaveh, Azar, and Talatahari (2008) applied a heuristic approach called ant colony optimization to stiffness design of spatial trusses under stress constraints. Lemonge, Barbosa, and Fonseca (2009) optimized shape and cross-sectional areas of latticed domes using a genetic algorithm with constraints on the number of different cross-sections to reduce construction cost.

In the design process of spatial frames, similar to regular building frames, cross-sectional properties are often selected from the list or catalog of standard sections. Therefore, the optimization problem is formulated as a combinatorial problem, for which the heuristic approaches presented in Secs. 1.12, 5.4, and Appendix A.3 can be effectively used for optimizing real-world structures.

Since the designer's preference plays a key role in the shape design of a spatial structure, its design problem can be formulated as a multiobjective optimization problem considering the geometrical and mechanical properties as the objective functions. Ohsaki, Nakamura, and Kohiyama (1997) presented an optimization approach to the design of a double-layer dome truss considering the properties of the surface and the curves formed by the members. Ohsaki and Hayashi (2000) developed an approach to shape optimization of ribbed shells considering the fairness metrics of the surface (Roulier and

Rando 1994). Their method was extended to a multiobjective programming approach for roundness of shape and mechanical efficiency of the structure (Ohsaki, Ogawa, and Tateishi 2003; Fujita and Ohsaki 2009).

Another unique aspect of designing spatial frames is that the safety against buckling often turns out to be the most critical design requirement. However, optimization against buckling is out of scope of this book, because it was extensively reviewed in the previous book by the author (Ohsaki and Ikeda 2007).

6.2   Seismic optimization of spatial trusses

6.2.1   Introduction

The concept called performance-based design has been proposed in countries prone to seismic risk to ensure servicability of structures and life safety, for specified load levels (Mahin, Malley, and Hamburger 2002). Among the many types of structures, spatial structures are very important as the evacuation facilities at an event of seismic disaster. The key aspects of seismic design of spatial structures are summarized as follows:

1. Vertical motions are not negligible in comparison to horizontal motions for evaluation of seismic responses. Furthermore, vertical responses due to horizontal motions should be considered for a curved roof structure, as shown in Sec. 6.2.3. The constraints should be given for the vertical accelerations of the roof so that the hanging nonstructural components and devices, including ceiling and lights, are not damaged during a severe earthquake.

2. Several modes may dominate in the seismic response, in contrast to a building frame in which only the lowest mode dominates. It is important to incorporate the effect of higher modes especially in acceleration responses.

3. The effects of spatial variation of seismic motions should be considered for long-span structures, e.g., bridges spanning between supports in different soil conditions.

In the following, illustrative examples are presented for sensitivity analysis of eigenvalues and eigenmodes as well as optimization of spatial trusses under constraints on seismic responses.

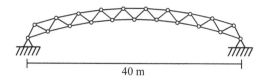

FIGURE 6.1: A 39-bar arch-type truss.

(a) First mode   (b) Second mode

FIGURE 6.2: Eigenmodes of the 39-bar arch-type truss with $\phi = 40$ deg.

6.2.2 Design sensitivity analysis

Let $\mathbf{K}(\mathbf{A})$ and $\mathbf{M}(\mathbf{A})$ denote the $n \times n$ stiffness matrix and mass matrix, respectively, of a truss, which are functions of the vector $\mathbf{A} = (A_1, \ldots, A_m)^{\top}$ of the cross-sectional areas of m members. The mass matrix consists of the structural (member) mass and the nonstructural (nodal) mass. The rth eigenvalue and eigenmode are denoted by Ω_r and $\mathbf{\Phi}_r$, respectively. The sensitivity coefficients of Ω_r with respect to A_i are given as (see Sec. 2.3)

$$\frac{\partial \Omega_r}{\partial A_i} = \mathbf{\Phi}_r^{\top} \left(\frac{\partial \mathbf{K}}{\partial A_i} - \Omega_r \frac{\partial \mathbf{M}}{\partial A_i} \right) \mathbf{\Phi}_r \qquad (6.1)$$

Because $\mathbf{K}$ and $\mathbf{M}$ of a truss are linear functions of A_i, the matrices $\partial \mathbf{K}/\partial A_i$ and $\partial \mathbf{M}/\partial A_i$ are constant. Furthermore, if the structural mass is negligibly small compared with the nonstructural mass, then Ω_r is generally a nondecreasing function of A_i, because $\partial \mathbf{K}/\partial A_i$ is positive semidefinite, except for the case where there exists a member that has a nonzero mass density and no deformation in the eigenmode. The sensitivity coefficient of the rth natural frequency f_r is computed from those of the eigenvalues as

$$\frac{\partial f_r}{\partial A_i} = \frac{1}{8\pi^2 f_r} \frac{\partial \Omega_r}{\partial A_i} \qquad (6.2)$$

Example 6.1
Consider, as an illustrative example, a 39-bar arch-type truss, as shown in Fig. 6.1, where the span length is 40 m, the open angle ϕ of the lower circle is 40 deg, and the difference between the radii of the lower and upper circles is 2 m. The lengths of the lower chords, upper chords, and diagonals are the same, respectively. The material of the member is steel, where the mass density is 7.86×10^{-6} kg/mm^3 and the elastic modulus is 210.0 kN/mm^2. The concentrated mass of 1.0×10^5 kg exists at each node.

FIGURE 6.3: Sensitivity coefficients of f_1 of the 39-bar arch-type truss.

If the cross-sectional areas are 0.01 m^2 for all members, the two lowest frequencies (Hz) are $f_1 = 2.0621$ and $f_2 = 2.2258$, and the corresponding eigenmodes are as shown in the solid lines in Fig. 6.2, while the dotted lines indicate the undeformed shape. Because the ratio of the rise to the span of the arch is relatively small, the first mode is symmetric with respect to the vertical center axis, while the second mode is antisymmetric. Note that the second mode is excited by a horizontal motion; therefore, as is seen from the mode shape in Fig. 6.2(b), the vertical response is induced by the horizontal input at the nodes except the lower node on the center line.

The sensitivity coefficients $\partial f_1/\partial A_i$ of f_1 with respect to the cross-sectional areas are plotted in Fig. 6.3, where the width of each member is proportional to the sensitivity coefficient, which is positive for all members; i.e., an increase of the cross-sectional area of any member results in an increase of f_1. For example, the sensitivity coefficient with respect to the cross-sectional area of member 1, which is the lower chord connected to the support, is 1.2364×10^{-5} mm^{-2}; i.e., an increase of 1 mm^2 of A_1 leads to an increase of f_1 of 1.2364×10^{-5} Hz. This way, by carrying out design sensitivity analysis, the designer can obtain valuable information on the effect of design modification in view of increasing the lowest natural frequency.

6.2.3 Optimization against seismic excitations

Optimal solutions are found for an arch-type truss and a cylindrical double-layer grid subjected to seismic excitations. The response spectrum by Newmark and Hall (1982) is used, and the modal responses are combined using the complete quadratic combination (CQC) method (Wilson, Der Kiureghian, and Bayo 1982) (see Appendix A.7 for details of the response spectrum approach). The seismic motions in the horizontal and vertical directions are simultaneously considered, where the level of the vertical motion is 2/3 of those for the horizontal motions; i.e., the values of the parameters for maximum acceleration, velocity, and displacement of the ground motion in (A.154) in Appendix A.7 are $C_A = 2.01$ m/s^2, $C_V = 0.25$ m/s, and $C_D = 0.1875$ m for horizontal motion, and $C_A = 1.34$ m/s^2, $C_V = 0.1667$ m/s, and $C_D = 0.125$ m for vertical motion. Responses to vertical and horizontal motions evaluated using the CQC method are further combined using the square-root-of-sum-of-square (SRSS) method as presented by Semby and Der Kiureghian (1985), because these two directions coincide with the principal directions of the struc-

(a) Cross-sectional area   (b) First eigenmode

FIGURE 6.4: Optimal solution for $\phi = 40$ deg.

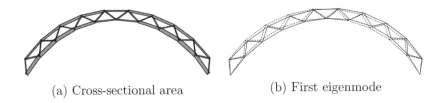

(a) Cross-sectional area   (b) First eigenmode

FIGURE 6.5: Optimal solution for $\phi = 120$ deg.

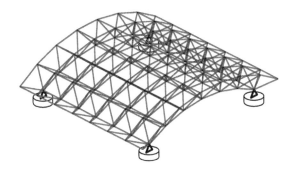

FIGURE 6.6: A 7×7 double-layer cylindrical grid.

ture. The sensitivity coefficients of the modal responses can also be obtained analytically, as shown in Appendix A.7. The cross-sectional areas are assumed to be continuous variables, and sequential quadratic programming is used for optimization.

Optimal solutions are first found for the arch-type truss in Fig. 6.1. The upper bound for the absolute value of stress is 105 N/mm², and the lower-bound cross-sectional area is 500 mm² for all members. The optimal solutions for the open angle $\phi = 40$ deg and 120 deg are shown in Figs. 6.4(a) and 6.5(a), respectively, where the width of each member is proportional to its cross-sectional area, and the first eigenmodes are shown in Figs. 6.4(b) and 6.5(b), respectively.

As is seen, the optimal truss for $\phi = 40$ deg has an antisymmetric first eigenmode, whereas it is symmetric if all the members have the same cross-sectional area, as shown in Fig. 6.2(a). The two lowest frequencies of the optimal truss with $\phi = 40$ deg are 2.1397 Hz and 2.6687 Hz, which are moderately close, and correspond to antisymmetric and symmetric modes, respectively. There-

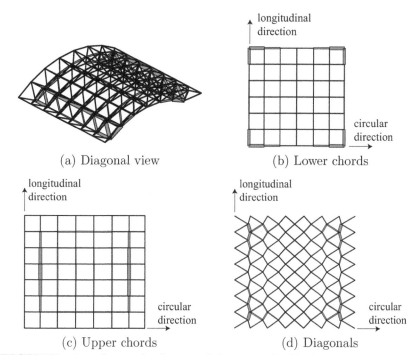

(a) Diagonal view

(b) Lower chords

(c) Upper chords

(d) Diagonals

FIGURE 6.7:   Optimal solution of the 7 × 7 double-layer cylindrical grid.

fore, due to the contribution of the second eigenmode to the responses against vertical motion, the optimal cross-sectional areas have large values around the center, because the second eigenmode has a large curvature change assuming that the arch-type truss represents a continuous arch. On the other hand, the antisymmetric first mode dominates in response of the optimal truss with $\phi = 120$ deg; hence, the cross-sectional areas have large values around the regions between the center and the supports.

We next find the optimal solutions for a 7 × 7 double-layer cylindrical grid, as shown in Fig. 6.6, which has pin supports at the four lower corners and lumped mass of 1000 kg at each node. The members are pin-jointed at the nodes. The span length between the supports is 12 m for both the longitudinal and circular directions. The open angle of the lower cylinder is 80 deg, and the distance between the lower and upper cylinders is 2 m. The lower and upper chords in each direction and the diagonals have the same lengths, respectively.

The material parameters, the lower-bound cross-sectional areas, and the upper-bound stress are the same as those of the arch-type truss. The truss is subjected to three-directional seismic motions. The optimal cross-sectional areas are shown in Figs. 6.7(a)–(d). As is seen, the upper chords in the longitudinal direction between the supports, and the lower chords and the diagonals near the supports have large cross-sectional areas. The lowest three frequen-

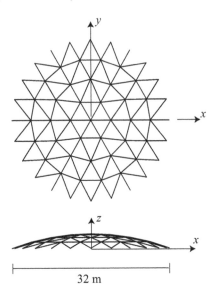

FIGURE 6.8: A single-layer spatial frame.

cies (Hz) are 2.7964, 2.8743, and 3.1496, which are very close. The first mode is symmetric, the second mode is antisymmetric with respect to the plane that is perpendicular to the circles, and the third mode is antisymmetric with respect to the plane that is parallel to the circles. These modes are excited by vertical motion and horizontal motions in two directions, respectively.

6.3   Heuristic approaches to optimization of a spatial frame

In the design process of spatial frames, the cross-sectional properties are often selected from the list or catalog of standard sections. Therefore, the optimization problem is formulated as a combinatorial problem, for which the heuristic approaches presented in Secs. 1.12, 5.4, and Appendix A.3 are effectively used for spatial trusses and frames. In this section, we summarize the results by Ohsaki (2005b) and investigate the performance of single-point-search heuristics to the design of a spatial frame with discrete cross-sectional properties.

Consider a 132-bar single-layer spatial frame, as shown in Fig. 6.8, subjected to static loads. All the nodes are on a sphere with open angle 40 deg and span length 32 m. The members in the longitudinal (meridian) and circumferential directions, respectively, have the same lengths. The members

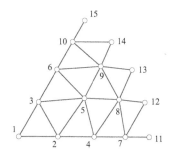

FIGURE 6.9: Node numbers of the single-layer spatial frame.

TABLE 6.1: Nodal coordinates (mm) of the single-layer spatial frame.

Node number	x	y	z
1	0.0	0.0	2821.24
2	4077.22	0.0	2643.22
3	2038.60	3530.98	2643.22
4	8123.42	0.0	2110.52
5	7035.08	4061.70	2110.52
6	4061.70	7035.08	2110.52
7	12,107.78	0.0	1227.22
8	11,377.60	4141.10	1227.22
9	9275.10	7782.74	1227.22
10	6053.88	10,485.64	1227.22
11	16000.0	0.0	0.0
12	15,454.82	4141.10	0.0
13	13,856.40	8000.0	0.0
14	11,313.70	11,313.70	0.0
15	8000.0	13,826.40	0.0

are rigidly jointed at the nodes and pin-jointed at the supports along the lowest circle. The node numbers and nodal coordinates of one of the six equal parts are shown in Fig. 6.9 and Table 6.1, respectively. The elastic modulus is 200 kN/mm^2, and the weight density is 7.7×10^{-5} N/mm^2. A concentrated load of 40 kN is applied in the negative z-direction at each node, and the self-weight of members is added to the nodal loads.

The members have cylindrical cross-sections with the external and internal radii denoted by R and r, respectively. For simplicity, we assume the relation between R and r as

$$r = 0.96R \qquad (6.3)$$

Therefore, all the cross-sectional properties, including the cross-sectional area and the second moment of inertia, are functions of R. The members are

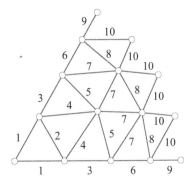

FIGURE 6.10: Member groups of the single-layer spatial frame.

TABLE 6.2: Optimal solutions using various approaches (Part 1).

Member group	Continuous	Greedy (Type 1)	Enumeration	Greedy (Type 2)	Stingy
1	184.2	4	4	4	4
2	60.0	1	1	1	1
3	156.3	3	3	4	3
4	116.2	2	2	1	2
5	109.2	2	2	3	3
6	153.0	3	3	4	4
7	126.9	3	3	3	3
8	117.9	2	2	1	1
9	60.0	2	1	2	2
10	144.5	3	3	3	3
V (m^3)	2.504	3.831	3.766	3.996	4.005
Maximum stress ratio	1.000	0.9957	0.9746	0.9409	0.8961
No. of analyses	1233	1237	489,331	16	14

classified into 10 groups, as shown in Fig. 6.10, and the external radius of the members in group i is denoted by R_i. A list $\mathcal{R}$ is given as follows, from which R_i of each group is to be selected:

$$\mathcal{R} = \{60, 120, 180, 240\} \tag{6.4}$$

where the unit of length is mm.

The objective function to be minimized is the total structural volume V. Let σ^A denote the stress at a member end due to axial force. The maximum stresses at the edge of a member end due to bending around the mutually perpendicular two axes are denoted by σ_1^B and σ_2^B. Because each member has a circular cylindrical section, the maximum absolute value $\sigma^{\max}$ of the stress is estimated by

$$\sigma^{\max} = |\sigma^{\mathrm{A}}| + \sqrt{(\sigma_1^{\mathrm{B}})^2 + (\sigma_2^{\mathrm{B}})^2} \tag{6.5}$$

The upper bound 50 N/mm^2 is given for $\sigma^{\max}$ of each member. In the following, the ratio of $\sigma^{\max}$ to the upper-bound stress is called the *stress ratio*.

An optimal solution is first found by considering R_i as a continuous variable. Optimization is carried out using IDESIGN Ver. 3.5 (Arora and Tseng 1987), where sequential quadratic programming is used. The lower and upper bounds for R_i are 60 mm and 240 mm, respectively, which are equal to the smallest and largest discrete values in the list $\mathcal{R}$. The optimal solution is listed in the second column of Table 6.2. Note that the number of analyses before reaching the optimal solution is 1233, which is very large, because a finite difference approach is used for design sensitivity analysis. The maximum value of the stress ratio of the members is 1.0 for all groups, which means that the optimal solution is fully stressed.

The nearest value of R_i from the continuous optimal solution is selected from the list $\mathcal{R}$ for each group; i.e., $R_1, \ldots, R_{10}$ are 180, 60, 180, 120, 120, 160, 120, 120, 60, 160. Since the solution obtained this way does not satisfy the stress constraints, the value of R_i corresponding to the maximum stress ratio among all groups is increased consecutively until stress constraints are satisfied in all members. This method is denoted as the greedy method (Type 1), and its results are listed in the third column of Table 6.2, where the numbers 1, 2, 3, and 4 correspond to $R_i = 60$, 120, 180, and 240, respectively. The number of analyses is 4 for the iterative correction using the greedy method to which is added 1233 for sequential quadratic programming to result in the total number of analyses $1233 + 4 = 1237$.

To confirm the accuracy of the heuristic methods, the global optimal solution has been found by enumerating all $4^{10} = 1,048,574$ solutions. The result is listed in the fourth column of Table 6.2. Note that the total number of analyses is less than 4^{10}, because structural analysis is not carried out in the enumeration process if the total structural volume is less than that of the current upper-bound solution satisfying stress constraints in all groups. As is seen, the simple greedy method (Type 1) can reach a good approximate solution.

Next we consider a greedy method (Type 2) starting with $R_i = 60$ mm, which is the smallest value in the list $\mathcal{R}$, for all groups. The result is listed in the fifth column of Table 6.2. On the other hand, if we use the stingy method starting with the initial solution $R_i = 240$ mm for all groups and reducing the cross-section consecutively, the result is as shown in the sixth column of Table 6.2. In this case, the stingy method reaches a better solution than the greedy method (Type 2). The number of analyses is very small for both methods.

The result of simulated annealing (SA) is listed in the second column of Table 6.3, where the quadratic exterior penalty function approach described

TABLE 6.3:   Optimal solutions using various approaches (Part 2).

Member group	SA	Random search	TS	Enumeration near continuous solution	Greedy (Type 3)
1	4	4	4	4	4
2	2	1	1	1	1
3	3	3	3	3	4
4	2	2	2	2	2
5	2	2	2	2	3
6	3	3	3	3	4
7	3	3	3	3	3
8	2	2	2	2	1
9	1	2	1	1	2
10	3	3	3	3	3
V (m^3)	3.831	3.766	3.766	3.766	4.157
Maximum stress ratio	0.9980	0.9746	0.9746	0.8441	0.9746
No. of analyses	2000	2000	282	2256	1244

in Appendix A.2.2.4 is used for incorporating the stress constraints, and the penalty parameter is 10^{11}, which is also used in all the examples below that utilize the penalty function approach. The initial value of the temperature parameter is 1, which is multiplied by 0.99 at each iterative step. We tried 10 different random initial solutions with 200 steps for each case. Therefore, the total number of analyses is 2000. The result of a random search is listed in the third column of Table 6.3, where the number of analyses is also 2000. Note that the random search has better performance than SA for this example.

The result of a tabu search (TS) is shown in the fourth column of Table 6.3. The maximum number of steps is 50, and the length of the tabu list is also 50, which means that a solution selected once is not chosen again. A maximum of 20 neighborhood solutions are randomly searched at each iterative step by increasing or decreasing the rank of the section by 1 for each variable. As is seen, the global optimal solution has been found by TS with 282 analyses from a randomly generated initial solution. Note that the same optimal solution has been found from five different initial solutions. The fifth column of Table 6.3 shows the result of enumeration in the neighborhood of the optimal solution with a continuous variable (second column of Table 6.2). For example, $R_i = 180$ and 240 are selected as the candidates for group 1, because $R_1 = 184.2$ for the continuous solution. In this example, there exists a global optimal solution in the neighborhood of the continuous solution.

Finally, a greedy method (Type 3) was tried starting with the largest value of R_i in the list that does not exceed the continuous solution in the second column of Table 6.2. The results are listed in the sixth column of Table 6.3. In this case, no good solution was found by this approach.

TABLE 6.4: Comparison of performances of the heuristic approaches for various lists of available radii.

R^*	60	58	56	54	52	50
Greedy (Type 1)	3.831	3.831	3.836	3.567	3.557	3.186
Enumeration	<u>3.766</u>	<u>3.682</u>	<u>3.565</u>	<u>3.314</u>	<u>3.259</u>	<u>3.092</u>
Greedy (Type 2)	3.996	4.026	3.754	3.709	4.249	4.434
Stingy	4.005	3.743	3.621	4.634	4.297	5.092
SA	3.831	<u>3.682</u>	3.587	<u>3.514</u>	<u>3.259</u>	3.137
Random search	<u>3.766</u>	<u>3.682</u>	3.586	<u>3.314</u>	3.273	3.102
TS	<u>3.766</u>	<u>3.682</u>	<u>3.565</u>	3.478	3.389	<u>3.092</u>
Enumeration near continuous solution	<u>3.766</u>	3.728	3.709	3.721	3.450	3.102
Greedy (Type 3)	4.157	3.831	3.836	3.567	3.557	4.037

The optimization results presented above, however, strongly depend on the geometry of the structure, the available values of R_i, the load level, etc. Therefore, in the following, we parametrically vary the available values of R_i, and compare the performances of the heuristic approaches.

Let R^* denote the unit value of the available cross-sectional areas, and define the list $\mathcal{R}$ as

$$\mathcal{R} = \{R^*, 2R^*, 3R^*, 4R^*\} \tag{6.6}$$

Note that $R^* = 60$ corresponds to the previous example. The optimal objective values for $R^* = 60, 58, 56, 54, 52$, and 50 by each method presented above are listed in Table 6.4, where the underline indicates that the value of the optimal solution coincides with that of the global optimal solution obtained by enumeration. It is observed from Table 6.4 that SA, TS, and random search can reach the global solution or a good approximate solution. The greedy methods and the stingy method may sometimes find a solution that has a very large objective value. Furthermore, it often happens that the global solution does not exist in the neighborhood of the continuous solution.

6.4 Shape optimization considering the designer's preference

6.4.1 Introduction

In the practical design process of long-span trusses and frames, the stresses and displacements under static and dynamic design loads are required to be within the specified bounds defined by building codes. Another important aspect in structural design in civil and architectural design is that the most pre-

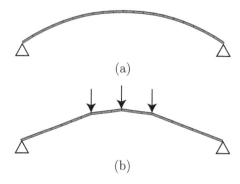

FIGURE 6.11:   Optimal shapes of an arch subjected to static loads; (a) self-weight, (b) concentrated loads.

ferred shape may be presented by the designer or architect prior to structural design considering mechanical properties. Therefore, one of the drawbacks in structural optimization under constraints on mechanical properties only is that the solution often turns out to be an unrealistic design that appears to be far from optimum from the geometrical and aesthetic point of view.

In this section, we consider the shape optimization problem in which optimal locations of the nodes are to be optimized while topology variation is not allowed (Lin, Che, and Yu 1982; Imai and Schmit 1982; Svanberg 1981; Saka 1980; Sadek 1986) (see Sec. 1.10 for simple examples of shape optimization). Smooth curves and surfaces are described efficiently with techniques developed in *computer aided geometric design* (CAGD) (Faux and Pratt 1979; Farin 1988; Rogers and Adams 1990; Barnhill 1994); see Appendix A.6. A method is presented for the shape design of a plane arch-type frame with nodal locations defined in terms of Bézier curves (Ohsaki, Nakamura, and Isshiki 1998). The deviation of the curve associated with the lower nodes from the shape preferred by the designer is considered as one of the objective functions. Compliance is also minimized as the performance measure of the stiffness.

Example 6.2
An arch is traditionally regarded as the ideal shape for carrying vertical loads. However, the optimal shape varies with the distribution of external loads. As an illustrative example, optimal shapes of pin-supported arches are found using a nonlinear programming algorithm. The arch is discretized to 20 beam elements, which have a cylindrical section with the same cross-sectional area and the fixed ratio 0.1 of the thickness to the external radius. Hence, the design variables are the cross-sectional area and the vertical coordinates of the 19 internal nodes between the elements, where the ratio of the rise (height of the center node) to the span is fixed at 0.15. Constraints are given for the maximum edge stresses evaluated at the nodes. The objective function

to be minimized is the total structural volume. A detailed description of the model is omitted here, because the purpose of this example is to illustrate the dependence of the optimal shape on the loading conditions.

The optimal shape under self-weight is shown in Fig. 6.11(a), which is close to the optimal shape of a hanging cable called a catenary if the vertical coordinates are reversed. Therefore, the arch in Fig. 6.11(a) is called a *catenary arch*. For the case in which the three concentrated loads are applied around the center, the optimal shape is as shown in Fig. 6.11(b). As is seen, the optimal shape strongly depends on the load distributions, and the trade-off between the shape preferred by the designer and the mechanically optimal shape should be considered in the practical shape optimization process.

6.4.2 Description of an arch-type frame using a Bézier curve

Let $\mathbf{P}(u) = (x(u), y(u))^\top$ denote a Bézier curve defined by the parameter $0 \le u \le 1$ in the two-dimensional space (x, y). The Bézier curve of order n^{C} is defined in terms of the control points $\mathbf{R}_i = (X_i, Y_i)^\top$ $(i = 0, 1, \ldots, n^{\mathrm{C}})$ as

$$
\begin{aligned}
\mathbf{P}(u) &= \begin{pmatrix} x(u) \\ y(u) \end{pmatrix} \\
&= \sum_{i=0}^{n^{\mathrm{C}}} \mathbf{R}_i B_i^{n^{\mathrm{C}}}(u) \\
&= \sum_{i=0}^{n^{\mathrm{C}}} \begin{pmatrix} X_i \\ Y_i \end{pmatrix} B_i^{n^{\mathrm{C}}}(u)
\end{aligned}
\tag{6.7}
$$

where $B_i^{n^{\mathrm{C}}}(u)$ is the Bernstein polynomial of order n^{C} (see Appendix A.6 for details of the Bézier curve).

Utilizing the *convex hull property*, the shape of the Bézier curve may be indirectly and interactively controlled by modifying the locations of the control points. Another important property of the Bézier curve is that $\mathbf{P}(0) = \mathbf{R}_0$ and $\mathbf{P}(1) = \mathbf{R}_{n^{\mathrm{C}}}$ are satisfied; i.e., the two ends of the curve coincide with the control points. Furthermore, the curve inscribes the control polygon at $\mathbf{R}_0$ and $\mathbf{R}_{n^{\mathrm{C}}}$.

It is straightforward to optimize the Bézier curves by considering the locations of the control points as design variables. For the design of a truss or a frame, however, the locations of the nodes need to be defined after the shape of the curve is given. Suppose $\mathbf{P}(u)$ in (6.7) defines the Bézier curve associated with the lower nodes. A sequence of points on a Bézier curve is specified by a set of parameters $\mathbf{u} = (u_1, \ldots, u_{n^{\mathrm{J}}})^\top$, where n^{J} is the number of nodes of the arch-type frame, including the upper and lower nodes as well as the supports. If the ith node is located on the lower curve, then its coordinate vector is computed as $\mathbf{P}(u_i)$.

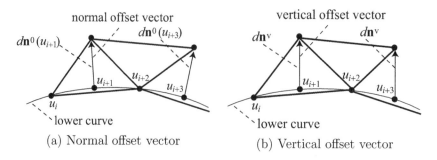

(a) Normal offset vector   (b) Vertical offset vector

FIGURE 6.12:   Process of generating an arch-type frame.

The locations of the upper nodes are defined in terms of the normal offset vector, as illustrated in Fig. 6.12(a), or the vertical offset vector in Fig. 6.12(b) (see Appendix A.6.3 for definitions of the adjoint curves, including the offset curves). Conversely, the upper nodes can be specified on a Bézier curve, and the lower nodes may be defined using the offset vector. Two approaches with the use of normal and constant vertical offset vectors were presented by Ohsaki, Nakamura, and Kohiyama (1997) for double-layer spatial trusses.

A vector $\mathbf{n}(u)$ that is normal to the lower Bézier curve $\mathbf{P}(u)$ is defined as

$$
\begin{aligned}
\mathbf{n}(u) &= \mathbf{T}\frac{\partial \mathbf{P}}{\partial u} \\
&= \mathbf{T}\sum_{i=0}^{n^{\mathrm{C}}} \mathbf{R}_i \frac{\partial B_i^{n^{\mathrm{C}}}(u)}{\partial u}
\end{aligned}
\tag{6.8}
$$

where $\mathbf{T}$ is the rotation matrix of the angle $\pi/2$. Then, the unit normal vector $\mathbf{n}^0$ is given as

$$
\mathbf{n}^0(u) = \frac{\mathbf{n}(u)}{\sqrt{\mathbf{n}^\top \mathbf{n}}}
\tag{6.9}
$$

The unit vector in the y-direction is defined as $\mathbf{n}^{\mathrm{v}} = (0,1)^\top$. Let d denote the prescribed length of the offset vector. Then the nodal coordinate vectors are computed as

$$
\begin{aligned}
&\text{Lower node}: \ \mathbf{P}(u_i) &&\text{(6.10a)}\\
&\text{Upper node}: \ \mathbf{P}(u_i) + d\mathbf{n}^0(u_i) \ \ \text{or} \ \ \mathbf{P}(u_i) + d\mathbf{n}^{\mathrm{v}} &&\text{(6.10b)}
\end{aligned}
$$

The process of generating an arch-type frame is illustrated in Fig. 6.12.

Examples of arch-type frames with vertical and normal offset vectors, respectively, as well as control points for the lower curve, are illustrated in Fig. 6.13, where the numbers of lower and upper nodes, except the two supports, are 17 and 18, respectively. Hence, the parameter values for the ith lower and upper nodes from the left are given as $i/18$ and $(2i-1)/36$, respectively.

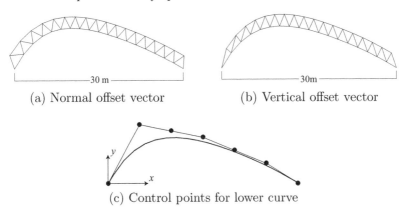

(a) Normal offset vector   (b) Vertical offset vector

(c) Control points for lower curve

FIGURE 6.13: An arch-type frame modeled using a Bézier curve of order 6.

Ohsaki, Nakamura, and Isshiki (1998) used a Bézier function for defining the cross-sectional areas. The Bézier function $A(u)$ of order n^C for the cross-sectional area is given as

$$A(u) = \sum_{i=0}^{n^C} R_i^A B_i^{n^C}(u) \tag{6.11}$$

where R_i^A $(i = 0, 1, \ldots, n^C)$ are the function values at the control points. Suppose member i connects nodes j and k, and let $u_i^A = (u_j + u_k)/2$. Then, the cross-sectional area of member i is given as $A_i = A(u_i^A)$. This way, smoothly distributing cross-sectional areas can be optimized by considering R_i^A as design variables. Note that different Bézier functions should be used for the lower chords, upper chords, and diagonals. In the following numerical examples, however, we assume that cross-sectional areas are fixed at the specified values.

6.4.3 Shape optimization incorporating the designer's preference

A multiobjective programming problem is formulated for obtaining the *trade-off designs* between the deviation from the desired shape and the mechanical performance that is defined by compliance under static loads. Let $\mathbf{K}$ and $\mathbf{U}$ denote the stiffness matrix and the displacement vector under specified loads, which are functions of $\mathbf{X}$ consisting of the variable coordinates of the control points. Note that the control points at the supports are usually fixed. Then, compliance $W(\mathbf{X})$ is defined using $\mathbf{K}(\mathbf{X})$ and $\mathbf{U}(\mathbf{X})$ as

$$W(\mathbf{X}) = \mathbf{U}(\mathbf{X})^\top \mathbf{K}(\mathbf{X}) \mathbf{U}(\mathbf{X}) \tag{6.12}$$

An upper bound W^{U} is given for the compliance so that the frame has enough stiffness against the external loads.

Suppose the shape of the lower curve preferred by the designer or architect is expressed using a Bézier curve as

$$\widetilde{\mathbf{P}}(u) = \sum_{i=0}^{n^{\mathrm{C}}} \widetilde{\mathbf{R}}_i B_i^{n^{\mathrm{C}}}(u) \tag{6.13}$$

In the following, a tilde indicates a specified value. Using (6.7), the deviation $G(\mathbf{X})$ of a Bézier curve $\mathbf{P}(u) = (x(u), y(u))^{\top}$ from $\widetilde{\mathbf{P}}(u) = (\widetilde{x}(u), \widetilde{y}(u))^{\top}$ is defined as follows:

$$
\begin{aligned}
G(\mathbf{X}) &= \int_0^1 (\mathbf{P}(u) - \widetilde{\mathbf{P}}(u))^{\top}(\mathbf{P}(u) - \widetilde{\mathbf{P}}(u)) \mathrm{d}u \\
&= \int_0^1 (x(u) - \widetilde{x}(u))^2 \mathrm{d}u + \int_0^1 (y(u) - \widetilde{y}(u))^2 \mathrm{d}u \\
&= \int_0^1 \left[\sum_{i=1}^{n^{\mathrm{C}}} (X_i - \widetilde{X}_i) B_i^{n^{\mathrm{C}}}(u) \right]^2 \mathrm{d}u \\
&\quad + \int_0^1 \left[\sum_{i=1}^{n^{\mathrm{C}}} (Y_i - \widetilde{Y}_i) B_i^{n^{\mathrm{C}}}(u) \right]^2 \mathrm{d}u \\
&= \sum_{i=1}^{n^{\mathrm{C}}} \sum_{j=1}^{n^{\mathrm{C}}} \left[(X_i - \widetilde{X}_i)(X_j - \widetilde{X}_j) \int_0^1 B_i^{n^{\mathrm{C}}}(u) B_j^{n^{\mathrm{C}}}(u) \mathrm{d}u \right] \\
&\quad + \sum_{i=1}^{n^{\mathrm{C}}} \sum_{j=1}^{n^{\mathrm{C}}} \left[(Y_i - \widetilde{Y}_i)(Y_j - \widetilde{Y}_j) \int_0^1 B_i^{n^{\mathrm{C}}}(u) B_j^{n^{\mathrm{C}}}(u) \mathrm{d}u \right]
\end{aligned}
\tag{6.14}
$$

It is seen from (6.14) that the integration of the basis functions is to be carried out independently of the design variables $\mathbf{X}$.

The multiobjective optimization problem is formulated as

$$\text{Minimize} \quad G(\mathbf{X}) \text{ and } W(\mathbf{X}) \tag{6.15a}$$

$$\text{subject to} \quad \mathbf{X}^{\mathrm{L}} \le \mathbf{X} \le \mathbf{X}^{\mathrm{U}} \tag{6.15b}$$

where $\mathbf{X}^{\mathrm{L}}$ and $\mathbf{X}^{\mathrm{U}}$ are the lower and upper bounds for $\mathbf{X}$, respectively. A set of Pareto optimal solutions is found using the constraint approach (see Appendix A.4 for details). First, we minimize the shape deviation $G(\mathbf{X})$ and the compliance $W(\mathbf{X})$ independently of side constraints (6.15b) on $\mathbf{X}$ by solving the single-objective optimization problem. Then appropriate upper bounds G^{U} and W^{U} for the constraint approach can be specified. For the region where $W(\mathbf{X})$ has a small value, $W(\mathbf{X})$ is minimized as follows under constraint on $G(\mathbf{X})$:

$$\text{Minimize} \quad W(\mathbf{X}) \tag{6.16a}$$

$$\text{subject to} \quad G(\mathbf{X}) \le G^{\mathrm{U}} \tag{6.16b}$$

$$\mathbf{X}^{\mathrm{L}} \le \mathbf{X} \le \mathbf{X}^{\mathrm{U}} \tag{6.16c}$$

In contrast, for the region where $G(\mathbf{X})$ has small values, $G(\mathbf{X})$ is minimized as follows under constraint on $W(\mathbf{X})$:

$$\text{Minimize} \quad G(\mathbf{X}) \tag{6.17a}$$

$$\text{subject to} \quad W(\mathbf{X}) \le W^{\mathrm{U}} \tag{6.17b}$$

$$\mathbf{X}^{\mathrm{L}} \le \mathbf{X} \le \mathbf{X}^{\mathrm{U}} \tag{6.17c}$$

6.4.4 Sensitivity analysis with respect to control points

Problems (6.16) and (6.17) can be solved using a gradient-based optimiza-tion algorithm. Therefore, in this section, the sensitivity coefficients with respect to the design variables X_i and Y_i are derived for the objective and constraint functions. The sensitivity coefficients of the static displacements are not presented here, because they can be found using the standard methods in Sec. 2.2.

If node k is a lower node, the sensitivity coefficients of its coordinates can be obtained directly from (6.7) as

$$\frac{\partial x(u_k)}{\partial X_i} = B_i^{n^{\mathrm{C}}}(u_k), \quad \frac{\partial y(u_k)}{\partial Y_i} = B_i^{n^{\mathrm{C}}}(u_k) \tag{6.18}$$

In the following, the argument u_k is omitted for simplicity. For the upper nodes, the sensitivity coefficients of the offset vector should be computed. Because the vertical offset vector is constant, we derive the sensitivity coeffi-cients of only the normal offset vector. By differentiating (6.8) and (6.9), and using (6.7), we obtain

$$\frac{\partial \mathbf{n}}{\partial X_i} = \mathbf{T} \frac{\partial B_i^{n^{\mathrm{C}}}}{\partial u} \tag{6.19a}$$

$$\frac{\partial \mathbf{n}^0}{\partial X_i} = \frac{1}{\sqrt{\mathbf{n}^{\mathrm{T}} \mathbf{n}}} \frac{\partial \mathbf{n}}{\partial X_i} + \frac{1}{\mathbf{n}^{\mathrm{T}} \mathbf{n}} \left(\mathbf{n}^{0\mathrm{T}} \frac{\partial \mathbf{n}}{\partial X_i} \right) \mathbf{n} \tag{6.19b}$$

For the shape deviation, the following equations are derived from (6.14):

$$\frac{\partial G}{\partial X_i} = 2 \sum_{j=1}^{n^{\mathrm{C}}} \left[(X_j - \tilde{X}_j) \int_0^1 B_i^{n^{\mathrm{C}}}(u) B_j^{n^{\mathrm{C}}}(u) \mathrm{d}u \right],$$

$$\frac{\partial G}{\partial Y_i} = 2 \sum_{j=1}^{n^{\mathrm{C}}} \left[(Y_j - \tilde{Y}_j) \int_0^1 B_i^{n^{\mathrm{C}}}(u) B_j^{n^{\mathrm{C}}}(u) \mathrm{d}u \right] \tag{6.20}$$

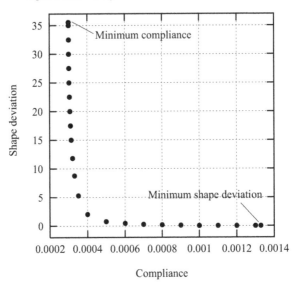

FIGURE 6.14:   Pareto optimal solutions for vertical offset vector, variable y-coordinates.

Note that the integration of basis functions has already been carried out in the process of evaluating $G(\mathbf{X})$.

6.4.5   Numerical examples

Consider a 71-bar plane arch-type frame shown in Fig. 6.13 defined using the Bézier curve of order 6; i.e., the number of control points for defining the lower curve is 7. Because the control points at both ends are fixed at the supports, the number of variable points is 5. The members are rigidly connected at the joints and pin-jointed at the two supports. The span length is 30 m, and the length d of the offset vector is 2 m.

The configuration in Fig. 6.13 is assumed to represent the shape preferred by the designer. Pareto optimal solutions are found for two cases where the y-directional coordinates and (x, y)-coordinates of the control points, respectively, are considered as design variables. Therefore, the numbers of design variables are 5 and 10 for these two cases, respectively. The vertical and normal offset vectors are considered for both cases. Optimal shapes are found by using SNOPT Ver. 7.2 (Gill, Murray, and Saunders 2002), where sequential quadratic programming is used.

The members consist of a circular tube section with external and internal radii of 50 mm and 46 mm, respectively. The arch is subjected to static loads that represent the self-weight. The elastic modulus is 210 kN/mm^2, and the weight density is 77.0 kN/m^3. The bounds for $\mathbf{X}$ are not given. In the fol-

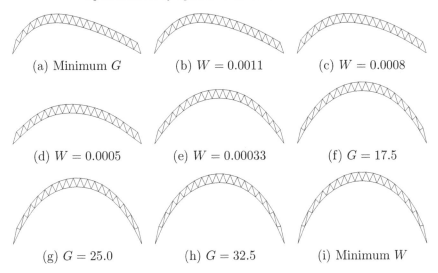

FIGURE 6.15: Optimal shapes for the vertical offset vector, variable y-coordinates.

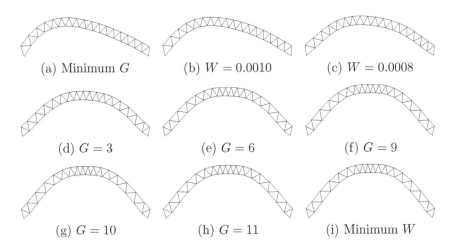

FIGURE 6.16: Optimal shapes for the normal offset vector, variable (x, y)-coordinates.

lowing, the units of length and force are m and kN, respectively, which are omitted for brevity.

The Pareto optimal solutions are first found for the vertical offset vector with only y-coordinates as variables. If the shape deviation G is minimized, the optimal value is 0 and the corresponding value of compliance W is 1.3283×10^{-3}. Since the preferred shape is defined using the Bézier curve of order 6, exactly the same shape as shown in Fig. 6.13 is obtained by minimizing G. On the other hand, if W is minimized, the optimal value of W is 3.0099×10^{-4}, and the corresponding value of G is 35.545.

We apply the constraint approach for generating the set of Pareto solutions. Problem (6.16) is solved first with the upper bound for W decreased as 0.0013, 0.0012, ..., 0.0004, 0.00035, 0.00033, 0.00032. Then Problem (6.17) is solved with the upper bound for G decreased as 35.0, 32.5, ..., 15.0. The Pareto optimal solutions are plotted in the objective function space in Fig. 6.14. As is seen, the Pareto solutions form a convex curve in the objective function space. Figs. 6.15(a)–(i) show the shapes of various Pareto solutions. We can see from these figures that a smaller compliance leads to a circular shape. However, the lengths of the members near the supports become very large if the compliance is minimized with y-directional variables.

If (x, y)-coordinates are considered as variables, the Pareto solutions in the objective function space are almost the same as those in Fig. 6.14. Therefore, no significant effect is found for considering x-coordinates of the control points as additional variables, if the vertical offset vector is used.

Pareto solutions are next generated with the normal offset vector and (x, y)-coordinates as variables. If the shape deviation G is minimized, the optimal value is 0 and the corresponding value of compliance W is 1.2541×10^{-3}, which is a little smaller than the case in which the vertical offset vector is considered with y-directional variables. On the other hand, if W is minimized, the optimal value of W is 5.4461×10^{-4}, and the corresponding value of G is 11.956, which shows that the minimum compliance is larger than that for the vertical offset vector even if the (x, y)-coordinates are considered as variables. The shapes of Pareto solutions are shown in Figs. 6.16(a)–(i). As is seen, the upper and lower chords around the center, as well as the members near the supports, become shorter as a result of minimization of compliance with variable (x, y)-coordinates.

6.5   Shape optimization of a single-layer latticed shell

6.5.1   Introduction

In Sec. 6.4, we presented an approach to shape optimization of plane arch-type frames considering the designer's preference. The smoothness of the surface, as well as the mesh pattern, is also important for designing latticed shells with non-standard shapes. Complex surfaces defined using parametric surfaces are often called free-form surfaces (e.g., Guillet, Noël, and Léon 1996), and the shapes of shell roofs defined using parametric surfaces are often called *free-form shells* (Kimura and Ohmori 2008).

Ramm, Bletzinger, and Reitinger (1993) used Bézier surfaces for shape opti-

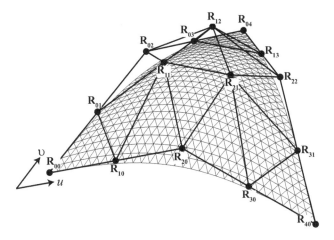

FIGURE 6.17: Relation between the control net and nodal locations of the surface defined using a triangular Bézier patch.

mization of shells. Ohsaki and Hayashi (2000) extended the fairness metrics of surface (Rando and Roulier 1991; Roulier and Rando 1994) and defined some roundness metrics for optimization of ribbed shells. Ohsaki, Nakamura, and Kohiyama (1997) carried out shape optimization of a double-layer roof truss, where the triangular Bézier patch is utilized for modeling smooth surfaces.

In this section, a formulation of the optimization problem and optimization results are presented for single-layer latticed shells modeled using a triangular Bézier patch. Pareto optimal solutions considering global properties of the surface defined by the variance of member lengths and compliance against static loads, respectively, representing geometrical and mechanical properties, can be successfully generated without loss of smoothness of the surface by modifying the control points of the parametric surface.

6.5.2 Description of a latticed shell and formulation of the optimization problem

The surface of the single-layer latticed shell is defined using the Bézier triangle, or triangular Bézier patch, as shown in Fig. 6.17, where the thick lines represent the control net, and the thin lines are members of the frame. The points $\mathbf{R}_{\alpha\beta}$ ($\alpha = 0, \ldots, n^C$; $\beta = 0, \ldots, n^C - \alpha$) are the control points, where n^C is the order of the surface that is 3 in Fig. 6.17 (see Appendix A.6 for details of the triangular Bézier patch).

Let u and v denote the parameters for the surface, which are combined to a vector $\mathbf{u} = (u, v)^\top$. The parameter vector of the point at which the kth node is located is denoted by $\mathbf{u}_k = (u_k, v_k)^\top$; i.e., the coordinates of the kth node are defined as $\mathbf{P}_k = \mathbf{P}(\mathbf{u}_k)$. The thin lines in Fig. 6.17 are obtained from the

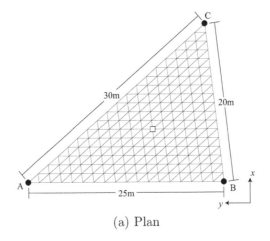

(a) Plan

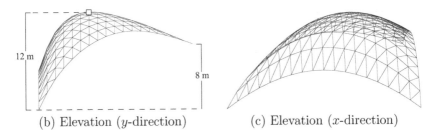

(b) Elevation (y-direction)   (c) Elevation (x-direction)

FIGURE 6.18:   A single-layer latticed dome.

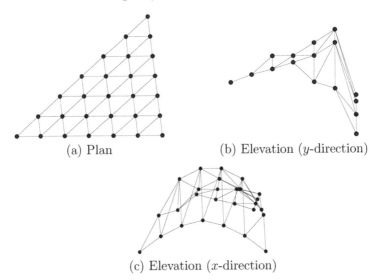

(a) Plan   (b) Elevation (y-direction)

(c) Elevation (x-direction)

FIGURE 6.19:   Bézier patch of the single-layer latticed dome.

uniformly spaced parameters 0.0, 0.05, 0.1, ..., 0.9, 0.95, 1.0 for u_k and v_k. This way, a single-layer frame with triangular mesh can be generated.

An example of a single-layer latticed dome is shown in Fig. 6.18, and the locations of the control points are plotted in Fig. 6.19. The frame is supported at the three corners A, B, and C, at which the control points exist owing to the basic property of the Bézier surface. Therefore, the locations of the supports can easily be fixed during shape optimization with the coordinates of the control points as design variables.

Let $\mathbf{K}$ and $\mathbf{U}$ denote the stiffness matrix and the displacement vector under the specified load vector, respectively. The vector consisting of the variable coordinates of the control points is denoted by $\mathbf{X}$. Then $\mathbf{K}$ and $\mathbf{U}$ are functions of $\mathbf{X}$. The compliance $W(\mathbf{X})$ defined as $W(\mathbf{X}) = \mathbf{U}^{\top}(\mathbf{X})\mathbf{K}(\mathbf{X})\mathbf{U}(\mathbf{X})$ is to be minimized in the following optimization problem so that the frame has sufficient stiffness against the external loads.

Let $L_i(\mathbf{X})$ denote the length of the ith member of the frame consisting of m members. The average value of $L_i(\mathbf{X})$ among all members is denoted by $L^{\mathrm{ave}}(\mathbf{X})$. The variance $D(\mathbf{X})$, defined as follows, of the member lengths is to be minimized so as to improve the regularity of the frame, which leads to efficiency in aesthetic and constructional aspects:

$$D(\mathbf{X}) = \sum_{i=1}^{m}(L^{\mathrm{ave}}(\mathbf{X}) - L_i(\mathbf{X}))^2 \qquad (6.21)$$

Hence, we have two objective functions to be minimized, and the optimization problem turns out to be a multiobjective programming (MOP) problem (see Appendix A.4 for basic properties and methodologies of MOP).

In order to assure regularity and smoothness of the surface and frame, upper and lower bounds are to be given for the locations of the control points and the nodes of the frame. Let $\mathbf{P}(\mathbf{X})$ denote the vector consisting of all components of $\mathbf{P}_k(\mathbf{X})$ $(k = 1, \ldots, n^{\mathrm{p}})$, where n^{p} is the number of nodes of the frame. Then the geometrical constraints are given as

$$H_i^{\mathrm{s}}(\mathbf{X}) \leq 0, \quad (i = 1, \ldots, n^{\mathrm{s}}) \qquad (6.22\mathrm{a})$$
$$H_i^{\mathrm{f}}(\mathbf{P}(\mathbf{X})) \leq 0, \quad (i = 1, \ldots, n^{\mathrm{f}}) \qquad (6.22\mathrm{b})$$

where n^{s} and n^{f} are the numbers of geometrical constraints for the surface and the frame, respectively, and the simple bound constraints for $\mathbf{X}$ and $\mathbf{P}(\mathbf{X})$ are assumed to be included. Then the MOP problem is formulated as

$$\text{Minimize} \quad W(\mathbf{X}) \text{ and } D(\mathbf{X}) \qquad (6.23\mathrm{a})$$
$$\text{subject to} \quad H_i^{\mathrm{s}}(\mathbf{X}) \leq 0, \quad (i = 1, \ldots, n^{\mathrm{s}}) \qquad (6.23\mathrm{b})$$
$$H_i^{\mathrm{f}}(\mathbf{P}(\mathbf{X})) \leq 0, \quad (i = 1, \ldots, n^{\mathrm{f}}) \qquad (6.23\mathrm{c})$$

In the following examples, Pareto optimal solutions are obtained using the constraint approach, and each single-objective optimization problem is solved utilizing a gradient-based approach. Therefore, equations are derived below for calculating sensitivity coefficients of the nodal coordinates with respect to the coordinates of the control points. The sensitivity coefficients of the compliance can be found from those of nodal coordinates directly using the formulations in Secs. 2.2 and 2.7.

The jth component ($j = 1, 2, 3$) of $\mathbf{R}_{\alpha\beta}$ and $\mathbf{P}_k$ in the three-dimensional space is denoted by $R_{\alpha\beta j}$ and P_{kj}, respectively. The following expression is derived from (A.136a) and (A.136b) in Appendix A.6:

$$\frac{\partial P_{kj}}{\partial R_{\alpha\beta j}} = \frac{n^{C}!}{\alpha!\beta!(n^{C} - \alpha - \beta)!} u_k^{\alpha} v_k^{\beta} (1 - u_k - v_k)^{n^{C} - \alpha - \beta},$$

$$(j = 1, 2, 3; \ k = 1, \ldots, n^{P})$$

$$(6.24)$$

Note that $\partial P_{kj}/\partial R_{\alpha\beta i} = 0$ for $j \neq i$. The sensitivity of the length L_k of the kth member connecting nodes r and s with respect to $R_{\alpha\beta j}$ is calculated from

$$\frac{\partial L_k}{\partial R_{\alpha\beta j}} = \frac{\partial L_k}{\partial P_{rj}} \frac{\partial P_{rj}}{\partial R_{\alpha\beta j}} + \frac{\partial L_k}{\partial P_{sj}} \frac{\partial P_{sj}}{\partial R_{\alpha\beta j}} \qquad (6.25)$$

6.5.3 Numerical examples

Optimal shapes are found for a single-layer latticed frame, as shown in Fig. 6.18, which has three supports and is modeled using a triangular Bézier surface of order 6; hence, the number of control points is 28. The height of the supports A, B, and C are 0, 0, and 8 m, respectively. The height of the node indicated by a blank square in Fig. 6.18 is constrained to be 12 m. The locations of the control points at the supports are fixed. Therefore, we have 25 free control points. The parameters u and v are divided into 18 uniformly spaced intervals between 0 and 1 to define the nodal locations; i.e., $u_k, v_k = 0, 1/18, \ldots, 17/18, 1$. Optimum designs are found using SNOPT Ver. 7.2 (Gill, Murray, and Saunders 2002).

The initial locations of the control points are illustrated in Fig. 6.19. The frame is subjected to static loads that represent the self-weight. The members consist of a circular tube section with external and internal radii of 50 mm and 46 mm, respectively. The elastic modulus is 210 kN/mm^2, Poisson's ratio for computing the stiffness for uniform torsion is 0.3, and the weight density is 77.0 kN/m^3. The bounds for $\mathbf{X}$ are not given. In the following, the units of length and force are m and kN, respectively, which are omitted for brevity.

Consider first the case in which only the z-coordinates of the free control points are considered as design variables; i.e., the number of variables is 25. We use the constraint approach in a similar manner as in Sec. 6.4 (see Appendix A.4 for details of the constraint approach). Single-objective problems are first solved for minimizing the compliance and the variance of member lengths, respectively, to find the admissible upper-bound values for the constraint approach. If the compliance is minimized, the optimal value is

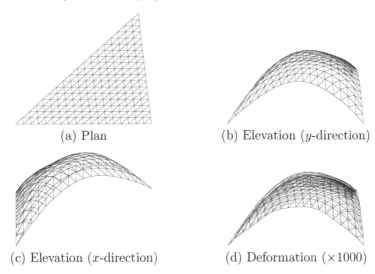

(a) Plan

(b) Elevation (y-direction)

(c) Elevation (x-direction)

(d) Deformation ($\times 1000$)

FIGURE 6.20: Optimal shape for minimizing compliance; variable z-coordinates.

$W = 0.01522$, and the variance of the lengths is $D = 59.74$. The optimal shape and deformation scaled by a factor of 1000 are plotted in Fig. 6.20, where the dotted lines in Fig. 6.20(d) represent the undeformed shape in Fig. 6.20(b). As is seen, the smooth and round shape is obtained by minimizing W, and the deformation is very small even with the scale factor of 1000. On the other hand, if the variance of the lengths is minimized, the optimal value is $D = 23.39$, and the compliance is $W = 10.51$, which is very large. The optimal shape and deformation multiplied by 2 are plotted in Fig. 6.21. As is seen, the smoothness of the surface is deteriorated by minimizing deviation of member lengths. In fact, the principal curvature has different signs at the center and a corner.

In order to find the Pareto optimal solutions, the upper bound of W for the minimization problem of D is reduced from 0.12 to 0.03 with the decrement 0.01. Then, W is minimized under constraint on D, where the upper bound of D is reduced from 59 to 32 with the decrement 3. The Pareto optimal solutions, which form a convex curve in the objective function space, are plotted in Fig. 6.22. The solution for $D = 32$ is shown in Fig. 6.23 with the deformation scaled by 500. This way, an intermediate shape between Figs. 6.20 and 6.21 that has moderately small deformation can be successfully found using the constraint approach.

Next we consider (x, y, z)-coordinates of the free control points as design variables; i.e., the number of design variables is $25 \times 3 = 75$. In this case, the variance of the member lengths can be minimized to $D = 7.513 \times 10^{-8}$, which is almost equal to zero, and the compliance of this solution has a very large

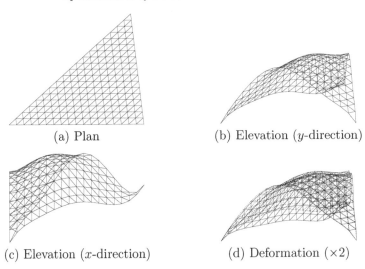

(a) Plan   (b) Elevation (y-direction)

(c) Elevation (x-direction)   (d) Deformation ($\times 2$)

FIGURE 6.21:   Optimal shape for minimizing deviation of lengths; variable z-coordinates.

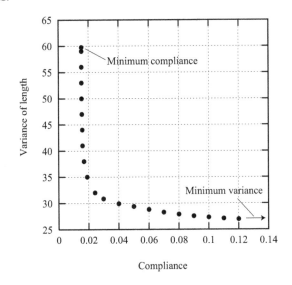

FIGURE 6.22:   Pareto optimal solutions; variable z-coordinates.

value, 10.94. The optimal solution that minimizes D and its deformation in real scale are shown in Fig. 6.24. As is seen, a cylindrical surface is generated by minimizing D, which results in a very large deformation. In contrast, the problem of minimizing the compliance did not converge. The constraint approach is also used here for finding the Pareto optimal solutions. The

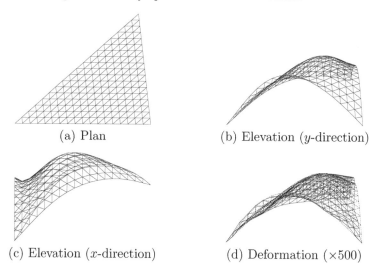

(a) Plan   (b) Elevation (y-direction)

(c) Elevation (x-direction)   (d) Deformation ($\times 500$)

FIGURE 6.23: Optimal shape for $D = 32$; variable z-coordinates.

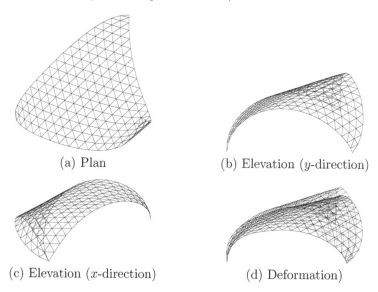

(a) Plan   (b) Elevation (y-direction)

(c) Elevation (x-direction)   (d) Deformation

FIGURE 6.24: Optimal shape for minimizing deviation of lengths; variable (x, y, z)-coordinates.

solution corresponding to $W = 0.02$ and its deformation scaled by 500 are shown in Fig. 6.25. This way, a moderately cylindrical shape with small deformation can be found by constraining W to a small value.

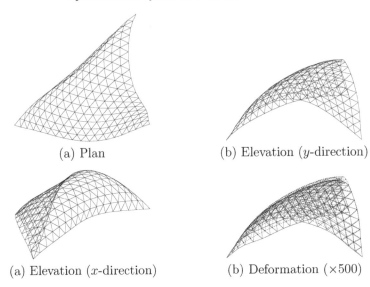

(a) Plan　　　　　　　　　　(b) Elevation (y-direction)

(a) Elevation (x-direction)　　　　(b) Deformation ($\times 500$)

FIGURE 6.25:　Optimal shape for $W = 0.02$; variable (x, y, z)-coordinates.

6.6　Configuration optimization of an arch-type truss with local geometrical constraints

6.6.1　Direct assignments of geometrical constraints

So far, we considered shape optimization under constraints on global geometrical measures, e.g., deviation from target shape and variance of member lengths. In this section, a shape optimization method with direct assignment of local geometrical constraints is presented for an arch-type truss modeled using a Bézier curve, which is optimized utilizing a genetic algorithm (GA) (Ohsaki and Kato 1997). We also optimize the topology in addition to the geometry; i.e., *optimal configuration* is to be found.

Consider an arch-type truss with triangular units (Warren truss), as shown in Fig. 6.26, where the lower nodes exist along the Bézier curve $\mathbf{P}_{\mathrm{L}}(u) = (P_{\mathrm{L}}^x(u), P_{\mathrm{L}}^y(u))^\top$. The parameter u $(0 \leq u \leq 1)$ is discretized to $n^{\mathrm{D}}+1$ points to define the nodes and supports on the lower curve:

$$u_{\mathrm{L}}^i = \frac{i}{n^{\mathrm{D}}}, \quad (i = 0, 1, \ldots, n^{\mathrm{D}}) \tag{6.26}$$

where $i = 0$ and n^{D} correspond to the two supports.

Let d denote the distance between the lower and upper curves. Then, the upper curve is given by the offset curve with normal distance d in a similar

FIGURE 6.26: A Warren arch truss.

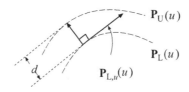

FIGURE 6.27: Definition of the upper curve using the normal offset vector with length d.

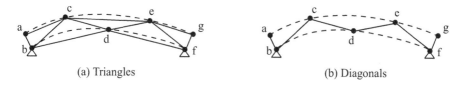

FIGURE 6.28: Feasible topology of a Warren truss.

manner as in Sec. 6.4:

$$\mathbf{P}_U(u) = \begin{pmatrix} P_L^x(u) \\ P_L^y(u) \end{pmatrix} + \frac{d}{\sqrt{(P_{L,u}^x)^2 + (P_{L,u}^y)^2}} \begin{pmatrix} 0 & -1 \\ 1 & 0 \end{pmatrix} \begin{pmatrix} P_{L,u}^x \\ P_{L,u}^y \end{pmatrix} \qquad (6.27)$$

where $(\,\cdot\,)_{,u}$ denotes differentiation with respect to u; see Appendix A.6.3 for details of the offset curve. Fig. 6.27 illustrates the relation between the upper and lower curves. The parameter values u_U^i for defining the upper nodes of a Warren truss with additional nodes above the supports are defined as

$$u_U^i = \frac{1}{2}(u_L^{i-1} + u_L^i), \quad (i = 1, \dots, n^D)$$
$$u_U^0 = u_L^0, \quad u_U^{n^P+1} = u_L^{n^P+1} \qquad (6.28)$$

Then, the location of an upper node is given as shown in Fig. 6.27 by incorporating $u = u_U^i$ into $\mathbf{P}_U(u)$ in (6.27).

We next define the geometrical parameters r_j and a_k for constraining the local properties. For example, for the truss in Fig. 6.28(a), there exist five triangles (a,b,c), (b,c,d), (c,d,e), (d,e,f), and (e,f,g). The sequence of diagonals is shown in Fig. 6.28(b). Let $\mathbf{P}_a, \dots, \mathbf{P}_g$ denote the location vectors of points 'a',...,'g'. Suppose that these vectors are in the three-dimensional space with

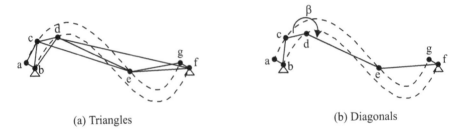

(a) Triangles   (b) Diagonals

FIGURE 6.29: Infeasible topology that does not correspond to a Warren truss.

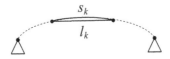

FIGURE 6.30: Definition of the arc of the lower curve formed by a lower chord.

the third component equal to 0. For the third triangle (c,d,e), for example, define a vector

$$\mathbf{R}_3 = \frac{1}{2}(\mathbf{P}_d - \mathbf{P}_c) \times (\mathbf{P}_e - \mathbf{P}_d) \qquad (6.29)$$

The vectors $\mathbf{R}_1$, $\mathbf{R}_2$, $\mathbf{R}_4$, and $\mathbf{R}_5$ are defined similarly using the vector product between the direction vectors of the diagonals. Then, define r_j as the third component of $\mathbf{R}_j$, which is equal to the signed area of the triangle formed by a pair of diagonals and a lower or upper chord. The regularity of the Warren truss is assured through assignment of the constraint on r_j. To preserve the configuration of the Warren truss, $r_j < 0$ should be satisfied if the triangle contains a lower chord, and $r_j > 0$ if the triangle contains an upper chord. The truss in Fig. 6.28 satisfies these conditions.

However, for the truss in Fig. 6.29, $r_3 < 0$ for the triangle (c,d,e) containing the upper chord (c,e). Accordingly, the angle β in Fig. 6.29(b) exceeds π, which leads to an infeasible configuration as a Warren truss. Note that the regularity of the truss is maintained by assigning stricter nonzero lower and upper bounds for r_j.

Furthermore, the distance between the lower curve and the lower chord can be restricted using the length l_k of the kth lower chord and the arc length s_k of the Bézier curve associated with the kth lower chord, as shown in Fig. 6.30. The parameter a_k is then defined as

$$a_k = \frac{s_k - l_k}{s_k} \qquad (6.30)$$

A lower chord closely follows the curve if a_k is close to 1.

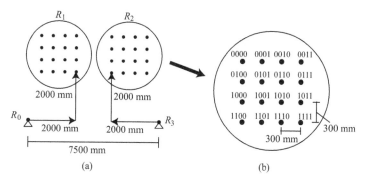

FIGURE 6.31: Assignment of control points.

6.6.2 Optimization using a genetic algorithm

Optimal configuration is found using a genetic algorithm. The objective function to be minimized is the compliance W; see Secs. 1.8 and 6.4 for the definition of compliance. We use the Bézier curve of order 3, for simplicity, considering symmetry conditions; see Appendix A.6 for details. Note that the locations of the supports are fixed. Therefore, we have only one control point with a variable location.

For a Warren truss as in Fig. 6.26, we have n^{D} lower chords, $n^{\mathrm{D}} - 1$ upper chords, and $2n^{\mathrm{D}}$ diagonals. Therefore, there are $2n^{\mathrm{D}} - 1$ triangle units, and the constraints are given as

$$r^{\mathrm{L}} \leq r_k \leq r^{\mathrm{U}}, \quad (k = 1, \ldots, 2n^{\mathrm{D}} - 1) \tag{6.31a}$$
$$a_k \leq a^{\mathrm{U}}, \quad (i = 1, \ldots, n^{\mathrm{D}}) \tag{6.31b}$$

where the superscripts $(\cdot)^{\mathrm{L}}$ and $(\cdot)^{\mathrm{U}}$ indicate lower and upper bounds, respectively. Note again that r^{L} and r^{U} have positive values if the triangle contains an upper chord, and they are negative if the triangle contains a lower chord.

A simple genetic algorithm is used for optimization. Fig. 6.31 illustrates the coding scheme for modeling 16 candidate locations of the control points R_1 and R_2, respectively, using a 4-bit string. A binary string of $n^{\mathrm{D}} - 1$ bits is used for representing by 1 and 0 the existence and nonexistence, respectively, of the lower node at each candidate point on the lower curve. Fig. 6.32 illustrates the various topologies of a truss with two lower nodes defined from four candidate nodes excluding the supports; i.e., $N^{\mathrm{D}} = 5$.

An exterior penalty function approach, presented in Appendix A.2.2.4, is used for incorporating the constraints into the objective function. The penalty

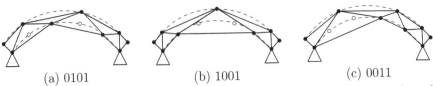

FIGURE 6.32: Various topologies of a truss with two lower nodes selected from four candidate nodes represented by 4-bit string.

function $d_r(r_k)$ for the constraint (6.31a) is defined as

$$
\begin{cases}
d_r(r_k) = 0 & \text{for} \quad r^L \le r_k \le r^U \\[2mm]
d_r(r_k) = \tau \left(\dfrac{r^L - r_k}{r^U - r^L} \right)^2 & \text{for} \quad r_k < r^L \\[2mm]
d_r(r_k) = \tau \left(\dfrac{r_k - r^U}{r^U - r^L} \right)^2 & \text{for} \quad r^U < r_k
\end{cases}
\tag{6.32}
$$

where τ is the penalty parameter that should be appropriately assigned in view of the magnitudes of the objective function W and the penalty terms.

Let n^P denote the size of the population (number of individuals in a generation). The binary string of the jth individual of the ith generation is denoted by X_i^j ($j = 1, \ldots, n^P$), and let $D_r^0(X_i^j)$ be defined as

$$
D_r^0(X_i^j) = \sum_{k=1}^{2n^D - 1} d_r(r_k)
\tag{6.33}
$$

Then, the total penalty $D_r(X_i^j)$ for constraint (6.31a) is given as

$$
\begin{cases}
D_r(X_i^j) = D_r^0(X_i^j) & \text{for} \quad D_r^0(X_i^j) \le D^* \\
D_r(X_i^j) = D^* + \alpha(D_r^0(X_i^j) - D^*) & \text{for} \quad D_r^0(X_i^j) > D^*
\end{cases}
\tag{6.34}
$$

where D^* and α (< 1) are the parameters, which are given so that the individual with a large violation of constraints does not have too large a penalty term.

Various definitions of penalty functions have been proposed for GAs. For instance, Chen and Chen (1997) defined the penalty parameter $\tau(k^g)$, which depends on the generation number k^g, as

$$
\tau(k^g) = \tau_1[1 + 0.2(k^g - 1)]
\tag{6.35}
$$

where τ_1 is the specified initial value. This way, the penalty parameter linearly increases as the generation proceeds. A linear penalty with a threshold value can also be used, as suggested by Hajela and Lee (1995).

The penalty function $D_a(X_i^j)$ for constraint (6.31b) is defined similarly, and the performance measure $C(X_i^j)$ for X_i^j is obtained by adding the penalty terms to the compliance $W(X_i^j)$ as

$$
C(X_i^j) = W(X_i^j) + D_r(X_i^j) + D_a(X_i^j)
\tag{6.36}
$$

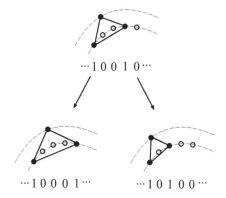

FIGURE 6.33: Illustration of the mutation process.

FIGURE 6.34: Optimal configuration and stresses with geometrical constraints.

Since GA maximizes the fitness function, the objective function $C(X_i^j)$ to be minimized can be transformed to $C_0 - C(X_i^j)$, with C_0 being a sufficiently large value, or to $1/C(X_i^j)$. However, in order to avoid ambiguity in the transformation of the fitness function, a ranking strategy with linear scaling is used here; i.e., the individual with rank N has the following fitness value $f(N)$:

$$f(N) = f^{\mathrm{w}} + (f^{\mathrm{b}} - f^{\mathrm{w}})\frac{n^{\mathrm{P}} - R}{n^{\mathrm{P}} - 1} \qquad (6.37)$$

where f^{w} and f^{b} are the parameters for defining the maximum and minimum fitness values, respectively. The two-point crossover is used with probability P^{c}. Mutation is defined such that a lower node moves to the right or left with the probability P^{m}. If the number of candidate lower nodes is far larger than that of the existing nodes in the optimal solutions, it is unlikely that there exists a node at one of the two neighboring candidate nodes. Therefore, the number of lower nodes is likely to be preserved and local search can be carried out by mutation, as illustrated in Fig. 6.33. An elitist strategy is used, where only the best individual remains in the next generation without carrying out mutation or crossover.

FIGURE 6.35: Optimal configuration and stresses without geometrical constraints.

6.6.3 Numerical examples

Optimization is carried out for a symmetric truss subjected to static loads. The depth d is 0.5 m and the span length is 7.5 m. The 16 candidate curves are generated from the location of the control point R_1 coded with a string of 4 bits as plotted in Fig. 6.31, where the location of R_2 is determined using the symmetry condition. The symmetrically located 100 candidate nodes are described by the string of 50 bits. Therefore, the total number of bits for each individual (solution) is $4 + 50 = 54$.

The concentrated loads corresponding to the nonstructural mass are applied at the upper nodes, where the distributed mass is 100.0 kg/m for the curve associated with the upper nodes, and the acceleration of gravity is 9.8 m/s^2. The self-weight is also considered with the mass density 7.86×10^3 kg/m^3. The cross-sectional areas are 5.0×10^{-4} m^2 for all members, and the elastic modulus is 205.8 kN/mm^2.

The bounds for the constraints are $(r^L, r^U) = (0.4, 1.4)$ for a triangle containing an upper chord, $(r^L, r^U) = (-1.4, -0.4)$ for a triangle containing a lower chord, and $a^U = 0.05$. The parameters for the penalty function are $D^* = 30.0$, $\tau = 8000.0$, and $\alpha = 0.1$. The probabilities for crossover and mutation are $P^c = 0.6$ and $P^m = 0.02$, respectively, and the size of the population is $n^P = 50$. The fitness parameters f^b and f^w for scaling are given so that the fitness of the best solution is 8/3 of that of the worst solution.

The optimal solution with $W = 138.08$ Nm and nine lower chords is found, as shown in Fig. 6.34, where the width of each member is proportional to the absolute value of the stress. If the geometrical constraints are not considered, the optimal solution has only three lower chords, as shown in Fig. 6.35, and the compliance is 113.00 Nm, which is smaller than that for the optimal solution with geometrical constraints. This way, the regularity of the truss can be preserved by sacrificing stiffness against static loads.

6.7 Seismic design for spatially varying ground motions

6.7.1 Introduction

Due to the increasing demand for constructing long-span dome structures, the spatial variation of seismic motions has become an important factor in the field of civil and architectural engineering. The simplest among the effects of spatial variation is the *wave passage effect*, as illustrated in Fig. 6.36, due to the delay of the wave being transmitted to the supports. This effect has been extensively investigated since the 1970s, mainly for the design and construction of bridges. There are other causes of spatial variation of seismic motions, i.e., the *incoherency effect* due to reflections and refractions of the wave in the heterogeneous medium, and the *local effect* or *site response effect* due to differences in soil conditions near the supports.

The response of a structure to spatially varying seismic motions may be computed by using time-history analysis against different forced displacements at each support (Clough and Penzien 1975; Price and Eberhard 1998), or a frequency domain analysis (DebChaudhury and Gazis 1988; Harichandran, Hawwari, and Swedian 1996). It will be convenient, however, in design practice, if a response spectrum approach can be used, because the characteristics of the ground motions are usually given in the form of a design response spectrum. Der Kiureghian and Neuenhofer (1992) presented a response spectrum approach incorporating the wave passage effect and incoherency effect that is modeled using coherency functions between the ground motions at the supports. The correlations between the modal responses are also considered in a similar manner as the complete quadratic combination (CQC) method (Wilson, Der Kiureghian, and Bayo 1982). Hao and Duan (1995) showed that incoherency of the support motions leads to torsional responses of frames. Zembaty (1996) carried out a parametric study to determine the effect of incoherency parameters on structural responses.

In this section, the method of Ohsaki (2001b) is summarized to carry out optimization and postoptimal analysis of long-span structures considering the incoherency effect, the wave passage effect, and the local amplification effect of the ground motions at the supports. It is shown that the second-order sensitivity coefficients of the optimal solution with respect to the parameters characterizing the spatial variation of the seismic motions can be easily obtained if the first-order parametric sensitivity coefficients vanish.

6.7.2 Response to spatially varying ground motions

The total degrees of freedom (DOFs) of the internal nodes and supports of a structure are classified into unconstrained DOFs (UDOFs) and support DOFs (SDOFs). Let $\mathbf{x}(t)$ and $\mathbf{u}(t)$ denote the vectors of absolute displacements

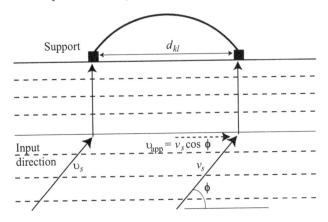

FIGURE 6.36:   Illustration of the wave passage effect of ground motion.

corresponding to UDOF and SDOF, respectively, which are functions of time t. The stiffness matrix, mass matrix, and damping matrix, denoted by $\mathbf{K}, \mathbf{M}$, and $\mathbf{C}$, respectively, are divided into the components corresponding to UDOF and SDOF that are indicated by the subscripts $(\cdot)_x$ and $(\cdot)_u$, respectively. Then, the equations of motion are written as (Clough and Penzien 1975):

$$\begin{pmatrix} \mathbf{M}_x & \mathbf{M}_{xu} \\ \mathbf{M}_{xu}^\top & \mathbf{M}_u \end{pmatrix} \begin{pmatrix} \ddot{\mathbf{x}}(t) \\ \ddot{\mathbf{u}}(t) \end{pmatrix} + \begin{pmatrix} \mathbf{C}_x & \mathbf{C}_{xu} \\ \mathbf{C}_{xu}^\top & \mathbf{C}_u \end{pmatrix} \begin{pmatrix} \dot{\mathbf{x}}(t) \\ \dot{\mathbf{u}}(t) \end{pmatrix}$$
$$+ \begin{pmatrix} \mathbf{K}_x & \mathbf{K}_{xu} \\ \mathbf{K}_{xu}^\top & \mathbf{K}_u \end{pmatrix} \begin{pmatrix} \mathbf{x}(t) \\ \mathbf{u}(t) \end{pmatrix} = \begin{pmatrix} \mathbf{0} \\ \mathbf{f}(t) \end{pmatrix}$$
$$(6.38)$$

where a dot indicates differentiation with respect to t, and $\mathbf{f}$ is the vector of reaction forces due to the displacements $\mathbf{u}$ at the support, which are regarded as forced displacements to the structure.

Let $\mathbf{x}^s(t)$ and $\mathbf{x}^d(t)$ denote the pseudostatic and dynamic components of $\mathbf{x}(t)$; i.e.,

$$\mathbf{x} = \mathbf{x}^s(t) + \mathbf{x}^d(t) \tag{6.39}$$

and $\mathbf{x}^s(t)$ is defined as

$$\mathbf{x}^s(t) = -\mathbf{K}_x^{-1}\mathbf{K}_{xu}\mathbf{u}(t) \tag{6.40}$$

Note that the inverse of $\mathbf{K}_x$ is not actually computed in the process of structural analysis; i.e., $\mathbf{x}^s(t)$ is computed from $\mathbf{K}_x\mathbf{x}^s(t) = \mathbf{K}_{xu}\mathbf{u}(t)$. The vector $\mathbf{x}^d$ is expressed as a sum of the modal components:

$$\mathbf{x}^d(t) = \sum_{i=1}^{n} y_i(t)\boldsymbol{\phi}_i = \boldsymbol{\Phi}\mathbf{y}(t) \tag{6.41}$$

where n is the number of UDOFs, and $\boldsymbol{\phi}_i$ is the ith eigenmode of the structure with fixed SDOF.

Let $s_{ki}(t)$ denote the modal response to the input $u_k(t)$ at the kth SDOF. A vector $\mathbf{r}_k$ is defined as

$$\mathbf{r}_k = -\mathbf{K}_{\mathrm{x}}^{-1}\mathbf{K}_{\mathrm{xu}}\mathbf{i}_k \qquad (6.42)$$

where $\mathbf{i}_k$ is a vector whose kth component is 1 and the remaining components are 0. Suppose the representative response $z(t)$, e.g., strain of a member, is defined as

$$z(t) = \mathbf{q}_1^\top \mathbf{x}(t) + \mathbf{q}_2^\top \mathbf{u}(t) \qquad (6.43)$$

which is rewritten as

$$z(t) = \sum_{k=1}^{s}\sum_{i=1}^{n} b_{ki}s_{ki}(t) + \sum_{k=1}^{s} a_k u_k(t) \qquad (6.44\mathrm{a})$$

$$a_k = \mathbf{q}_1^\top \mathbf{r}_k + \mathbf{q}_2^\top \mathbf{i}_k, \quad (k = 1,\dots,s) \qquad (6.44\mathrm{b})$$

$$b_{ki} = \mathbf{q}_1^\top \boldsymbol{\phi}_i \beta_{ki}, \quad (k = 1,\dots,s;\ i = 1,\dots,n) \qquad (6.44\mathrm{c})$$

where s is the number of SDOFs, and β_{ki} is defined as follows as the participation factor of the ith mode to the input at the kth SDOF:

$$\beta_{ki} = \boldsymbol{\phi}_i^\top (\mathbf{M}_{\mathrm{x}}\mathbf{r}_k + \mathbf{M}_{\mathrm{xu}}\mathbf{i}_k) \qquad (6.45)$$

Note that the eigenmodes are normalized with respect to $\mathbf{M}_{\mathrm{x}}$ as $\boldsymbol{\phi}_i^\top \mathbf{M}_{\mathrm{x}}\boldsymbol{\phi}_i = 1$.

Assuming that u_k is a zero-mean jointly stationary process, the mean square γ_z^2 of $z(t)$ is given as

$$
\begin{aligned}
\gamma_z^2 = {} & \sum_{k=1}^{s}\sum_{l=1}^{s} a_k a_l \rho_{u_k u_l} \gamma_{u_k} \gamma_{u_l} + 2\sum_{k=1}^{s}\sum_{l=1}^{s}\sum_{j=1}^{n} a_k b_{lj} \rho_{u_k s_{lj}} \gamma_{u_k} \gamma_{s_{lj}} \\
& + \sum_{k=1}^{s}\sum_{l=1}^{s}\sum_{i=1}^{n}\sum_{j=1}^{n} b_{ki} b_{lj} \rho_{s_{ki}s_{lj}} \gamma_{s_{ki}} \gamma_{s_{lj}}
\end{aligned}
\qquad (6.46)
$$

where γ_{u_k}, $\gamma_{s_{ki}}$, $\rho_{u_k u_l}$, $\rho_{u_k s_{lj}}$, and $\rho_{s_{ki}s_{lj}}$ are coefficients. For example, the coefficient $\gamma_{s_{ki}}$ is given as

$$\gamma_{s_{ki}} = \left(\int_{-\infty}^{\infty} |H_i(\mathrm{i}\omega)|^2 G_{\ddot{u}_k \ddot{u}_k}(\omega)\,\mathrm{d}\omega\right)^{\frac{1}{2}} \qquad (6.47)$$

where ω is the circular frequency, $\mathrm{i} = \sqrt{-1}$, $G_{\ddot{u}_k \ddot{u}_k}$ is the power spectrum of the input acceleration $\ddot{u}_k$, and $H_i(\mathrm{i}\omega)$ is the transfer function of the ith mode that is a function of the circular frequency ω; see Der Kiureghian and Neuenhofer (1992) for details of other coefficients. Note that the first, third, and second terms in (6.46) represent the pseudostatic response, dynamic response, and their combination, respectively.

Consider a response to the horizontal component of a ground motion. The coherency between the ground accelerations $\ddot{u}_k$ and $\ddot{u}_l$ at kth and lth SDOFs is given as (Luco and Wong 1986; Zerva 1990; Abrahamson, Schneider, and Stepp 1991).

$$\gamma_{kl}(\mathrm{i}\omega) = \exp\left[-\left(\frac{\alpha\omega d_{kl}}{v_\mathrm{s}}\right)^2\right]\exp\left(\mathrm{i}\frac{\eta\omega d_{kl}^\mathrm{L}}{v_\mathrm{app}}\right) \tag{6.48}$$

where α is the incoherency factor, d_{kl} is the horizontal distance between the kth and lth SDOFs, as illustrated in Fig. 6.36, d_{kl}^L is the projected distance of d_{kl} to the horizontal plane, v_s is the velocity of shear wave, and v_app is the apparent velocity of the shear wave that is given by the angle ψ of the wave direction to the horizontal plane as $v_\mathrm{app} = v_\mathrm{s}/\cos\psi$.

In (6.48), the first and second exponential terms represent the incoherency effect and the wave passage effect, respectively. Note that η is an auxiliary parameter for indicating incorporation and non-incorporation of the wave passage effect with $\eta = 1$ and 0, respectively. If $\gamma_{kl}(\mathrm{i}\omega)$ and $G_{\ddot{u}_k\ddot{u}_k}(\omega)$ are given, then $G_{u_ku_k}(\omega)$, $G_{u_ku_l}(\mathrm{i}\omega)$, $G_{\ddot{u}_k\ddot{u}_l}(\mathrm{i}\omega)$, and $G_{u_k\ddot{u}_k}(\mathrm{i}\omega)$ are easily computed by using the standard formulas of frequency domain analysis, and the integrations are carried out to find the mean square response of $z(t)$ defined by (6.46).

The ith natural circular frequency and damping ratio are denoted by ω_i and h_i, respectively. Let z^max denote the maximum absolute value of $z(t)$. The ratio of the maximum absolute value to the standard deviation of a zero-mean random process is called the *peak factor*. Assuming that the peak factors for the input quantities and the responses are the same (Der Kiureghian 1980), the mean maximum response $\mathrm{E}(z^\mathrm{max})$ of z^max is rewritten from (6.46) as

$$\begin{aligned}
\mathrm{E}(z^\mathrm{max}) \simeq \Bigg[&\sum_{k=1}^{s}\sum_{l=1}^{s} a_k a_l \rho_{u_k u_l} u_k^\mathrm{max} u_l^\mathrm{max} \\
&+ 2\sum_{k=1}^{s}\sum_{l=1}^{s}\sum_{j=1}^{n} a_k b_{lj}\rho_{u_k s_{lj}} u_k^\mathrm{max} S_{\mathrm{D}l}(\omega_j, h_j) \\
&+ \sum_{k=1}^{s}\sum_{l=1}^{s}\sum_{i=1}^{n}\sum_{j=1}^{n} b_{ki}b_{lj}\rho_{s_{ki}s_{lj}} S_{\mathrm{D}k}(\omega_i, h_i) S_{\mathrm{D}l}(\omega_j, h_j) \Bigg]^{\frac{1}{2}}
\end{aligned} \tag{6.49}$$

where u_k^max and $S_{\mathrm{D}k}(\omega, h)$ are the specified maximum displacement and the displacement response spectrum of the kth SDOF. The value of $S_{\mathrm{D}k}(\omega, h)$ varies among the supports due to local soil conditions that characterize the site responses (Der Kiureghian 1996).

The power spectrum $G_{\ddot{u}_k\ddot{u}_k}(\omega)$ of the input acceleration is defined in terms of the response spectrum. In the numerical examples, the following definition is used (Der Kiureghian and Neuenhofer 1991, 1992):

$$G_{\ddot{u}_k\ddot{u}_k}(\omega) = \frac{\omega^{c_k+2}}{\omega^{c_k} + \omega_{(\mathrm{f})k}^{c_k}}\left(\frac{2h\omega}{\pi} + \frac{4}{\pi\tau}\right)\left(\frac{S_{\mathrm{D}k}(\omega, h)}{\nu(\omega)}\right)^2 \tag{6.50}$$

where τ is the duration of the motion, $\nu(\omega)$ is the peak factor of the response to white noise, which, together with $\omega_{(f)k}$, are calculated to represent a finite power for the pseudostatic input (Der Kiureghian and Neuenhofer 1991; Ohsaki, Tagawa, and Kato 2000; Ohsaki 2001b).

6.7.3 Problem formulation and design sensitivity analysis

Consider a spatial frame with m members. Let L_i and A_i denote the length and cross-sectional area of the ith member, respectively. The nodal locations and topology of the structure are fixed. All the cross-sectional properties of each member are assumed to be defined by the cross-sectional area; i.e., the design variables are the cross-sectional areas $\mathbf{A} = (A_1, \ldots, A_m)^\top$. The objective function is the total structural volume, and constraints are given for the representative stresses, e.g., the maximum stress at the two edges of the two ends of each member, as illustrated in Fig. 5.2 in Sec. 5.1; i.e., there are four points at which the stress is constrained in each member.

Let $\sigma_{i,j}^{\mathrm{y}}(\mathbf{A})$ denote the representative mean maximum stress at the jth point of the ith member computed using (6.49), where a_k and b_{lk} are to be defined appropriately in view of the geometry and topology of the frame. The stress due to static loads, including self-weight, is denoted by $\sigma_{i,j}^{\mathrm{w}}(\mathbf{A})$. The optimization problem for minimizing the total structural volume V is formulated as

$$\text{Minimize} \quad V(\mathbf{A}) = \sum_{i=1}^{m} A_i L_i \tag{6.51a}$$

$$\text{subject to} \quad \sigma_{i,j}(\mathbf{A}) = \sigma_{i,j}^{\mathrm{y}}(\mathbf{A}) + |\sigma_{i,j}^{\mathrm{w}}(\mathbf{A})| \leq \sigma_{i,j}^{\mathrm{U}},$$
$$(i = 1, \ldots, m;\ j = 1, \ldots, 4) \tag{6.51b}$$

$$A_i^{\mathrm{L}} \leq A_i \tag{6.51c}$$

where A_i^{L} is the lower bound for A_i, and $\sigma_{i,j}^{\mathrm{U}}$ is the upper bound for $\sigma_{i,j}(\mathbf{A})$.

Problem (6.51) is a nonlinear programming problem, which is solved in the following examples using a gradient-based approach. Therefore, the design sensitivity coefficients of the objective and constraint functions with respect to A_i are needed. Because the sensitivity coefficients of static stresses are obtained using a well-established method in Sec. 2.2, only the formulations for $\sigma_{i,j}^{\mathrm{y}}(\mathbf{A})$ are presented below.

It is straightforward to differentiate (6.40)–(6.42) with respect to A_i; e.g., (6.42) is first rewritten as

$$\mathbf{K}_{\mathrm{x}} \mathbf{r}_k = -\mathbf{K}_{\mathrm{xu}} \mathbf{i}_k \tag{6.52}$$

which is differentiated with respect to A_i as

$$\mathbf{K}_{\mathrm{x}} \frac{\partial \mathbf{r}_k}{\partial A_i} = -\frac{\partial \mathbf{K}_{\mathrm{x}}}{\partial A_i} \mathbf{r}_k - \frac{\partial \mathbf{K}_{\mathrm{xu}}}{\partial A_i} \mathbf{i}_k \tag{6.53}$$

Therefore, the right-hand side of (6.53) can be conceived as a static load vector, and we can assume $\mathbf{K}_x$ has been already factorized when solving (6.52) for $\mathbf{r}_k$. Hence, the inverse matrix of $\mathbf{K}_x$ or its sensitivity coefficient is not needed to compute $\partial \mathbf{r}_k / \partial A_i$. Then, using (6.44b) and (6.44c), the sensitivity coefficients of a_k and b_{ki} are obtained as

$$\frac{\partial a_k}{\partial A_i} = \frac{\partial \mathbf{q}_1^\mathsf{T}}{\partial A_i} \mathbf{r}_k + \mathbf{q}_1^\mathsf{T} \frac{\partial \mathbf{r}_k}{\partial A_i} + \frac{\partial \mathbf{q}_2^\mathsf{T}}{\partial A_i} \mathbf{i}_k \tag{6.54a}$$

$$\frac{\partial b_{ki}}{\partial A_i} = \frac{\partial \mathbf{q}_1^\mathsf{T}}{\partial A_i} \boldsymbol{\phi}_i \beta_{ki} + \mathbf{q}_1^\mathsf{T} \frac{\partial \boldsymbol{\phi}_i}{\partial A_i} \beta_{ki} + \mathbf{q}_1^\mathsf{T} \boldsymbol{\phi}_i \frac{\partial \beta_{ki}}{\partial A_i} \tag{6.54b}$$

Note that the sensitivity coefficients of the eigenvalues and eigenmodes are computed utilizing the standard formulas shown in Sec. 2.3. For the participation factor,

$$\frac{\partial \beta_{ki}}{\partial A_i} = \frac{\partial \boldsymbol{\phi}_i^\mathsf{T}}{\partial A_i} (\mathbf{M}_x \mathbf{r}_k + \mathbf{M}_{xu} \mathbf{i}_k)$$

$$+ \boldsymbol{\phi}_i^\mathsf{T} \left(\frac{\partial \mathbf{M}_x}{\partial A_i} \mathbf{r}_k + \mathbf{M}_x \frac{\partial \mathbf{r}_k}{\partial A_i} + \boldsymbol{\phi}_i^\mathsf{T} \frac{\partial \mathbf{M}_{xu}}{\partial A_i} \mathbf{i}_k \right) \tag{6.55}$$

is derived from (6.45), and the sensitivity coefficients of S_{Dk} are found as described in Appendix A.7.

The sensitivity coefficients of the transfer function are derived using $\Omega_j = \omega_j^2$ as

$$\frac{\partial H_j(\pm i\omega)}{\partial A_i} = \frac{\partial}{\partial A_i} \left(\frac{1}{\omega_j^2 - \omega^2 \pm 2ih_j\omega_j\omega} \right)$$

$$= -(H_j(\pm i\omega))^2 \left(\frac{\partial \Omega_j}{\partial A_i} \pm 2ih_j \frac{\partial \omega_j}{\partial A_i} \omega \right) \tag{6.56}$$

$$= -(H_j(\pm i\omega))^2 \frac{\partial \Omega_j}{\partial A_i} \left(1 \pm i\frac{h_j\omega}{\omega_j} \right)$$

Note that $G_{\ddot{u}_k \ddot{u}_k}(\omega)$, $G_{u_k \ddot{u}_l}(i\omega)$, $G_{u_k u_l}(i\omega)$, and γ_{u_k} do not depend on A_i. From (6.47), we obtain

$$\frac{\partial \gamma_{s_{lj}}}{\partial A_i} = \frac{\partial}{\partial A_i} \left[\left(\int_{-\infty}^{\infty} \frac{1}{(\omega_j^2 - \omega^2)^2 + 4h_j^2\omega_j^2\omega^2} G_{\ddot{u}_l \ddot{u}_l}(\omega) d\omega \right)^{\frac{1}{2}} \right]$$

$$= \frac{1}{2\gamma_{s_{lj}}} \int_{-\infty}^{\infty} \frac{\partial}{\partial \Omega_j} \left[\frac{1}{(\Omega_j - \omega^2)^2 + 4h_j^2\Omega_j\omega^2} \right] \frac{\partial \Omega_j}{\partial A_i} G_{\ddot{u}_l \ddot{u}_l}(\omega) d\omega \tag{6.57}$$

$$= -\frac{1}{2\gamma_{s_{lj}}} \frac{\partial \Omega_j}{\partial A_i} \int_{-\infty}^{\infty} \left\{ \frac{2(\Omega_j - \omega^2) + 4h_j^2\omega^2}{[(\Omega_j - \omega^2)^2 + 4h_j^2\Omega_j\omega^2]^2} \right\} G_{\ddot{u}_l \ddot{u}_l}(\omega) d\omega$$

Other terms are differentiated similarly. Because the terms of partial differentiation are not included in the integrand, as is seen from (6.57), integration should be carried out only once, and the increase in the number of design variables does not lead to a rapid increase in computational cost.

6.7.4 Postoptimal analysis

Since the solution of the optimization problem (6.51) depends on the parameters that define the spatial variation of the seismic motion, it is practically important to investigate the sensitivity of the optimal solution and/or the optimal objective value with respect to such parameters. In the following examples, α, η, and the maximum ground displacements are taken as the parameters, and the sensitivity coefficients of optimal solutions with respect to the parameters are computed using the parametric programming approach described in Appendix A.5.

Consider, in general, an optimization problem

$$\text{Find} \qquad \widehat{C}(\mathbf{p}) = \min_{\mathbf{A}} C(\mathbf{A}, \mathbf{p}) \tag{6.58a}$$

$$\text{subject to} \quad H_i(\mathbf{A}, \mathbf{p}) \leq 0, \quad (i = 1, \ldots, n^{\mathrm{I}}) \tag{6.58b}$$

where $\mathbf{A}$ and $\mathbf{p} = (p_1, \ldots, p_{n^{\mathrm{P}}})^{\top}$ are the vectors of variables and parameters, respectively, n^{I} is the number of inequality constraints, including the bound constraints for $\mathbf{A}$, and $\widehat{C}(\mathbf{p})$ is the optimal objective value, which is conceived as a function of the parameter vector $\mathbf{p}$. In the following, a function of $\mathbf{p}$ is indicated with a hat as $\widehat{(\cdot)}$.

Suppose an optimal solution has been found using an appropriate method of nonlinear programming, where the Lagrange multiplier λ_i for the constraint $H_i(\mathbf{A}, \mathbf{p}) \leq 0$ has also been found as the result of optimization. The sensitivity of the optimal objective value with respect to p_j is found in a similar manner as (A.122) in Appendix A.5. The second-order sensitivity coefficients of the optimal objective value with respect to the parameters are given in (A.124). Note that the sensitivity coefficients $\partial \widehat{A}_i / \partial p_k$ and $\partial \widehat{\lambda}_i / \partial p_k$ of the variables and multipliers are generally needed for computing the second-order sensitivity coefficients $\partial^2 \widehat{C} / \partial p_j \partial p_k$ of the optimal objective value. However, it is not practically acceptable to compute those values for a complex optimization problem as considered in this section, because the Hessian of each constraint function with respect to the variables is needed for computing $\partial \widehat{A}_i / \partial p_k$ and $\partial \widehat{\lambda}_i / \partial p_k$.

Consider a case where all the variables of the optimal solution are even functions of a parameter p_k. In this case, the first-order parametric sensitivity coefficients $\partial \widehat{A}_i / \partial p_k$ and $\partial \widehat{\lambda}_i / \partial p_k$ vanish at $p_k = 0$ and the second-order sensitivity coefficients of the optimal objective value with respect to the parameters are obtained from

$$\frac{\partial^2 \widehat{C}}{\partial p_j \partial p_k} = \frac{\partial^2 C}{\partial p_j \partial p_k} + \sum_{i=1}^{n^{\mathrm{I}}} \lambda_i \frac{\partial^2 H_i}{\partial p_j \partial p_k} \tag{6.59}$$

Therefore, for a symmetric structure subjected to ground motions with symmetry or antisymmetry properties, the first-order sensitivity coefficients vanish and the second-order coefficients give useful information on the characteristics of the optimal solutions.

The entire process of analysis, sensitivity analysis, optimization, and postoptimal analysis is summarized as follows:

Step 1: Define the geometry and material properties of the structure, and set a_k and b_{ki} in (6.44) for each response quantity to be constrained.

Step 2: Assign the parameters c_k, S_{Dk}, α, η, v_s, d_{kl}, and d_{kl}^L, and compute ν and $\omega_{(f)k}$; see Ohsaki, Tagawa, and Kato (2000) and Ohsaki (2001b) for details.

Step 3: Initialize the design variable vector $\mathbf{A}$ and optimize the structure as follows:

   3.1 Analysis:

   (a) Carry out eigenvalue analysis of free vibration of the structure with fixed supports.

   (b) Compute $\gamma_{kl}(i\omega)$ and $G_{\ddot{u}_k \ddot{u}_k}(\omega)$ from (6.48) and (6.50), respectively.

   (c) Evaluate $G_{\ddot{u}_k \ddot{u}_l}(i\omega)$, $G_{u_k \ddot{u}_k}(i\omega)$, etc. from $G_{\ddot{u}_k \ddot{u}_k}(\omega)$ and $H_i(i\omega)$ using the standard formulas of frequency domain analysis.

   (d) Calculate $\mathrm{E}(z^{\max})$ for each response quantity from (6.49).

   3.2 Compute the design sensitivity coefficients of $\mathrm{E}(z^{\max})$ as described in Sec. 6.7.3.

   3.3 Modify $\mathbf{A}$ in accordance with the optimization algorithm and go to Step 3.1 if not converged.

Step 4: Carry out postoptimal analysis.

6.7.5   Numerical examples

Consider a rigidly jointed 39-bar arch-type plane frame, as shown in Fig. 6.37. The span length is 100.0 m and the lower nodes are located along a circle with an open angle of 50 deg. The upper nodes are also on a circle, and the difference between the radii of the two circles with the same center is 7.5 m. Note that the upper chords, lower chords, and diagonals have the same lengths, respectively. The frame is pin-supported at the two ends.

The frame is subjected to a set of horizontal seismic motions compatible with the response spectrum of Newmark and Hall (1982); see Appendix A.7 for the definition and parameters of the spectrum. In the definition in Appendix A.7, however, the maximum response displacement at $\Omega_r = 0$ does not agree with the maximum displacement C_{D} of the ground motion. Therefore,

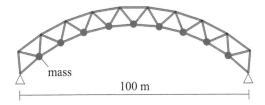

FIGURE 6.37: A 39-bar arch-type plane frame.

FIGURE 6.38: Optimal cross-sectional areas without spatial variation of seismic motion.

$S_{\mathrm{D}}^{(5)}(\Omega_r, h_r)$ is replaced by $S_{\mathrm{D}}^{(5)}(\Omega_r) = C_{\mathrm{D}}(1.0 + \kappa\Omega_r)$, and κ is defined to satisfy $S_{\mathrm{D}}^{(4)}(\Omega_r) = S_{\mathrm{D}}^{(5)}(\Omega_r)$ at the prescribed eigenvalue. The value of $u_k^{\max}$ for the kth SDOF is defined by using the scaling parameter μ_k to incorporate the local amplification effect as $u_k^{\max} = \mu_k C_{\mathrm{D}}$. Note that $S_{\mathrm{D}k}$ is also multiplied by μ_k.

Each member of the frame is assumed to be made of sandwich sections with $I_i = r^2 A_i$, where I_i is the second moment of inertia, and $r = 0.5$ m is the distance between the flange and the member axis that is considered as a constant. The elastic modulus is 205.8 kN/mm^2. The upper-bound stress is given as $\sigma_{i,j}^{\mathrm{U}} = 205.8$ N/mm^2 for all four points of members, and the lower-bound cross-sectional area is 10.0 mm^2, which is not active at the optimal solution in the following examples. The mass density of the members is 7.86×10^3 kg/m^3. A nonstructural mass of 2.0×10^4 kg is located at each lower node, and its rotational inertia is neglected. The loads in the vertical direction representing the weights of the members and the nonstructural masses are applied at the nodes to calculate the static stresses $\sigma_{i,j}^{\mathrm{W}}$.

The parameters of soil condition and the direction of seismic wave are $v_{\mathrm{s}} = 400.0$ m/s and $v_{\mathrm{app}} = 2000.0$ m/s. The modal damping ratio of the structure is 0.02, which is assumed to be independent of the frequency, for simplicity. The duration τ of the motion is 25.0 sec. In this case, the value of $\omega_{(\mathrm{f})k}$ to satisfy finite power against the pseudostatic input corresponding to the given response spectrum is 0.258. The parameter c_k in (6.50) is 3.0. The library DOT Ver. 5.0 (VR&D 1999) is used for optimization. The 16-point Gaussian quadrature is used for integration for the coefficients $\gamma_{s_{ki}}$, etc., in the frequency domain.

Fig. 6.38 shows the optimal cross-sectional areas for the case without the

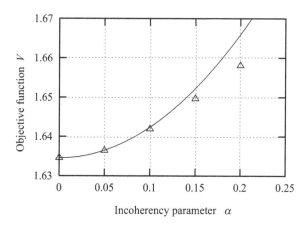

FIGURE 6.39: Optimal objective values for $\alpha = 0.0, 0.01, 0.1, 0.15, 0.2$ (triangular mark), and second-order approximation at $\alpha = 0.0$ (solid line).

effects of spatial variation of seismic motions; i.e., $\alpha = \eta = 0$, where the width of each member is proportional to the cross-sectional area. As is seen, the lower chords near the supports and the upper chords around the center have large cross-sectional areas. The optimal objective value $\widehat{V}$ is equal to 1.63454 m³. The triangles in Fig. 6.39 are the values of $\widehat{V}$ for $\alpha = 0.0, 0.05, 0.1, 0.15$, and 0.2. The solid line shows the second-order approximation at $\alpha = 0.0$, where $\partial^2 \widehat{V}/\partial \alpha^2 = 1.57076$ m³. Note that $\widehat{V}$ is an even function of α; i.e., $\partial \widehat{V}/\partial \alpha = 0$ at $\alpha = 0$, which is obvious from the absence of a linear term and the existence of quadratic term of α in (6.48). It may be observed from Fig. 6.39 that the optimal objective values are successfully approximated as a quadratic function of α.

The optimum design for $\alpha = 1.0$, shown in Fig. 6.40, has larger cross-sectional areas in the upper chords around the center than those for $\alpha = 0$, because the difference in the movements of two supports causes pseudostatic deformation, which leads to bending deformation of the arch around the center. The optimal objective value for $\alpha = 1.0$ is 1.71497 m³. The value of $\partial \widehat{V}/\partial \alpha$ at $\alpha = 0.5$ is 0.485797 m³, while the parametric sensitivity coefficient obtained by the central difference method with $\Delta \alpha = 0.01$ is 0.48 m³, which agrees with good accuracy with the analytical result. CPU time for postoptimal analysis is only about 0.2% of that for optimization.

The optimal cross-sectional areas considering the wave passage effect only, i.e., $\alpha = 0$ and $\eta = 1$, are almost the same as those in Fig. 6.40, where the optimal objective value is 1.69317 m³. Note that η is an auxiliary integer parameter for indicating incorporation of the wave passage effect. Therefore, variation of η indirectly corresponds to variation of v_{app}, as observed from the term η/v_{app} in (6.48). The parametric sensitivity coefficient with respect to η at $\eta = 1.0$ is 5.14221×10^{-2} m³, and the sensitivity coefficient obtained by

FIGURE 6.40: Optimal cross-sectional areas considering the incoherency effect ($\alpha = 1.0$).

the central finite difference approach with $\Delta\eta = 0.1$ is 5.14×10^{-2} m³, which agrees with good accuracy with the analytical result.

Finally, optimum designs are found considering the difference in the amplification of the seismic motions by local soil. Let μ_1 and μ_2 denote the scaling factors of the maximum ground displacements and the response spectra corresponding to the horizontal displacements of the two supports. The distribution of optimal cross-sectional areas for $(\mu_1, \mu_2) = (1.2, 0.8)$ is similar to that in Fig. 6.40, and the optimal objective value is 1.77694 m³. Because $\widehat{V}$ is an even function of $\mu_1 - \mu_2$, and an odd function of $\mu_1 + \mu_2$, the sensitivity coefficients with respect to μ_1 and μ_2 are the same and equal to 0.493577 m³, whereas the coefficient by the central finite difference approach with $\Delta\mu_1 = \Delta\mu_2 = 0.05$ is 0.5107 m³, which is slightly different from the analytical result. However, the local effect can be successfully approximated using parametric sensitivity coefficients.

6.8 Substructure approach to seismic optimization

6.8.1 Introduction

Spatial structures, such as long-span dome structures and bridges, are often built on supporting (boundary) structure. Therefore, the flexibility of the lower supporting structure should be taken into account when evaluating the seismic responses of the upper roof structure.

A response spectrum approach, e.g., the complete quadratic combination (CQC) method, may be effectively used for reducing the computational cost; see Appendix A.7. However, when only the upper structure is modified, the computational cost for eigenvalue analysis is still large if the whole structure is to be analyzed and the number of degrees of freedom (DOFs) of the lower structure is much larger than that of the upper structure. If the damping properties are modeled based on Rayleigh damping, then a difficulty also arises for the case where the upper and lower structures have different damping properties, e.g., a steel roof supported by a reinforced concrete structure.

Consider the process of carrying out seismic response analysis for upper and lower structures independently. In this case, the response of the lower structure can be conceived as an input to the upper structure, and different motions may be applied to the connections between the two substructures. This situation is similar to the seismic response analysis considering spatial variation of seismic motions that results in quasistatic forced deformation of the structure; see Sec. 6.7 for details.

The dynamic analysis method utilizing division of the structure into several domains is called the substructure approach (Meirovitchand and Hale 1981). If the substructures have hierarchical relations, the substructure supported by the *primary structure* (PS) is called the *secondary structure* (SS), which includes nonstructural equipment, a tower-type structure on a building, etc. The dynamic response of the PS can be conceived as the input to the SS, and an SS should be designed to prevent resonance with the PS. Hence, various approaches have been presented for evaluating the seismic response of the SS. The floor response spectra can be used if the weight of the SS is sufficiently small compared with that of the PS, and accordingly, the interaction in inertia between the SS and the PS can be neglected (Villaverde 1997).

Gupta (1997) developed a method of frequency domain analysis that rigorously considers the interaction between the PS and the SS that is supported at a single point by the PS. Dey and Gupta (1998) extended this method to SS with multiple supports, assuming one-to-one correspondence between the DOFs of the SS and the PS at the connections.

In this section, a general and computationally efficient approach by Ohsaki (2003b) is summarized for the seismic design of the SS while the properties of the PS are fixed. The method is applied to the optimum design of arch-type trusses supported by column-type trusses.

6.8.2   Frequency domain analysis for a secondary structure

Consider an arch-type truss, as shown in Fig. 6.41, supported by column-type trusses, which are conceived as the SS and the PS, respectively. Fig. 6.42 illustrates the connection between the PS and the SS, where the solid lines are the connecting members.

Let N^{P} and N^{s} denote the numbers of DOFs of the PS and the SS, respectively. In the following, the superscripts $(\cdot)^{\mathrm{P}}$ and $(\cdot)^{\mathrm{s}}$ are used for the values corresponding to the PS and the SS, respectively, for which the displacement vectors are denoted by $\mathbf{u}^{\mathrm{P}} = (u_1^{\mathrm{P}}, \ldots, u_{N^{\mathrm{P}}}^{\mathrm{P}})^{\top}$ and $\mathbf{u}^{\mathrm{s}} = (u_1^{\mathrm{s}}, \ldots, u_{N^{\mathrm{s}}}^{\mathrm{s}})^{\top}$. Note that the first n^{P} components of $\mathbf{u}^{\mathrm{P}}$ are the displacements of the nodes of the connecting members on the PS (node 'a' in Fig. 6.42). Similarly, the first n^{s} components of $\mathbf{u}^{\mathrm{s}}$ are the displacements of the nodes of the connecting members on the SS (nodes 'b' and 'c' in Fig. 6.42). Therefore, $n^{\mathrm{P}} = 2$ and $n^{\mathrm{s}} = 4$ for the connection in Fig. 6.42.

Let $\mathbf{d}_i$ denote the displacement vector of the connecting nodes in the SS that is induced by a unit rigid-body translation of the connecting members in

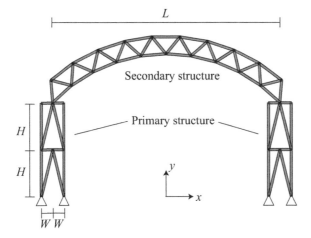

FIGURE 6.41: An arch-type truss supported by column-type trusses.

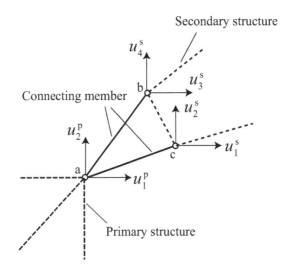

FIGURE 6.42: Connection between the PS and the SS.

the direction of $\mathbf{u}^{\mathrm{p}}$. For the connection in Fig. 6.42, the unit translation in the direction of displacement u_1^{p} leads to the unit translations in displacements u_1^{s} and u_3^{s}; i.e., $\mathbf{d}_1 = (1, 0, 1, 0)^{\top}$. The $n^{\mathrm{s}} \times n^{\mathrm{p}}$ matrix of which the ith column is $\mathbf{d}_i$ is denoted by $\mathbf{D}$.

The rth eigenmode of free vibration of the PS without the SS is denoted by $\mathbf{\Phi}^{\mathrm{p}r} = (\Phi_1^{\mathrm{p}r}, \ldots, \Phi_{N^{\mathrm{p}}}^{\mathrm{p}r})^{\top}$. Similarly, the eigenmode of the SS, including the connecting members with fixed supports at the connecting nodes in the PS, is denoted by $\mathbf{\Phi}^{\mathrm{s}r} = (\Phi_1^{\mathrm{s}r}, \ldots, \Phi_{N^{\mathrm{s}}}^{\mathrm{s}r})^{\top}$; i.e., for the example in Fig. 6.42, node 'a' is to be fixed. Let $H_r^{\mathrm{p}}(\omega)$ denote the transfer function of the rth eigenmode of PS. In the following, the circular frequency ω in the argument indicates a function in the frequency domain.

Let $\mathbf{u}^{\mathrm{sc}}$ denote the vector consisting of the first n^{s} components of $\mathbf{u}^{\mathrm{s}}$. The first n^{s} columns of the damping and stiffness matrices of the SS are denoted by $\mathbf{C}^{\mathrm{s}}$ and $\mathbf{K}^{\mathrm{s}}$, respectively. The first n^{s} columns of the mass matrix of the SS after removing the connecting members is denoted by $\mathbf{M}^{\mathrm{s}}$. Then, an $N^{\mathrm{s}} \times n^{\mathrm{s}}$ matrix $\mathbf{R}$ is defined as

$$\mathbf{R}(\omega) = \omega^2 \mathbf{M}^{\mathrm{s}} - \mathrm{i}\omega \mathbf{C}^{\mathrm{s}} - \mathbf{K}^{\mathrm{s}} \tag{6.60}$$

where i is the imaginary unit. The interaction force between the PS and the SS can be evaluated from the deformation of the connecting members and $\mathbf{R}(\omega)$.

The Fourier transformation of the seismic acceleration is denoted by $\ddot{z}(\omega)$. The participation factor of the rth eigenmode corresponding to the specified input direction is denoted by β_r^{p}. The transfer function of the rth mode of the SS to the seismic motion at the fixed base is denoted by $H_r^{\mathrm{s}}(\omega)$. The $n^{\mathrm{p}} \times n^{\mathrm{p}}$ stiffness matrix of the connecting members is denoted by $\mathbf{K}^{\mathrm{c}}$.

Then the Fourier transformation of the displacements $u_i^{\mathrm{s}}(\omega)$ $(i = 1, \ldots, N^{\mathrm{s}})$ of the SS can be expressed by $\ddot{z}(\omega)$ after solving the following set of N^{s} linear equations:

$$u_i^{\mathrm{s}}(\omega) - \sum_{k=1}^{n^{\mathrm{p}}} \left(\sum_{l=1}^{N^{\mathrm{s}}} \sum_{t=1}^{N^{\mathrm{s}}} \sum_{j=1}^{n^{\mathrm{s}}} \sum_{m=1}^{n^{\mathrm{p}}} \sum_{r=1}^{N^{\mathrm{p}}} \Phi_i^{\mathrm{s}l} \Phi_t^{\mathrm{s}l} \Phi_m^{\mathrm{p}r} \Phi_k^{\mathrm{p}r} \right.$$

$$\left. H_l^{\mathrm{s}}(\omega) H_r^{\mathrm{p}}(\omega) R_{tj}(\omega) D_{jm} K_{mk}^{\mathrm{c}} \right) u_k^{\mathrm{s}}(\omega)$$

$$= -\left[\sum_{l=1}^{N^{\mathrm{s}}} \sum_{t=1}^{N^{\mathrm{s}}} \Phi_i^{\mathrm{s}l} \Phi_t^{\mathrm{s}l} H_l^{\mathrm{s}}(\omega) \right. \tag{6.61}$$

$$\left. \left(M_{tt}^{\mathrm{s}} r_t + \sum_{j=1}^{n^{\mathrm{s}}} \sum_{r=1}^{N^{\mathrm{p}}} \sum_{m=1}^{n^{\mathrm{p}}} R_{tj}(\omega) D_{jm} \Phi_m^{\mathrm{p}r} H_r^{\mathrm{p}}(\omega) \beta_r^{\mathrm{p}} \right) \right] \ddot{z}(\omega),$$

$$(i = 1, \ldots, N^{\mathrm{s}})$$

where the components of matrices are indicated using subscripts. This way, the transfer function $h_i^{\mathrm{s}}(\omega) = u_i^{\mathrm{s}}(\omega)/\ddot{z}(\omega)$ of u_i^{s} can be found for seismic motion $\ddot{z}(\omega)$. The power spectrum $G_{u_i^{\mathrm{s}} u_i^{\mathrm{s}}}(\omega)$ of u_i^{s} is obtained by multiplying $(h_i^{\mathrm{s}}(\omega))^2$ by $G_{\ddot{z}\ddot{z}}(\omega)$ for the seismic acceleration. The mean square response can be found by integrating $G_{u_i^{\mathrm{s}} u_i^{\mathrm{s}}}(\omega)$ in an appropriate frequency domain. The specified peak factor is multiplied by the mean square response to obtain the maximum response in a similar manner as in Sec. 6.7.

In the process of the seismic design of structures, a response spectrum, instead of a power spectrum, is usually given. Therefore, for example, (6.50) in Sec. 6.7 can be used for converting the displacement response spectrum $S_D(\omega, h)$ to the power spectrum (Der Kiureghian and Neuenhofer 1992).

6.8.3 Optimization problem

The cross-sectional areas $\mathbf{A} = (A_1, \ldots, A_m)^\top$ of the SS are optimized for a given design of the PS, where m is the number of members in the SS. The objective function to be minimized is the total structural volume $V(\mathbf{A})$ of the SS. The direction of seismic motion, the design response spectrum $S_D(\omega, h)$, and the modal damping ratios are given. Constraints are assigned for the mean-maximum response strain ε_i (> 0) of the ith member of SS. Let ε^U denote the upper bound for ε_i. The optimization problem is formulated as

$$\text{Minimize} \quad V(\mathbf{A}) \tag{6.62a}$$

$$\text{subject to} \quad \varepsilon_i(\mathbf{A}) \leq \varepsilon^U, \quad (i = 1, \cdots, m) \tag{6.62b}$$

$$A_i \geq A_i^L, \quad (i = 1, \ldots, m) \tag{6.62c}$$

where A_i^L is the lower bound for A_i.

Optimal cross-sectional areas are found using the following algorithm:

Step 1: Specify the properties of the PS and carry out eigenvalue analysis to find the natural circular frequencies ω_r^P, eigenmodes $\mathbf{\Phi}_r^P$, and participation factors β_r^P of the PS. Define the damping ratio h_r^P and compute the modal transfer function $H_r^P(\omega)$ of the PS.

Step 2: Specify the duration of the seismic motion and the peak factors for the seismic motion and white noise, and obtain $G_{\ddot{z}\ddot{z}}(\omega)$ from $S_D(\omega, h)$ using (6.50).

Step 3: Define the damping ratios of the SS, and assign the initial values of A_i.

Step 4: Compute the natural circular frequencies ω_r^s, eigenmodes $\mathbf{\Phi}_r^s$, and modal transfer function $H_r^s(\omega)$ of the SS with fixed support.

Step 5: Solve (6.61) for the transfer function of the responses of the SS against the specified seismic motion. Multiply the peak factor by the mean-square response strains to obtain the mean-maximum response strain ε_i.

Step 6: Modify $\mathbf{A}$ according to the optimization algorithm, and go to Step 4 if not converged.

This way, the eigenvalue analysis of the PS should be carried out only once at the beginning of the optimization process. Hence, this algorithm is very effective if the number of degrees of freedom of the PS is much larger than that of the SS.

FIGURE 6.43: Optimal cross-sectional areas for $\alpha = 1.0$ and $\varepsilon^U = 0.0005$.

6.8.4 Numerical examples

Optimum designs are found for an arch-type truss supported by column-type trusses, as shown in Fig. 6.41, where $L = 20$ m, $W = 1$ m, $H = 4$ m, and the open angle of the lower circle of the arch is 50 deg. The upper and lower circles have the same center, and the difference in their radii is 1.5 m. The lower chords, upper chords, and diagonals have the same lengths, respectively.

The elastic modulus is 210.0 kN/mm^2 for both the PS and the SS. Although the method in this section can be used for the case where the PS and the SS have different damping properties, the stiffness-proportional damping with a damping ratio of 0.02 for the lowest eigenmode is used for both the PS and the SS for comparison with the CQC method (Wilson, Der Kiureghian, and Bayo 1982); see Appendix A.7. Optimization is carried out using IDESIGN Ver.3.5 (Arora and Tseng 1987). A finite difference approach is used for sensitivity analysis.

The cross-sectional areas of the PS are defined by a parameter α as 0.05α m^2 to investigate the dependence of the optimal solution of the SS on the stiffness of the PS. The lower bound for A_i of the SS is 0.01 m^2. Concentrated masses of 1.0×10^4 kg and 1.0×10^5 kg are located at the nodes of the SS and the PS, respectively. The member mass is not included for brevity.

Only the horizontal motions are considered, and the displacement response spectrum by Newmark and Hall (1982) is used for defining the set of strong seismic motions (see Appendix A.7 for the definition of the parameter values for the spectrum). The duration of motion is 25.0 s, and the peak factors for both the seismic motion and white noise are 3.0 (Der Kiureghian 1980; Der Kiureghian and Neuenhofer 1992).

The optimal cross-sectional areas for $\alpha = 1.0$ and $\varepsilon^U = 0.0005$ are as shown in Fig. 6.43, where the width of each member is proportional to its cross-sectional area. The fundamental eigenmode of the SS is antisymmetric with respect to the vertical center axis; hence, the cross-sectional areas of the members around the center are relatively small, because the deformation corresponding to the first mode is relatively small in this region.

The relations between ε^U and the optimal value of V for various values of α are plotted in Fig. 6.44. As is seen, V is a decreasing function of ε^U. For a larger value of α, the axial forces near the connections between the PS and the SS have larger values, and V increases accordingly. Let T_1^s and T_1^p denote the fundamental natural periods of the SS and the PS, respectively,

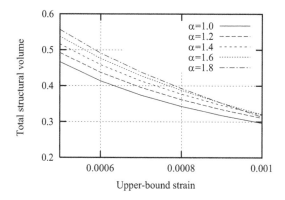

FIGURE 6.44: Relation between the upper-bound strain and the optimal total structural volume (m^3).

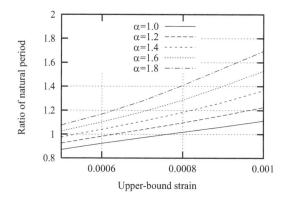

FIGURE 6.45: Relation between the upper-bound strain and the ratios of the fundamental periods of the optimal solutions.

with fixed supports. The relation between ε^U and the ratio T_1^s/T_1^p of the optimal solutions is plotted in Fig. 6.45. As is seen, the ratio decreases as ε^U is decreased, because the stiffness of the SS becomes larger. Note that there are some optimal solutions with $T_1^s/T_1^p = 1$, which is usually to be prevented. This fact indicates that the properties of the responses cannot be discussed based on the lowest eigenmodes only; i.e., the effects of the higher modes should be incorporated.

Next, we consider asymmetric input to the arch-type truss through modification of the cross-sectional areas A^R of the members in the right column-type truss for the case $\alpha = 1.0$ and $\varepsilon^U = 0.0005$. The cross-sectional areas of the arch are linked to preserve the symmetry property. The values of V (m^3) for $A^R = 0.06$ and 0.04 m^2 are 0.54250 and 0.56672, respectively, which are 16%

FIGURE 6.46: Optimal cross-sectional areas for $\alpha = 1$, $\varepsilon^U = 0.0005$, and $A^R = 0.06$ m^2.

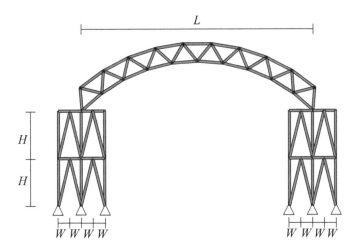

FIGURE 6.47: An arch-type truss supported by column-type trusses ($S = 2$).

and 21% larger than that of the symmetric case with $A^R = 0.05$ m^2. Therefore, larger cross-sectional areas are required for asymmetric cases to satisfy the strain constraints, because the quasistatic component of the response increases as a result of the difference of the motions at the two connections. The optimal cross-sectional areas for $A^R = 0.06$ m^2 are as shown in Fig. 6.46. As is seen, the cross-sectional areas of the upper and lower chords around the center have larger values than those in Fig. 6.43 due to the bending deformation of the arch corresponding to the quasistatic component.

To investigate the relation between the DOFs of the PS and the computational cost, optimal solutions are found for different numbers of spans S of the PS. The geometry for $S = 2$ is as shown in Fig. 6.47. The results are compared with those by the completely quadratic combination (CQC) method. Computation was carried out using a PC with AMD Athron 1.0GHz. The optimization results are shown in Table 6.5, where the number of steps indicates the number of design modifications in the optimization process. The parameters are $\alpha = 1.0$ and $\varepsilon^U = 0.0005$. It is seen from Table 6.5 that the mean CPU time for an optimization process (Steps 4–6) does not depend on

TABLE 6.5: Number of steps, total CPU time (s), and mean CPU time (s) for the optimization process of the PS with various numbers of span S.

S	DOF of PS	Substructure			CQC method		
		Steps	Total CPU	Mean CPU	Steps	Total CPU	Mean CPU
1	24	9	25.4	2.44	12	10.1	0.84
2	40	12	35.9	2.74	9	11.3	1.26
3	56	9	27.3	2.69	11	22.1	2.01
4	72	9	28.0	2.76	17	50.9	2.99
5	86	8	25.7	2.82	19	81.6	4.29

S if the substructure method is used, while it is almost proportional to S if the CQC method is used; hence, the substructure method is very effective if the number of DOFs of the PS is very large compared with that of the SS.

Appendix

A.1 Mathematical preliminaries

A.1.1 Positive definite matrix and convex functions

A symmetric $n \times n$ matrix $\mathbf{B}$ is said to be *positive semidefinite* if

$$\mathbf{b}^\top \mathbf{B} \mathbf{b} \geq 0 \qquad (A.1)$$

for any nonzero n-vector $\mathbf{b}$. The matrix $\mathbf{B}$ is *positive definite* if (A.1) is satisfied with strict inequality. All the eigenvalues of a positive definite matrix are positive.

A set $\mathcal{C}$ of n-vector $\mathbf{x}$ is said to be *convex* if

$$\mathbf{x} = \alpha \mathbf{x}_1 + (1 - \alpha)\mathbf{x}_2 \in \mathcal{C} \qquad (A.2)$$

for any parameter value $0 \leq \alpha \leq 1$ and any pair of n-vectors $\mathbf{x}_1 \in \mathcal{C}$ and $\mathbf{x}_2 \in \mathcal{C}$.

A function $F(\mathbf{x})$ defined in a convex set $\mathcal{C}$ of n-vector $\mathbf{x}$ is said to be convex if

$$F(\mathbf{x}) \leq F(\alpha \mathbf{x}_1 + (1 - \alpha)\mathbf{x}_2) \qquad (A.3)$$

for any parameter value $0 \leq \alpha \leq 1$ and any pair of n-vectors $\mathbf{x}_1 \in \mathcal{C}$ and $\mathbf{x}_2 \in \mathcal{C}$. Furthermore, $F(\mathbf{x})$ is *strictly convex* if (A.3) is satisfied with strict inequality.

Let $F_1(\mathbf{x}), \ldots, F_p(\mathbf{x})$ denote convex functions defined in the convex set $\mathcal{C}$. Then the *pointwise maximum*

$$F^{\max}(\mathbf{x}) = \max\{F_1(\mathbf{x}), \ldots, F_p(\mathbf{x})\} \qquad (A.4)$$

is also a convex function (Boyd and Vandenberghe 2004), as illustrated in Fig. A.1.

A function $F(\mathbf{x})$ defined in a convex set $\mathcal{C}$ of n-vector $\mathbf{x}$ is said to be *quasiconvex* if

$$F(\mathbf{x}_2) \leq F(\mathbf{x}_1) \implies F(\alpha \mathbf{x}_1 + (1 - \alpha)\mathbf{x}_2) \leq F(\mathbf{x}_1) \qquad (A.5)$$

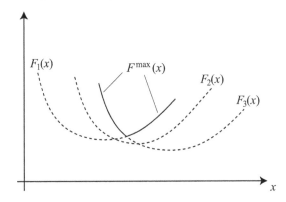

FIGURE A.1:   An example of a pointwise maximum of convex functions.

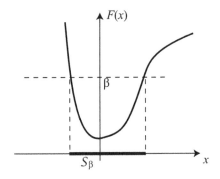

FIGURE A.2:   An example of a quasiconvex function.

holds for any parameter value $0 \leq \alpha \leq 1$. Also, if the *level set* $\mathcal{S}_\beta$ defined as

$$\mathcal{S}_\beta = \{\mathbf{x} \in \mathcal{C} \mid F(\mathbf{x}) \leq \beta\} \tag{A.6}$$

is a convex set, then $F(\mathbf{x})$ is quasiconvex, as illustrated in Fig. A.2. Note that a convex function is quasiconvex.

A function $F(\mathbf{x})$ defined in a convex set $\mathcal{C}$ of n-vector $\mathbf{x}$ is said to be *pseudoconvex* if

$$(\mathbf{x}_1 - \mathbf{x}_2)^\top \nabla F(\mathbf{x}_2) \leq 0 \Longrightarrow F(\mathbf{x}_1) \leq F(\mathbf{x}_2) \tag{A.7}$$

Note that a convex differentiable function is pseudoconvex.

A.1.2   Rayleigh's principle

Let $\mathbf{B}$ denote an $n \times n$ symmetric positive definite matrix, and $\mathbf{C}$ be an $n \times n$ symmetric matrix. A generalized eigenvalue problem is formulated as

$$\mathbf{C}\Phi_i = \Omega_i \mathbf{B}\Phi_i, \quad (i = 1, \ldots, n) \tag{A.8}$$

where Ω_i and $\mathbf{\Phi}_i$ are the ith eigenvalue and eigenvector, respectively, and Ω_i is ordered as

$$\Omega_1 \leq \Omega_2 \leq \cdots \leq \Omega_n \tag{A.9}$$

The ortho-normalization condition is given as

$$\mathbf{\Phi}_i^\top \mathbf{B} \mathbf{\Phi}_j = \delta_{ij}, \quad (i = 1, \ldots, n) \tag{A.10}$$

where δ_{ij} is the Kronecker delta. From (A.8) and (A.10), we obtain

$$\mathbf{\Phi}_i^\top \mathbf{C} \mathbf{\Phi}_j = \delta_{ij} \Omega_i, \quad (i, j = 1, \ldots, n) \tag{A.11}$$

The Rayleigh quotient $R(\mathbf{b})$ is defined for a nonzero n-vector $\mathbf{b}$ as

$$R(\mathbf{b}) = \frac{\mathbf{b}^\top \mathbf{C} \mathbf{b}}{\mathbf{b}^\top \mathbf{B} \mathbf{b}} \tag{A.12}$$

Suppose $\mathbf{b}$ is written as a linear combination of the eigenvectors as

$$\mathbf{b} = \sum_{i=1}^n c_i \mathbf{\Phi}_i \tag{A.13}$$

with the coefficients c_i satisfying

$$\sum_{i=1}^n c_i^2 \neq 1 \tag{A.14}$$

From (A.10)–(A.14), we obtain

$$R(\mathbf{b}) = \frac{\displaystyle\sum_{i=1}^n c_i^2 \Omega_i}{\displaystyle\sum_{i=1}^n c_i^2} \tag{A.15}$$

Hence, from (A.9), (A.14), and (A.15), Rayleigh's principle states that the following inequalities are satisfied for any nonzero vector $\mathbf{b}$:

$$\Omega_1 \leq R(\mathbf{b}) \leq \Omega_n \tag{A.16}$$

The equalities $R(\mathbf{b}) = \Omega_1$ and $R(\mathbf{b}) = \Omega_n$ are satisfied for $\mathbf{b} = \mathbf{\Phi}_1$ and $\mathbf{b} = \mathbf{\Phi}_n$, respectively, and the maximum and minimum eigenvalues Ω_n and Ω_1 are obtained, respectively, by maximizing and minimizing $R(\mathbf{b})$ with respect to $\mathbf{b}$.

Example A.1
Let $\mathbf{B}$ and $\mathbf{C}$ be given as

$$\mathbf{B} = \begin{pmatrix} 1 & 0 \\ 0 & 1 \end{pmatrix}, \quad \mathbf{C} = \begin{pmatrix} 5 & -2 \\ -2 & 2 \end{pmatrix} \tag{A.17}$$

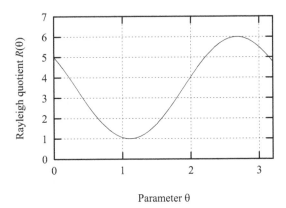

Parameter θ

FIGURE A.3:   Variation of Rayleigh quotient $R(\theta)$ with respect to parameter θ.

Then we have

$$\boldsymbol{\Phi}_1 = \frac{1}{\sqrt{5}} \begin{pmatrix} 1 \\ 2 \end{pmatrix}, \quad \boldsymbol{\Phi}_2 = \frac{1}{\sqrt{5}} \begin{pmatrix} 2 \\ -1 \end{pmatrix}, \quad \Omega_1 = 1, \quad \Omega_2 = 6 \tag{A.18}$$

Suppose $\mathbf{b} = (b_1, b_2)^\top$ is parameterized by θ as

$$b_1 = \cos\theta, \quad b_2 = \sin\theta \tag{A.19}$$

Then the Rayleigh quotient $R(\theta)$ is obtained as

$$R(\theta) = 5\cos^2\theta - 4\sin\theta\cos\theta + 2\sin^2\theta \tag{A.20}$$

which is plotted in Fig. A.3. As is seen, $R(\theta)$ has maximum value 6 and minimum value 1, which are equal to Ω_2 and Ω_1, respectively.

A.1.3   Singular value decomposition

Let $\mathbf{D}$ denote an $n \times m$ rectangular matrix, and suppose $n \le m$ for simplicity. The eigenvalues of the $m \times m$ symmetric matrix $\mathbf{D}^\top \mathbf{D}$ are denoted by Ω_i $(i = 1, \ldots, m)$, which are nonnegative because $\mathbf{D}^\top \mathbf{D}$ is positive semidefinite. The number r of nonzero eigenvalues is equal to the rank r of $\mathbf{D}$. Then the *singular value decomposition* (SVD) of $\mathbf{D}$ is written as (Horn and Johnson 1990; Atkinson 1989)

$$\mathbf{D} = \mathbf{S}\boldsymbol{\Omega}\mathbf{R}^\top \tag{A.21}$$

where

$$\boldsymbol{\Omega} = \begin{pmatrix} \operatorname{diag}(\omega_1, \ldots, \omega_r) & \mathbf{O} \\ \mathbf{O} & \mathbf{O} \end{pmatrix} \tag{A.22}$$

Here, $\mathbf{O}$ is a null matrix of appropriate size, and the diagonal terms $\omega_i = \sqrt{\Omega_i}$ of the $n \times m$ rectangular matrix $\mathbf{\Omega}$ are called singular values of $\mathbf{D}$, which are defined in nonincreasing order as $\omega_1 \geq \ldots, \geq \omega_r > 0$.

The $m \times m$ matrix $\mathbf{R}$ and $n \times n$ matrix $\mathbf{S}$ are orthogonal matrices satisfying

$$\mathbf{R}^\top \mathbf{R} = \mathbf{R}\mathbf{R}^\top = \mathbf{I}_m, \quad \mathbf{S}^\top \mathbf{S} = \mathbf{S}\mathbf{S}^\top = \mathbf{I}_n \tag{A.23}$$

where $\mathbf{I}_m$ and $\mathbf{I}_n$ are the $m \times m$ and $n \times n$ identity matrices, respectively.

By premultiplying $\mathbf{S}^\top$ and postmultiplying $\mathbf{R}$ on both sides of (A.21) and using (A.23), we obtain

$$\mathbf{S}^\top \mathbf{D} = \mathbf{\Omega}\mathbf{R}^\top, \quad \mathbf{D}\mathbf{R} = \mathbf{S}\mathbf{\Omega} \tag{A.24}$$

Hence, the column vectors of $\mathbf{S}$ and $\mathbf{R}$ are called left and right singular vectors, respectively. For the case $r < m$, the column vectors $\mathbf{R}_i$ $(i = r + 1, \ldots, m)$ of $\mathbf{R}$ correspond to the zero singular value as

$$\mathbf{D}\mathbf{R}_i = \mathbf{0} \tag{A.25}$$

A.1.4 Directional derivative and subgradient

The directional derivative $F'(\mathbf{x}^0, \mathbf{s})$ of the function $F(\mathbf{x})$ at $\mathbf{x} = \mathbf{x}^0$ in the direction of $\mathbf{s}$ is defined as

$$F'(\mathbf{x}^0, \mathbf{s}) = \lim_{t \to 0+} \frac{F(\mathbf{x}^0 + t\mathbf{s}) - F(\mathbf{x}^0)}{t} \tag{A.26}$$

Let $\mathcal{X}$ denote a nonempty open convex set of $\mathbf{x}$, and suppose $F(\mathbf{x})$ is a convex function of $\mathbf{x}$ in $\mathcal{X}$. The subgradient $\boldsymbol{\zeta}$ of $F(\mathbf{x})$ at $\mathbf{x} = \mathbf{x}^0$ is defined as a vector satisfying (Rocakfellar 1970; Ekeland and Témam 1999)

$$F(\mathbf{x}) \geq F(\mathbf{x}^0) + \boldsymbol{\zeta}^\top (\mathbf{x} - \mathbf{x}^0) \text{ for all } \mathbf{x} \in \mathcal{X} \tag{A.27}$$

Although a subgradient is originally defined for a convex function, it can also be defined for a concave function as (Floudas 1995)

$$F(\mathbf{x}) \leq F(\mathbf{x}^0) + \boldsymbol{\zeta}^\top (\mathbf{x} - \mathbf{x}^0) \text{ for all } \mathbf{x} \in \mathcal{X} \tag{A.28}$$

The set of all subgradients of $F(\mathbf{x})$ is called the *subdifferential* of $F(\mathbf{x})$, which is denoted by $\partial F(\mathbf{x})$.

A.2 Optimization methods

A.2.1 Classification of optimization problems

A problem of minimizing or maximizing an objective function under some constraints on the functions of the variables is called a *mathematical program-*

ming problem or *optimization problem*. The solution methods for mathematical programming problems are called *mathematical programming* (Arora 2004; Haftka, Gürdal, and Kamat 1990; Luenberger 2003).

Although most of the structural optimization problems presented in this book can be solved by simply using optimization libraries or software packages, it is necessary to have a good knowledge of optimization algorithms and the classification of problems to select an appropriate method and to understand the output data and messages from the program if it could not produce a reasonable optimal solution. The optimization algorithms are classified into mathematical programming and heuristics, including genetic algorithms and simulated annealing. We focus on mathematical programming approaches in this section (see Sec. A.3 for details of heuristic approaches).

Let $\mathbf{x} = (x_1, \ldots, x_m)^\top$ denote the variable vector, where m is the number of variables. In the following, a vector is assumed to be a column vector, and its component is indicated by a subscript. The objective function to be minimized is denoted by $F(\mathbf{x})$. The equality constraints and inequality constraints are given as $G_i(\mathbf{x}) = 0$ $(i = 1, \ldots, n^{\mathrm{E}})$ and $H_i(\mathbf{x}) \leq 0$ $(i = 1, \ldots, n^{\mathrm{I}})$, which can be simply written as $\mathbf{G}(\mathbf{x}) = \mathbf{0}$ and $\mathbf{H}(\mathbf{x}) \leq \mathbf{0}$, respectively. Throughout the book, an inequality $\mathbf{a} \leq \mathbf{b}$ for vectors $\mathbf{a}$ and $\mathbf{b}$ of the same size means that inequality is satisfied for each pair of components of the vectors.

A mathematical programming problem is generally formulated as

$$\text{Minimize} \quad F(\mathbf{x}) \tag{A.29a}$$

$$\text{subject to} \quad G_i(\mathbf{x}) = 0, \quad (i = 1, \ldots, n^{\mathrm{E}}) \tag{A.29b}$$

$$H_i(\mathbf{x}) \leq 0, \quad (i = 1, \ldots, n^{\mathrm{I}}) \tag{A.29c}$$

$$\mathbf{x} \in \mathcal{X} \tag{A.29d}$$

where $\mathcal{X}$ is the admissible region of $\mathbf{x}$, which also distinguishes the types of the variables, i.e., real, integer, 0–1, etc. The constraints defining the range of variables as

$$x_i^{\mathrm{L}} \leq x_i \leq x_i^{\mathrm{U}}, \quad (i = 1, \ldots, m) \tag{A.30}$$

with lower bound x_i^{L} and upper bound x_i^{U} are called *side constraints*, *bound constraints*, or *box constraints*, and are generally handled separately from the general inequality constraints (A.29c) in the optimization algorithm.

A solution is said to be *feasible* if it satisfies all the constraints (A.29b), (A.29c), and (A.29d). The region of feasible solutions is called the *feasible region*. The feasible solution that minimizes the objective function $F(\mathbf{x})$ is called the *global optimal solution* or simply the *optimal solution*. The value of the objective function at the optimal solution is called the *optimal objective value* or the *optimal value*. A solution that minimizes $F(\mathbf{x})$ among its neighborhood feasible solutions is called the *local optimal solution*. The inequality constraint $H_i(\mathbf{x}) \leq 0$ is said to be *active* if it is satisfied with equality; otherwise, it is *inactive*.

An optimization problem is called a *linear programming* (LP) problem if all the objective and constraint functions are linear functions of real variables. In contrast, the problem is called a *nonlinear programming* (NLP) problem if at least one of the functions is nonlinear. A solution method for NLP is called *nonlinear programming*. Contrary to LP, which has general solution methods called the simplex method and the interior-point method, there is no general method that is effective for all types of NLP problems. Therefore, good knowledge is needed to select the most appropriate algorithm for the problem at hand. Since we mainly utilize NLP in this book, details of LP may be consulted in textbooks, e.g., Mangasarian (1969), Peressini, Sullivan, and Uhl (1988), and Luenberger (2003).

A.2.2 Nonlinear programming

A.2.2.1 Unconstrained optimization problem

An NLP problem without constraint is called an unconstrained NLP problem. We first define the *gradient* $\nabla F(\mathbf{x})$ of the objective function $F(\mathbf{x})$ with respect to the variables $\mathbf{x} = (x_1, \ldots, x_m)^\top$ as

$$\nabla F(\mathbf{x}) = \left(\frac{\partial F}{\partial x_1}, \ldots, \frac{\partial F}{\partial x_m} \right)^\top \tag{A.31}$$

Note that $\nabla F(\mathbf{x})$ corresponds to the direction of maximum increment of $F(\mathbf{x})$ for a specified value of $\Delta \mathbf{x}^\top \Delta \mathbf{x}$ for the increment $\Delta \mathbf{x}$ of $\mathbf{x}$. Therefore, the objective function $F(\mathbf{x})$ is most efficiently decreased if the solution is modified in the direction $-\nabla F(\mathbf{x})$. The optimization method of updating $\mathbf{x}$ as follows in the direction $-\nabla F(\mathbf{x})$ is called the *steepest descent method*:

$$\mathbf{x}^{(k+1)} = \mathbf{x}^{(k)} - \tau \nabla F(\mathbf{x}^{(k)}) \tag{A.32}$$

where k is the iteration counter, and τ is the parameter defining the magnitude of modification of $\mathbf{x}$.

In this approach, however, the estimated point overshoots the minimum point in the direction $-\nabla F(\mathbf{x})$ if τ is too large. In contrast, the computational cost increases if a smaller value is given for τ. Therefore, only the direction is defined by $-\nabla F(\mathbf{x})$, and the minimum point in this direction is found by the process called *line search*.

Example A.2

Consider the following function defined with respect to two variables x_1 and x_2 (Katoh, Ohsaki, and Tani 2002):

$$F(\mathbf{x}) = \frac{1}{2}(x_1 - 3)^2 + (x_2 - 2)^2 \tag{A.33}$$

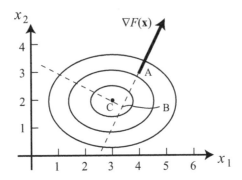

FIGURE A.4: Contour lines and gradient of a two-variable function.

The ellipses in Fig. A.4 indicate the contour lines of $F(\mathbf{x})$. The gradient of $F(\mathbf{x})$ is given as

$$\nabla F(\mathbf{x}) = \begin{pmatrix} x_1 - 3 \\ 2(x_2 - 2) \end{pmatrix} \tag{A.34}$$

For example, at point 'A' with $(x_1, x_2) = (4, 3)$, $F(\mathbf{x}) = 3/2$ and $\nabla F(\mathbf{x}) = (1, 2)^\top$, as shown in Fig. A.4.

Suppose point 'A' in Fig. A.4 is given as the initial solution $\mathbf{x}^{(0)}$. Then the variables are modified in the direction $-\nabla F(\mathbf{x}^{(0)}) = (-1, -2)^\top$. Because the objective function is a quadratic function of $\mathbf{x}$, the value of F along the line $\mathbf{x} = \mathbf{x}^{(0)} - \tau \nabla F(\mathbf{x}^{(0)})$ is a quadratic function of τ as

$$F = \frac{1}{2}(1 - \tau)^2 + (1 - 2\tau)^2 \tag{A.35}$$

From the stationary condition of F in (A.35) with respect to τ, we obtain $\tau = 5/9$, from which $\mathbf{x}^{(1)} = (31/9, 17/9)^\top$ is derived. Hence, $\mathbf{x}$ is updated to $\mathbf{x}^{(1)}$, indicated by point 'B' in Fig. A.4, where the gradient is $\nabla F(\mathbf{x}) = (4/9, -2/9)^\top$. By carrying out the line search again, we move to $\mathbf{x}^{(2)} = (83/27, 56/27)$, which is indicated by point 'C'. This way, the point close to the optimal solution $(3, 2)$ has been obtained with two steps of iteration for this small example with ellipsoidal contour lines of $F(\mathbf{x})$. However, if the contour lines have irregular shapes, convergence is very slow; furthermore, a local optimal solution may be found for a nonconvex objective function.

Since the objective function is not usually a quadratic function, a more general approach is needed for the line search. Among several algorithms, including equal-interval search and golden section search, the simplest approach is the bi-section search as follows:

Step 1: Suppose the solution $\mathbf{x}^{(k)}$ is obtained at the kth iteration of the steepest descent method. Initialize the parameter as $\tau^{[0]} = \tau_0$ with the specified value τ_0, and set the iteration counter of line search $j = 0$ and $\mathbf{x}^{[0]} = \mathbf{x}^{(k)}$.

Step 2: Evaluate $F(\mathbf{x}^{[j+1]})$ at $\mathbf{x}^{[j+1]} = \mathbf{x}^{[j]} - \tau^{[j]} \nabla F(\mathbf{x}^{(k)})$.

Step 3: If $|F(\mathbf{x}^{[j+1]}) - F(\mathbf{x}^{[j]})|$ is smaller than the specified value, then terminate the line search, let $\mathbf{x}^{(k+1)} \leftarrow \mathbf{x}^{[j]}$, and go to the next step of the steepest descent method.

Step 4: If $F(\mathbf{x}^{[j+1]}) < F(\mathbf{x}^{[j]})$, let $\tau^{[j+1]} = \tau^{[j]}$; otherwise let $\tau^{[j+1]} = -\tau^{[j]}/2$. Update the iteration counter as $j \leftarrow j + 1$ and go to Step 2.

The optimality of the solution can be verified using the stationary condition

$$\nabla F(\mathbf{x}) = \mathbf{0} \tag{A.36}$$

and the positive semidefiniteness of the Hessian $\mathbf{D}$, for which the (i, j)-component D_{ij} is defined as

$$D_{ij} = \frac{\partial^2 F}{\partial x_i \partial x_j} \tag{A.37}$$

A.2.2.2 Constrained optimization problem: equality constraints

In structural design problems in various fields of engineering, design requirements are usually given in the form of constraints on responses, such as stresses and displacements under a specified design load. Since the responses are generally nonlinear functions of the design variables, constrained nonlinear programming problems are to be solved for optimizing structures.

For a problem with equality constraints only, the conditions for optimality are obtained using the Lagrange multiplier approach (Bersekas 1982). Let λ_j denote the Lagrange multiplier for the equality constraint $G_j(\mathbf{x}) = 0$, and define the Lagrangian as

$$\psi(\mathbf{x}, \boldsymbol{\lambda}) = F(\mathbf{x}) + \sum_{j=1}^{n^{\mathrm{E}}} \lambda_j G_j(\mathbf{x}) \tag{A.38}$$

where $\boldsymbol{\lambda} = (\lambda_1, \ldots, \lambda_{n^{\mathrm{E}}})^{\mathsf{T}}$. In the following, the argument $\mathbf{x}$ is omitted for brevity.

By differentiating $\psi(\mathbf{x}, \boldsymbol{\lambda})$ with respect to x_i, we obtain the following stationary conditions:

$$\frac{\partial \psi}{\partial x_i} = \frac{\partial F}{\partial x_i} + \sum_{j=1}^{n^{\mathrm{E}}} \lambda_j \frac{\partial G_j}{\partial x_i} = 0, \quad (i = 1, \ldots, m) \tag{A.39}$$

The unknown variables $x_1, \ldots, x_m$ and multipliers $\lambda_1, \ldots, \lambda_{n^{\mathrm{E}}}$ are obtained from n^{E} constraints $G_j(\mathbf{x}) = 0$ and m stationary conditions (A.39). Note

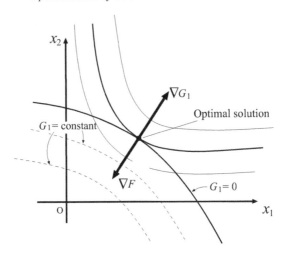

FIGURE A.5: Gradients of objective and constraint functions at the optimal the solution with a single equality constraint.

that the stationary conditions are necessary conditions for optimality, and global optimal solutions cannot always be obtained by the Lagrange multiplier method; i.e., the set of $n^E + m$ equations may have multiple solutions.

Eq. (A.39) can be written in a vector form as

$$\nabla\psi = \nabla F + \sum_{j=1}^{n^E} \lambda_j \nabla G_j = \mathbf{0} \qquad (A.40)$$

It is seen from (A.40) that ∇F is expressed as a linear combination of the gradients ∇G_j of the constraints at the optimal solution:

$$\nabla F = -\sum_{j=1}^{n^E} \lambda_j \nabla G_j \qquad (A.41)$$

As a simple case, if there exists only one constraint $G_1 = 0$ with respect to two variables, ∇F and ∇G_1 are in the same or the opposite direction, as illustrated in Fig. A.5, and the contour line of the constraint is tangential to that of the objective function at the optimal solution.

Example A.3

Consider the following quadratic programming (QP) problem (Katoh, Ohsaki, and Tani 2002):

$$\text{Minimize} \quad F(x_1, x_2) = \frac{1}{2}(x_1 - 3)^2 + (x_2 - 2)^2 \qquad (A.42a)$$

$$\text{subject to} \quad G_1(x_1, x_2) = x_1 + 3x_2 - 12 = 0 \qquad (A.42b)$$

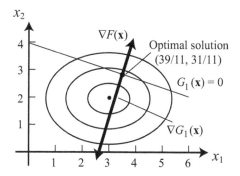

FIGURE A.6: Optimal solution of an equality constrained nonlinear programming problem.

The Lagrangian is given as

$$\psi(x_1, x_2, \lambda_1) = \frac{1}{2}(x_1 - 3)^2 + (x_2 - 2)^2 + \lambda_1(x_1 + 3x_2 - 12) \qquad (A.43)$$

The stationary conditions of the Lagrangian are obtained as

$$x_1 - 3 + \lambda_1 = 0, \quad 2x_2 - 4 + 3\lambda_1 = 0 \qquad (A.44)$$

from which and constraint (A.42b), we have the following optimal solution:

$$x_1 = \frac{39}{11}, \quad x_2 = \frac{31}{11}, \quad \lambda = -\frac{6}{11} \qquad (A.45)$$

In fact, the following relations hold at the optimal solution:

$$\nabla F = \frac{6}{11}\begin{pmatrix} 1 \\ 3 \end{pmatrix}, \quad \nabla G_1 = \begin{pmatrix} 1 \\ 3 \end{pmatrix}, \quad \nabla F = -\lambda_1 \nabla G_1 \qquad (A.46)$$

which is illustrated in Fig. A.6. This way, a QP problem with a convex quadratic objective function and linear equality constraints can be solved explicitly using the Lagrange multiplier approach.

A.2.2.3 Constrained optimization problem: inequality constraint

Consider an NLP problem with inequality constraints $H_i(\mathbf{x}) \leq 0$ ($i = 1, \ldots n^I$), where equality constraints are not considered for brevity. As an example, the feasible region is illustrated in Fig. A.7 for a problem with two variables x_1, x_2, and two constraints $H_1(\mathbf{x}) \leq 0$, $H_2(\mathbf{x}) \leq 0$. The Lagrangian is defined as

$$\psi(\mathbf{x}, \boldsymbol{\mu}) = F(\mathbf{x}) + \sum_{j=1}^{n^I} \mu_j H_j(\mathbf{x}) \qquad (A.47)$$

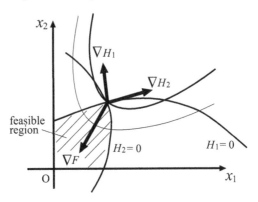

FIGURE A.7:   Gradients of objective and constraint functions at the optimal solution with inequality constraints.

where μ_j (≥ 0) is the Lagrange multiplier for the inequality constraint $H_j(\mathbf{x}) \leq 0$.

Roughly speaking, a solution is locally optimal if there is no feasible neighborhood solution that has a smaller objective value than the current solution. Therefore, F should not decrease in the direction in which all of the functions of the active constraints satisfying $H_j = 0$ decrease. On the other hand, the inactive constraint satisfying $H_j < 0$ does not have any effect on the optimality of the solution. Hence, as illustrated in Fig. A.7, the gradient ∇F of the objective function should be expressed as a non-positive linear combination of the gradients of the active constraints as

$$\nabla F = -\sum_{j=1}^{n^I} \mu_j \nabla H_j \qquad (A.48)$$

with

$$\begin{cases} \mu_j \geq 0 & \text{for } H_j = 0 \\ \mu_j = 0 & \text{for } H_j < 0 \end{cases} \qquad (A.49)$$

Eq. (A.48) is rewritten as

$$\nabla \psi = \nabla F + \sum_{j=1}^{n^I} \mu_j \nabla H_j = \mathbf{0} \qquad (A.50)$$

Eqs. (A.49) and (A.50) together with constraints $H_i \leq 0$ are called *optimality conditions* or *Karush-Kuhn-Tucker (KKT) conditions*, and (A.49) is called *complementarity conditions*. The solution satisfying the KKT conditions is called the KKT point in the variable space. Note that the KKT conditions are the first-order necessary conditions for local optimality.

Suppose the variable vector is modified by $\Delta\mathbf{x}$ from the solution satisfying the KKT conditions. Then, using the linear approximation

$$\Delta F = \nabla F^\top \Delta\mathbf{x}, \quad \Delta H_j = \nabla H_j^\top \Delta\mathbf{x} \tag{A.51}$$

the increment of the objective function is approximated with (A.48) as

$$\Delta F = -\sum_{j=1}^{n^{\mathrm{I}}} \mu_j \Delta H_j \tag{A.52}$$

Therefore, $\Delta F \geq 0$ is satisfied with first-order approximation for any increment $\Delta\mathbf{x}$ in the direction of the feasible region satisfying $\Delta H_j \leq 0$ for the active constraints.

The KKT conditions are sufficient conditions for global (local) optimality if all the objective and constraint functions are globally (locally) convex; i.e., the solution satisfying the KKT conditions is locally optimal if the Hessian of the Lagrangian is positive semidefinite. More general and rigorous sufficient conditions and second-order optimality conditions may be consulted in the textbooks, e.g., Floudas (1995).

If we have bound constraints $x_i^{\mathrm{L}} \leq x_i \leq x_i^{\mathrm{U}}$, the Lagrangian is reformulated as

$$\psi(\mathbf{x}, \boldsymbol{\mu}, \boldsymbol{\eta}, \boldsymbol{\kappa}) = F(\mathbf{x}) + \sum_{j=1}^{n^{\mathrm{I}}} \mu_j H_j(\mathbf{x}) + \sum_{i=1}^{m} \eta_i (x_i - x_i^{\mathrm{U}}) + \sum_{i=1}^{m} \kappa_i (x_i^{\mathrm{L}} - x_i) \tag{A.53}$$

where η_i and κ_i are the nonnegative Lagrange multipliers for the bound constraints. Then, from the stationary conditions and complementarity conditions, we have the following optimality conditions:

$$\begin{cases} Z_i = 0 & \text{for } x_i^{\mathrm{L}} < x_i < x_i^{\mathrm{U}} \\ Z_i \leq 0 & \text{for } x_i = x_i^{\mathrm{U}} \\ Z_i \geq 0 & \text{for } x_i = x_i^{\mathrm{L}} \end{cases} , \quad (i = 1, \ldots, m) \tag{A.54}$$

with the complementarity conditions (A.49), where

$$Z_i = \frac{\partial F}{\partial x_i} + \sum_{j=1}^{n^{\mathrm{I}}} \mu_j \frac{\partial H_j}{\partial x_i}, \quad (i = 1, \ldots, m) \tag{A.55}$$

For a problem with side constraints only, Z_i in (A.54) turns out to be the derivative of the objective function $F(\mathbf{x})$.

The optimality conditions can be directly solved to obtain the optimal solution (see, e.g., Sec. 1.7 for the *optimality criteria approach* for structural optimization). The optimality conditions are also used as the termination conditions for a gradient-based nonlinear programming approach.

A.2.2.4   Penalty function approach

The most convenient approach for incorporating the constraints into the optimization algorithm is the *penalty function approach*, or *penalty approach*, which simply increases the objective function value by adding the term of penalization for the violated constraints; thus the constrained problem is transformed into an unconstrained problem. This method is widely used for heuristic methods, e.g., genetic algorithms; see Sec. 6.6.

Consider, for simplicity, a minimization problem of an objective function under inequality constraints only:

$$\text{Minimize} \quad F(\mathbf{x}) \tag{A.56a}$$

$$\text{subject to} \quad H_i(\mathbf{x}) \leq 0, \quad (i = 1, \ldots, n^{\mathrm{I}}) \tag{A.56b}$$

A penalty function $R(H_j(\mathbf{x}))$ (> 0) is given for the inequality constraint $H_j \leq 0$. Let r denote the positive penalty parameter. Then the objective function $F(\mathbf{x})$ is transformed to $F^*(\mathbf{x}, r)$ as

$$F^*(\mathbf{x}, r) = F(\mathbf{x}) + r \sum_{j=1}^{n^{\mathrm{I}}} R(H_j(\mathbf{x})) \tag{A.57}$$

and the following unconstrained problem is to be solved:

$$\text{Minimize} \quad F^*(\mathbf{x}, r) \tag{A.58}$$

where the parameter r is fixed while optimizing with respect to the variable vector $\mathbf{x}$.

Penalty function approaches are classified into the *exterior penalty approach* and the *interior penalty approach* depending on the definition of the penalty function. There exist several definitions of exterior penalty functions, among which the simplest one is the quadratic function:

$$\begin{cases} R(H_j(\mathbf{x})) = (H_j(\mathbf{x}))^2 & \text{for} \ \ H_j(\mathbf{x}) > 0 \\ R(H_j(\mathbf{x})) = 0 & \text{for} \ \ H_j(\mathbf{x}) \leq 0 \end{cases} \tag{A.59}$$

Suppose the optimal solution of the original constrained problem (A.56) exists at the boundary of the feasible region defined by the inequality constraints; i.e., at least one inequality constraint is satisfied with equality at the optimal solution. Then, in the exterior penalty approach, the constraint is slightly violated at the optimal solution of Problem (A.56), and the amount of violation is large if r is small. Therefore, if the violation is larger than the specified tolerance, r is increased and an approximate solution is searched again by solving the unconstrained problem (A.58). Hence, a very large value should be assigned for r to obtain the solution with good accuracy. However, the Hessian of the problem becomes ill-conditioned, and the gradient of $F^*(\mathbf{x}, r)$ diverges if r is too large. Therefore, a method such as the *sequential unconstrained minimization technique* (SUMT) (Fiacco and Cormic 1968) can be used for gradually and systematically updating the parameter r.

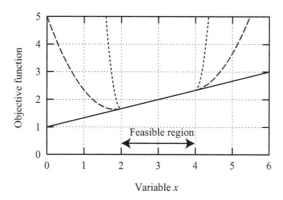

FIGURE A.8: Exterior penalty function; solid line: $F(x)$, dotted line: $r = 20$, dashed line: $r = 1$.

Example A.4

Fig. A.8 illustrates the problem of minimizing the objective function

$$F(x) = \frac{1}{3}x + 1 \tag{A.60}$$

under simple bound constraints

$$2 \le x \le 4 \tag{A.61}$$

The inequalities in (A.61) are converted to standard forms as

$$H_1(x) = -x + 2 \le 0 \tag{A.62a}$$
$$H_2(x) = x - 4 \le 0 \tag{A.62b}$$

Hence, the penalized objective function is formulated as

$$F^*(x, r) = \frac{1}{3}x + 1 + r(R(H_1) + R(H_2)) \tag{A.63}$$

Suppose $x < 2$, i.e., the constraint $H_1(x) \le 0$ is not satisfied. Then (A.63) is written as

$$F^*(x, r) = \frac{1}{3}x + 1 + r(-x + 2)^2 \tag{A.64}$$

From (A.64), the stationary condition of F^* with respect to x is obtained as

$$\frac{\mathrm{d}F^*}{\mathrm{d}x} = 2rx - 4r + \frac{1}{3} = 0 \tag{A.65}$$

Therefore, the approximate optimal solution $\tilde{x}$ that minimizes F^* is explicitly written as

$$\tilde{x} = 2 - \frac{1}{6r} \tag{A.66}$$

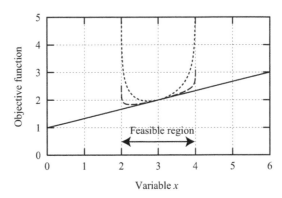

FIGURE A.9: Interior penalty function; solid line: $F(x)$, dotted line: $r = 0.6$, dashed line: $r = 0.1$.

Fig. A.8 shows the plots of $F^*(x, r)$ corresponding to two different values of r as well as the original objective function. As is seen from (A.66), the error $1/(6r)$ decreases as r is increased. Therefore, $\tilde{x}$ approaches the optimal solution $x = 2$ as r is increased. However, if r is too large, the differential coefficient $\mathrm{d}F^*/\mathrm{d}x$ diverges, and the convergence property of the optimization process for minimizing F^* will deteriorate if a gradient-based optimization algorithm is used. Another disadvantage of this approach is that the approximate optimal solution always exists outside the feasible region if there exist active constraints at the optimal solution.

In the interior penalty function approach, the following logarithmic barrier function is usually used:

$$F^*(\mathbf{x}, r) = F(\mathbf{x}) - r \sum_{j=1}^{n^{\mathrm{I}}} \log(-H_j(\mathbf{x})) \qquad \text{(A.67)}$$

where the penalty parameter r should be sufficiently small. Fig. A.9 illustrates the penalized objective functions for the problem defined by (A.60) and (A.61). As is seen, a smaller value of r leads to a better approximate optimal solution. The main advantage of this approach is that the solution always exists in the feasible region. However, this is also a disadvantage because $\log(-H_j(\mathbf{x}))$ is not defined for $H_j(\mathbf{x}) > 0$; hence, only the feasible solutions should be searched, which may be very difficult for a large-scale optimization problem.

There are various types of interior penalty function approaches, which are simply called the *interior point method*, for quadratic programming (QP) and semidefinite programming (SDP); see Appendix A.2.4 for a brief introduction to SDP. For these problems, the solution of minimizing the penalized objective function in (A.67) is traced parametrically with respect to r by solving the KKT conditions; see, e.g., Kojima, Shindoh, and Hara (1997) and Mehrotra (1992) for details.

A.2.2.5 Sequential quadratic programming

Sequential quadratic programming (SQP) is an extension of *sequential linear programming* (SLP), which solves linearized subproblems successively using the standard approaches of LP to find the optimal solution of a general NLP problem (Vanderplaats 1999; Gill, Murray, and Saunders 2002; Arora 2004).

Taylor's expansion of the objective function $F(\mathbf{x})$ as well as the constraint functions $G_j(\mathbf{x})$ and $H_j(\mathbf{x})$ at the solution $\mathbf{x}^{(k)}$ at the kth step of solving the original NLP problem (A.29) leads to

$$F(\mathbf{x}) = F(\mathbf{x}^{(k)}) + \nabla F(\mathbf{x}^{(k)})^\top (\mathbf{x} - \mathbf{x}^{(k)}) + \cdots \qquad (A.68a)$$

$$G_j(\mathbf{x}) = G_j(\mathbf{x}^{(k)}) + \nabla G_j(\mathbf{x}^{(k)})^\top (\mathbf{x} - \mathbf{x}^{(k)}) + \cdots \qquad (A.68b)$$

$$H_j(\mathbf{x}) = H_j(\mathbf{x}^{(k)}) + \nabla H_j(\mathbf{x}^{(k)})^\top (\mathbf{x} - \mathbf{x}^{(k)}) + \cdots \qquad (A.68c)$$

If we use only linear terms and define $\Delta\mathbf{x}^{(k)} = \mathbf{x}^{(k+1)} - \mathbf{x}^{(k)}$, the following LP subproblem of SLP is derived:

$$\text{Minimize} \quad F(\mathbf{x}) = F(\mathbf{x}^{(k)}) + \nabla F(\mathbf{x}^{(k)})^\top \Delta\mathbf{x}^{(k)} \qquad (A.69a)$$

$$\text{subject to} \quad \nabla G_j(\mathbf{x}^{(k)})^\top \Delta\mathbf{x}^{(k)} = -G_j(\mathbf{x}^{(k)}), \quad (i = 1, \dots n^{\mathrm{E}}) \qquad (A.69b)$$

$$\nabla H_j(\mathbf{x}^{(k)})^\top \Delta\mathbf{x}^{(k)} \leq -H_j(\mathbf{x}^{(k)}), \quad (i = 1, \dots n^{\mathrm{I}}) \qquad (A.69c)$$

which is solved for the variable vector $\Delta\mathbf{x}^{(k)} = (\Delta x_1^{(k)}, \dots, \Delta x_m^{(k)})^\top$ using the method of LP. Then, $\mathbf{x}$ is updated to $\mathbf{x}^{(k+1)} = \mathbf{x}^{(k)} + \Delta\mathbf{x}^{(k)}$, and subproblem (A.69) is solved successively.

However, in SLP, the error due to linearization may be very large if there is no upper bound for the magnitude of $\Delta\mathbf{x}^{(k)}$. Therefore, we usually assign the bound $\Delta\overline{x}_i^{(k)}$, called the *move limit* for the absolute value $|\Delta x_i^{(k)}|$ of each component of $\Delta\mathbf{x}^{(k)}$, which is proportionally reduced by a factor c (< 1) as $\Delta\overline{x}_i^{(k+1)} = c\Delta\overline{x}_i^{(k)}$. The algorithm converges if the norm of $\Delta\mathbf{x}^{(k)}$ becomes sufficiently small. However, there will be no feasible solution if the current solution is infeasible and the move limit is too strict. Furthermore, the solution converges to a non-optimal solution if the move limit is reduced too quickly. Hence, it is very difficult to find an appropriate value of the reduction factor c.

To alleviate this difficulty, SLP is extended to SQP, where a quadratic penalty is given for the magnitude of $\Delta\mathbf{x}^{(k)}$ as follows to formulate a quadratic programming subproblem:

$$\text{Minimize} \quad F(\mathbf{x}) = F(\mathbf{x}^{(k)}) + \nabla F(\mathbf{x}^{(k)})^\top \Delta \mathbf{x}^{(k)} + \frac{1}{2} \Delta \mathbf{x}^{(k)\top} \mathbf{D} \Delta \mathbf{x}^{(k)} \quad \text{(A.70a)}$$

$$\text{subject to} \quad \nabla G_j(\mathbf{x}^{(k)})^\top \Delta \mathbf{x}^{(k)} = -G_j(\mathbf{x}^{(k)}), \quad (i = 1, \ldots n^E) \qquad \text{(A.70b)}$$

$$\nabla H_j(\mathbf{x}^{(k)})^\top \Delta \mathbf{x}^{(k)} \leq -H_j(\mathbf{x}^{(k)}), \quad (i = 1, \ldots n^I) \qquad \text{(A.70c)}$$

Although $\mathbf{D}$ may be any $m \times m$ positive definite matrix in view of assigning a penalty for the large magnitude of solution update $\Delta \mathbf{x}^{(k)}$, an approximate Hessian of the Lagrangian of the original problem is preferred in order to assure convergence to a local optimum. The Hessian or its inverse can be appropriately computed using an algorithm of the quasi-Newton method.

After solving Problem (A.70) for $\Delta \mathbf{x}^{(k)}$, the solution may be simply updated to $\mathbf{x}^{(k)} + \Delta \mathbf{x}^{(k)}$; however, the best solution is usually searched in the direction of $\Delta \mathbf{x}^{(k)}$ using the technique of line search in Sec. A.2.2.1 to minimize a *descent function* or *merit function* defined by the objective function and the penalty for violation of the constraints.

In this book, we mainly use the software SNOPT Ver. 7.2 (Gill, Murray, and Saunders 2002), which consists of a major iteration which utilizes a quasi-Newton method called the Broyden–Fletcher–Goldfarb–Shanno (BFGS) method for approximately updating the Hessian, and a minor iteration for solving the QP subproblem using a reduced-gradient method. An augmented Lagrangian merit function is used for computing the step size of the line search.

A.2.2.6   Method of feasible directions

The simplest approach to an unconstrained NLP problem is to update the solution in the steepest descent direction of the objective function, as presented in Sec. A.2.2.1. For a constrained problem, the gradient of the objective function can be projected to the hyper-plane of the active constraints to obtain a feasible direction of solution update. However, because the constraints are generally nonlinear, the updated solution very often violates the constraints, and we cannot allow a large magnitude of solution update using this approach.

In the *method of feasible directions* (MFD), the search direction $\mathbf{S}^{(k)}$ at the kth step is directed into the feasible region, and the solution is updated as

$$\mathbf{x}^{(k+1)} = \mathbf{x}^{(k)} + \alpha \mathbf{S}^{(k)} \qquad \text{(A.71)}$$

where the scaling parameter α is found by line search (Ben-Israel, Ben-Tal, and Zolbec 1981; Vanderplaats 1999).

Suppose we have only inequality constraints, and let $\mathcal{J}$ denote the set of active constraints at the kth iteration. Then the *usable direction* $\mathbf{S}^{(k)}$ is characterized by

$$\nabla F(\mathbf{x}^{(k)})^\top \mathbf{S}^{(k)} \leq 0,$$
$$\nabla H_j(\mathbf{x}^{(k)})^\top \mathbf{S}^{(k)} \leq 0, \quad (j \in \mathcal{J}) \tag{A.72}$$

which means that $\mathbf{S}^{(k)}$ is directed into the inside of the feasible region and the objective function decreases as the solution is modified in the direction of $\mathbf{S}^{(k)}$. However, the best direction that minimizes $\nabla F(\mathbf{x}^{(k)})^\top \mathbf{S}^{(k)}$ in the feasible direction may be generally tangent to the boundary of the feasible region. Therefore, a *pushoff factor* θ is introduced, and the auxiliary variable β is maximized by solving the following LP problem:

$$\text{Minimize} \quad \beta \tag{A.73a}$$
$$\text{subject to} \quad \nabla F(\mathbf{x}^{(k)})^\top \mathbf{S}^{(k)} + \beta \leq 0 \tag{A.73b}$$
$$\nabla H_j(\mathbf{x}^{(k)})^\top \mathbf{S}^{(k)} + \theta\beta \leq 0, \quad (j \in \mathcal{J}) \tag{A.73c}$$
$$\mathbf{S}^{\mathrm{L}} \leq \mathbf{S}^{(k)} \leq \mathbf{S}^{\mathrm{U}} \tag{A.73d}$$

where the variables are $\mathbf{S}^{(k)}$ and β, and $\mathbf{S}^{\mathrm{L}}$ and $\mathbf{S}^{\mathrm{U}}$ are the lower and upper bounds for $\mathbf{S}^{(k)}$. This way, a feasible direction that reduces the objective function is obtained, and the best solution is searched in this direction using a line search.

A.2.2.7 Method of moving asymptote

The *method of moving asymptote* (MMA) developed by Svanberg (1987) is based on a successive convex approximation using the reciprocals of the variables. Therefore, the MMA may be regarded as an extension of the convex linearization method (CONLIN) (Fleury 1989a).

Consider an optimization problem with inequality constraints $H_i(\mathbf{x}) \leq 0$ $(i = 1, \ldots, n^{\mathrm{I}})$ and side constraints $\mathbf{x}^{\mathrm{L}} \leq \mathbf{x} \leq \mathbf{x}^{\mathrm{U}}$. Let $M_i^{\mathrm{L}(k)}$ and $M_i^{\mathrm{U}(k)}$ denote the *moving asymptotes* at the kth iteration given for x_i as

$$M_i^{\mathrm{L}(k)} < x_i^{(k)} < M_i^{\mathrm{U}(k)} \tag{A.74}$$

The variable x_i is converted to the reciprocal $1/(x_i - M_i^{\mathrm{L}(k)})$ or $1/(M_i^{\mathrm{U}(k)} - x_i)$ in view of the signs of the derivatives of the objective function and the constraint functions with respect to x_i. Let $\partial H_j^{(k)}/\partial x_i$ denote the value of $\partial H_j/\partial x_i$ at $\mathbf{x} = \mathbf{x}^{(k)}$. The approximation $H_j^{\mathrm{a}}(\mathbf{x})$ of $H_j(\mathbf{x})$ is defined with respect to the reciprocals as

$$H_j^{\mathrm{a}}(\mathbf{x}) = r_j^{(k)} + \sum_{i=1}^{m} \left(\frac{p_{ij}^{(k)}}{M_i^{\mathrm{U}(k)} - x_i} + \frac{q_{ij}^{(k)}}{x_i - M_i^{\mathrm{L}(k)}} \right) \tag{A.75}$$

where

$$
p_{ij}^{(k)} =
\begin{cases}
(M_i^{U(k)} - x_i)^2 \dfrac{\partial H_j^{(k)}}{\partial x_i} & \text{for } \dfrac{\partial H_j^{(k)}}{\partial x_i} > 0 \\[4mm]
0 & \text{for } \dfrac{\partial H_j^{(k)}}{\partial x_i} \le 0
\end{cases}
\tag{A.76}
$$

$$
q_{ij}^{(k)} =
\begin{cases}
0 & \text{for } \dfrac{\partial H_j^{(k)}}{\partial x_i} \ge 0 \\[4mm]
-(x_i - M_i^{L(k)})^2 \dfrac{\partial H_j}{\partial x_i} & \text{for } \dfrac{\partial H_j^{(k)}}{\partial x_i} < 0
\end{cases}
\tag{A.77}
$$

$$
r_j^{(k)} = H_j(\mathbf{x}^{(k)}) - \sum_{i=1}^{m} \left(\frac{p_{ij}^{(k)}}{M_i^{U(k)} - x_i^{(k)}} + \frac{q_{ij}^{(k)}}{x_i^{(k)} - M_i^{L(k)}} \right)
\tag{A.78}
$$

The objective function is expanded similarly with $p_{i0}^{(k)}$, $q_{i0}^{(k)}$, and $r_0^{(k)}$ defined by assuming $H_0(\mathbf{x}) = F(\mathbf{x})$ in (A.75)–(A.78). Then the following subproblem is to be solved:

$$
\text{Minimize} \quad r_0^{(k)} + \sum_{i=1}^{m} \left(\frac{p_{i0}^{(k)}}{M_i^{U(k)} - x_i} + \frac{q_{i0}^{(k)}}{x_i - M_i^{L(k)}} \right)
\tag{A.79a}
$$

$$
\text{subject to} \quad r_j^{(k)} + \sum_{i=1}^{m} \left(\frac{p_{ij}^{(k)}}{M_i^{U(k)} - x_i} + \frac{q_{ij}^{(k)}}{x_i - M_i^{L(k)}} \right) \le 0,
$$

$$
(j = 1, \dots, N^{\mathrm{I}})
\tag{A.79b}
$$

$$
\alpha_i^{\mathrm{L}} \le x_i - x_i^{(k)} \le \alpha_i^{\mathrm{U}}
\tag{A.79c}
$$

where α_i^{L} and α_i^{U} are the appropriately assigned move limits satisfying

$$
M_i^{L(k)} < \alpha_i^{\mathrm{L}} + x_i^{(k)} \le x_i \le \alpha_i^{\mathrm{U}} + x_i^{(k)} < M_i^{U(k)}
\tag{A.80}
$$

A.2.3   Dual problem

As a simple example, consider first an LP problem

$$
\text{Minimize} \quad \mathbf{c}^{\mathsf{T}} \mathbf{x}
\tag{A.81a}
$$

$$
\text{subject to} \quad \mathbf{Bx} - \mathbf{b} \ge 0
\tag{A.81b}
$$

$$
\mathbf{x} \ge 0
\tag{A.81c}
$$

where $\mathbf{c} = (c_1, \dots, c_m)^{\mathsf{T}}$ and $\mathbf{b} = (b_1, \dots, b_n)^{\mathsf{T}}$ are constant vectors, and $\mathbf{B}$ is an $n \times m$ constant matrix.

The *dual problem* of the *primal problem* (A.81) is formulated using the dual variables $\boldsymbol{\mu} = (\mu_1, \ldots, \mu_n)^{\top}$ as

$$\text{Maximize} \quad \mathbf{b}^{\top} \boldsymbol{\mu} \tag{A.82a}$$

$$\text{subject to} \quad \mathbf{B}^{\top} \boldsymbol{\mu} - \mathbf{c} \leq \mathbf{0} \tag{A.82b}$$

$$\boldsymbol{\mu} \geq \mathbf{0} \tag{A.82c}$$

where $\boldsymbol{\mu}$ coincides with the vector of Lagrange multipliers of the constraints (A.81b) of the primal problem.

For the feasible solutions $\mathbf{x}$ and $\boldsymbol{\mu}$ of problems (A.81) and (A.82), respectively, the following inequality holds:

$$\mathbf{c}^{\top} \mathbf{x} \geq \boldsymbol{\mu}^{\top} \mathbf{B} \mathbf{x} \geq \boldsymbol{\mu}^{\top} \mathbf{b} \tag{A.83}$$

where (A.81b) and (A.82b) have been used. Therefore, the optimal value of Problem (A.81) is not less than that of Problem (A.82). This condition is called *weak duality*. Furthermore, the equality in (A.83) is satisfied by the optimal solutions of (A.81) and (A.82), which is called *strong duality*.

Consider next the following NLP problem with inequality constraints only:

$$\text{Minimize} \quad F(\mathbf{x}) \tag{A.84a}$$

$$\text{subject to} \quad H_i(\mathbf{x}) \leq 0, \quad (i = 1, \ldots, n^{\mathrm{I}}) \tag{A.84b}$$

The Lagrangian for this problem is given as

$$\psi(\mathbf{x}, \boldsymbol{\mu}) = F(\mathbf{x}) + \sum_{j=1}^{n^{\mathrm{I}}} \mu_j H_j(\mathbf{x}) \tag{A.85}$$

There are several definitions of a dual problem for an NLP problem. For example, Wolf's dual problem is formulated as

$$\text{Maximize} \quad \psi(\mathbf{x}, \boldsymbol{\mu}) = F(\mathbf{x}) + \sum_{j=1}^{n^{\mathrm{I}}} \mu_j H_j(\mathbf{x}) \tag{A.86a}$$

$$\text{subject to} \quad \nabla F(\mathbf{x}) + \sum_{j=1}^{n^{\mathrm{I}}} \mu_j \nabla H_j(\mathbf{x}) = \mathbf{0} \tag{A.86b}$$

$$\mu_i \geq 0, \quad (i = 1, \ldots, n^{\mathrm{I}}) \tag{A.86c}$$

where the variables are $\mathbf{x}$ and $\boldsymbol{\mu}$.

Let $\mathbf{x}^{\mathrm{P}}$ and $(\mathbf{x}^{\mathrm{D}}, \boldsymbol{\mu})$ denote arbitrary feasible solutions of Problems (A.84) and (A.86), respectively. Then the following weak duality holds:

$$F(\mathbf{x}^{\mathrm{P}}) \geq \psi(\mathbf{x}^{\mathrm{D}}, \boldsymbol{\mu}) \tag{A.87}$$

Furthermore, if $F(\mathbf{x})$ and $H_i(\mathbf{x})$ are convex, and the primal problem (A.84) has an optimal solution $\tilde{\mathbf{x}}$, then there exists the optimal solution $(\tilde{\mathbf{x}}, \tilde{\boldsymbol{\mu}})$ of dual problem (A.86) satisfying the strong duality condition:

$$F(\tilde{\mathbf{x}}) = \psi(\tilde{\mathbf{x}}, \tilde{\boldsymbol{\mu}}) \tag{A.88}$$

Example A.5
As a small example, consider the following simple convex quadratic programming problem with two variables:

$$\text{Minimize} \quad x_1^2 + x_2^2 \tag{A.89a}$$
$$\text{subject to} \quad x_1 + x_2 + 1 \leq 0 \tag{A.89b}$$

The dual problem is formulated as

$$\text{Maximize} \quad x_1^2 + x_2^2 + \mu_1(x_1 + x_2 + 1) \tag{A.90a}$$
$$\text{subject to} \quad 2x_1 + \mu_1 = 0 \tag{A.90b}$$
$$2x_2 + \mu_1 = 0 \tag{A.90c}$$
$$\mu_1 \geq 0 \tag{A.90d}$$

From the constraints, we have $x_1 = x_2 = -\mu_1/2$, and the objective function is expressed with respect to μ_1 only as $-\mu_1^2/2 + \mu_1$. Therefore, the optimal solution of the dual problem is obtained as $x_1 = x_2 = -1/2$, $\mu_1 = 1$, which leads to the optimal objective value $1/2$, which is the same as that of the primal problem.

As another formulation, the *Lagrange dual problem* is defined as

$$\text{Maximize} \quad \Phi(\boldsymbol{\mu}) = \min_{\mathbf{x}} \psi(\mathbf{x}, \boldsymbol{\mu}) \tag{A.91a}$$
$$\text{subject to} \quad \boldsymbol{\mu} \geq \mathbf{0} \tag{A.91b}$$

For example, consider again the simple primal problem (A.89). The solution that minimizes ψ in the dual problem (A.91) is obtained from the stationary conditions (A.90b) and (A.90c) of ψ, which leads to $x_1 = x_2 = -\mu_1/2$. Then, the objective function is given as $\Phi(\mu_1) = -\mu_1^2/2 + \mu_1$. Hence, maximization of $\Phi(\mu_1)$ leads to $\mu_1 = 1$ and $\Phi(\mu_1) = 1/2$, which is equal to the optimal value of the primal problem; i.e, the strong duality is satisfied.

A.2.4   Semidefinite programming

An optimization problem that has constraints such that the variable matrix is positive semidefinite is called a *semidefinite programming* (SDP) problem. The objective function and other constraints are linear in the standard form of an SDP problem (Wolkowicz, Saigal, and Vandenberghe 2000; Ohsaki and

Kanno 2007). Since the constraints of positive semidefiniteness of matrices include linear and convex quadratic constraints, the SDP is an extension of linear programming and convex quadratic programming. Interior-point methods for linear and quadratic programming have been extended to solve SDPs (Kojima, Shindoh, and Hara 1997).

Let $\mathbf{X} \succeq \mathbf{O}$ indicate that the symmetric matrix $\mathbf{X}$ is positive semidefinite. The inner product $\mathbf{X} \bullet \mathbf{Y}$ of the $n \times n$ matrices $\mathbf{X} = (X_{ij})$ and $\mathbf{Y} = (Y_{ij})$ is defined as

$$\mathbf{X} \bullet \mathbf{Y} = \sum_{i=1}^{n} \sum_{j=1}^{n} X_{ij} Y_{ij} \tag{A.92}$$

The standard form of the SDP problem is given as

$$\text{Minimize} \quad \mathbf{C} \bullet \mathbf{X} \tag{A.93a}$$
$$\text{subject to} \quad \mathbf{A}_i \bullet \mathbf{X} = b_i, \quad (i = 1, \ldots, m) \tag{A.93b}$$
$$\mathbf{X} \succeq \mathbf{O} \tag{A.93c}$$

where $n \times n$ matrices $\mathbf{C}$ and $\mathbf{A}_i$, and vector $\mathbf{b} = (b_1, \ldots, b_m)^\top$ are constant, and $\mathbf{X}$ is the variable matrix.

The dual problem of Problem (A.93) is formulated as

$$\text{Minimize} \quad \mathbf{b}^\top \mathbf{y} \tag{A.94a}$$
$$\text{subject to} \quad \sum_{i=1}^{m} \mathbf{A}_i y_i + \mathbf{Z} = \mathbf{C} \tag{A.94b}$$
$$\mathbf{Z} \succeq \mathbf{O} \tag{A.94c}$$

where the vector $\mathbf{y} = (y_1, \ldots, y_m)^\top$ and $n \times n$ matrix $\mathbf{Z}$ are the variables.

The KKT conditions for the pair of primal and dual SDP problems are written as

$$\mathbf{X}\mathbf{Z} = \mathbf{O} \tag{A.95}$$

with the constraints (A.93b), (A.93c), (A.94b), and (A.94c) of the primal and dual problems. The algorithm for solving the KKT conditions incorporating the relaxation parameter for complementary conditions (A.95) and interior penalty function is generally called the *primal-dual interior-point method.*

Various structural optimization problems, including the robust optimization problem (Ben-Tal and Nemirovski 1997) and the truss topology optimization problem under frequency constraint (Ohsaki, Fujisawa, Katoh, and Kanno 1999), can be formulated as an SDP problem; see Sec. 3.9.

A.2.5 Combinatorial problem

If the variables can take only integer values, then the optimization problem is called an *integer programming* (IP) problem, which is formulated as

$$\text{Minimize} \quad F(\mathbf{x}) \tag{A.96a}$$

$$\text{subject to} \quad G_i(\mathbf{x}) = 0, \quad (i = 1, \ldots, n^{\mathrm{E}}) \tag{A.96b}$$

$$H_i(\mathbf{x}) \leq 0, \quad (i = 1, \ldots, n^{\mathrm{I}}) \tag{A.96c}$$

$$\mathbf{x} \in \mathbb{Z}^m \tag{A.96d}$$

where $\mathbb{Z}$ denotes the set of integers, and m is the number of variables. Since the problem of finding the optimal combination of the variables is formulated as an IP problem, it is equivalently called a *combinatorial optimization problem*. If some of the variables can take real values, then the problem is called a *mixed integer programming* (MIP) problem, which usually denotes a problem with linear objective and constraint functions. An MIP with nonlinear objective and/or constraint functions is called a *mixed integer nonlinear programming* (MINLP) problem, which often appears in structural optimization problems with discrete design variables and continuous state variables.

The most standard approach to IP is the *branch-and-bound method* (Horst, Pardalos, and Thoai 1995; Floudas 1995), where the original problem is successively divided into subproblems, and the upper and lower bounds of the optimal objective value are updated by solving the relaxed problem assuming that the variables can have real numbers. Note that all the objective and constraint functions of the relaxed problem should be convex in order to find the global solution of the original problem. In the process of branch-and-bound, the subproblem that remains to be solved is called the *active problem*. Since any IP can be reformulated as a 0–1 problem, we present the basic algorithm of IP, as follows, assuming the variables x_i can take either 0 or 1, and $F(\mathbf{x})$, $G_i(\mathbf{x})$, and $H_i(\mathbf{x})$ are linear functions of $\mathbf{x}$:

Step 0: Initialize the upper bound F^{U} of $F(\mathbf{x})$ as $F^{\mathrm{U}} = \infty$. Let the set $\mathcal{A}$ of the active problems consist of the original IP (A.96).

Step 1: Select a problem P from $\mathcal{A}$ and remove it from $\mathcal{A}$. Note that the number of branching processes can be reduced if a smaller initial value is given for F^{U} using, e.g., a heuristically obtained approximate optimal solution.

Step 2: Solve the relaxed LP problem $\overline{P}$ of P by relaxing the integer conditions of the variables to inequality constraints $0 \leq x_i \leq 1$, and select a variable j satisfying $0 < x_j < 1$ at the optimal solution of $\overline{P}$. Let P_0 and P_1 denote the subproblems of P by specifying $x_j = 0$ and 1, respectively. Solve the relaxed LPs $\overline{P}_0$ and $\overline{P}_1$, respectively, of P_0 and P_1.

Step 3: Let F_0 and F_1 denote the optimal objective values of $\overline{P}_0$ and $\overline{P}_1$, respectively. If $F_0 > F^U$, then set $x_j = 1$ and terminate P_0. If $F_1 > F^U$, then set $x_j = 0$ and terminate P_1.

Step 4: If the solution of $\overline{P}_0$ satisfies the integer conditions, then let $F^U = \min\{F_0, F^U\}$, and terminate P_0; otherwise, add P_0 to $\mathcal{A}$.

Step 5: If the solution of $\overline{P}_1$ satisfies the integer conditions, then let $F^U = \min\{F_1, F^U\}$, and terminate P_1; otherwise, add P_1 to $\mathcal{A}$.

Step 6: If $\mathcal{A} \neq \emptyset$, then go to Step 1; otherwise, output the best value of F^U, and terminate the process.

Example A.6

As a simple example, consider the following LP problem:

$$P: \text{Minimize} \quad x_1 + x_2 \tag{A.97a}$$
$$\text{subject to} \quad -2x_1 - x_2 + 4 \leq 0 \tag{A.97b}$$
$$-x_1 - 4x_2 + 4 \leq 0 \tag{A.97c}$$
$$x_1 \in \mathbb{Z}, \quad x_2 \in \mathbb{Z} \tag{A.97d}$$

which is solved using the branch-and-bound method as follows:

1. At the initial stage, let $F^U = \infty$. The set $\mathcal{A}$ of active problems consists of P, which is to be selected and solved.

2. The optimal solution of the relaxed LP of P, denoted by $\overline{P}$, without the integer constraint (A.97d) is found as $(x_1, x_2) = (12/7, 4/7)$.

3. We select x_1 for generating the two subproblems P_0 and P_1 of P with $x_1 = 1$ and $x_1 = 2$, respectively, and formulate the following relaxed problems $\overline{P}_0$ and $\overline{P}_1$, respectively, of P_0 and P_1:

$$\overline{P}_0: \text{Minimize} \quad 1 + x_2 \tag{A.98a}$$
$$\text{subject to} \quad -2 - x_2 + 4 \leq 0 \tag{A.98b}$$
$$-2 - 4x_2 + 4 \leq 0 \tag{A.98c}$$
$$x_2 \in \mathbb{R} \tag{A.98d}$$

$$\overline{P}_1: \text{Minimize} \quad 2 + x_2 \tag{A.99a}$$
$$\text{subject to} \quad -4 - x_2 + 4 \leq 0 \tag{A.99b}$$
$$-1 - 4x_2 + 4 \leq 0 \tag{A.99c}$$
$$x_2 \in \mathbb{R} \tag{A.99d}$$

where $\mathbb{R}$ is the set of real numbers.

4. The optimal values F_0 and F_1 of $\overline{P}_0$ and $\overline{P}_1$, respectively, are found as 3 and $5/2$, and the corresponding optimal solutions are $(x_1, x_2) = (1, 2)$ and $(2, 1/2)$.

5. Because the solution of $\overline{P}_0$ satisfies the integer conditions and $F_0 < F^U = \infty$, F^U is updated as $F^U = F_0 = 3$, and P_0 is terminated.

6. Because the solution of P_1 does not satisfy the integer conditions, $\overline{P}_1$ is added to the set $\mathcal{A}$ of the active problems.

7. Problem P_1 is selected from $\mathcal{A}$ to formulate subproblems P_{10} and P_{11} by fixing x_2 to 0 and 1, respectively. Since no variable is left, and P_{10} is infeasible, we can immediately compute the objective value of P_{11} as $F_1 = 3$, which is the same as the current upper bound F^U.

8. Since no problem is left in $\mathcal{A}$, optimal solutions are found to be $(x_1, x_2) = (1,2)$ and $(2,1)$ with the same objective value $F(\mathbf{x}) = 3$.

A.3   Heuristics

A.3.1   Introduction

Heuristic approaches (or *heuristics* for simplicity) have been developed to obtain approximate optimal solutions within reasonable computational cost, although there is no theoretical proof of convergence to the global optimal solution (Reeves 1995). Among many heuristic approaches, the most popular approach is the genetic algorithm (GA) (Goldberg 1989), which can be categorized as a multipoint search or population-based approach that has many solutions at each iterative step called generation. Although a GA generally requires a very large population size, some methods, e.g., micro-GA, scatter search (Laguna and Marti 2003), particle swarm optimization (Kennedy 1997), immune algorithm, and harmony search (Geem, Kim, and Loganathan 2001; Lee and Geem 2004), have been developed to find approximate optimal solutions with a relatively small population size. However, because the computational cost for evaluating the objective and/or constraint functions at each step is very large for structural optimization problems, a multipoint strategy may not be appropriate, especially for optimization of a structure with a large number of degrees of freedom. Therefore, single-point-search heuristics, including *simulated annealing* (SA) (Kirkpatrick, Gelatt, and Vecchi 1983; Aarts and Korst 1989; Cerny 1985), and *tabu search* (or taboo search, TS) (Glover 1989; Glover and Laguna 1997), may have advantages over the multipoint strategies.

A.3.2   Single-point-search heuristics

A.3.2.1   Basic algorithm

Single-point-search heuristics are based on *local search* (Aarts and Lenstra 1997), in which the solution is consecutively updated to a neighborhood solution if it improves ('reduces' for a minimization problem) the value of the objective function, where the neighborhood solutions are generated by modifying the value of one or several variables to neighboring values. Because it is not always possible to find a good approximate optimal solution by simple local searches, heuristic approaches have been proposed to improve the convergence properties.

The basic algorithm of a single-point-search heuristic approach can be stated as follows:

Step 1: Assign an initial solution.

Step 2: Carry out local search to select a candidate solution for the next step from the neighborhood solutions of the current solution.

Step 3: Accept or reject the candidate solution in accordance with the criteria defined by the specific algorithm.

Step 4: Go to Step 2 if not converged.

For the initial solution, we can assign either specified or randomly generated values. An optimal solution found by using another approach can be modified to be used as the initial solution; e.g., the nearest discrete solution from the optimal solution with continuous variables can be used. The neighborhood solutions are the set of solutions that can be reached from the current solution by the specified operation; e.g., for the truss topology optimization problem, the cross-sectional area of a randomly selected member can be increased or decreased by the specified value, or the locations of a pair of existing and non-existing members are exchanged, and so on.

Let $\mathbf{J}$ denote the vector of integer variables. Consider a minimization problem of the objective function $F(\mathbf{J})$ with inequality constraints $H_j(\mathbf{J}) \leq 0$ $(j = 1, \ldots, n^{\mathrm{I}})$. If inequality constraints cannot be directly handled in the algorithm, the following penalty function is used to evaluate the performance of the solution:

$$F^*(\mathbf{J}) = F(\mathbf{J}) + \sum_{j=1}^{n^{\mathrm{I}}} \alpha_j (\max\{H_j(\mathbf{J}), 0\})^2 \qquad (\text{A.100})$$

where α_j is a penalty coefficient, and no penalty is given if the constraint $H_j \leq 0$ is satisfied; see Appendix A.2.2.4 for details. Other penalty function approaches such as *dynamic penalty*, as well as the augmented Lagrangian approach, can also be used (Lagaros, Papadrakakis, and Kokossalakis 2002).

A.3.2.2   Greedy/stingy method

Heuristic approaches can be classified into deterministic and probabilistic approaches. The simplest deterministic approach is the *greedy method* described as follows for a constrained optimization problem:

Step 1: Assign an initial solution that does not satisfy the constraints; e.g., choose the smallest value for all variables for the case in which the constraint functions $H_j(\mathbf{J})$ are decreasing functions of $\mathbf{J}$.

Step 2: Move to a neighborhood solution which most efficiently improves the objective function and constraints, where a penalty function is used for the definition of the efficiency of the solution.

Step 3: Go to Step 2 if one of the constraints is not satisfied or the solution cannot be improved.

In contrast, an approach that starts with a solution satisfying all the constraints and reduces the objective value consecutively is called the *stingy method*, described as follows:

Step 1: Assign an initial solution that satisfies all the constraints; e.g., choose the largest value for all variables for the case in which the constraint functions $H_j(\mathbf{J})$ are decreasing functions of $\mathbf{J}$.

Step 2: Move to a neighborhood solution which most efficiently reduces the objective function.

Step 3: Go to Step 2 if one of the constraints is not satisfied.

Applications of greedy and stingy methods can be found in Secs. 1.12, 5.1, and 6.3.

A.3.2.3   Tabu search

The convergence property of the global optimal solution may be enhanced if many solutions are searched before moving to a neighborhood solution, or preferably, all neighborhood solutions may be searched to select the best neighborhood solution. A neighborhood solution that does not reduce (for a minimization problem) the objective value can also be selected to improve the possibility of reaching the global optimal solution. However, in this case, a so-called *cycling* or *loop* can occur where a set of neighboring solutions is chosen iteratively. TS has been developed to prevent cycling utilizing the *tabu list* containing the prohibited solutions that have already been searched.

The algorithm of TS is summarized follows, where the superscript $(\cdot)^{(k)}$ denotes a value at the kth iteration:

Step 1: Assign an initial solution $\mathbf{J}^{(0)}$, and initialize the tabu list $\mathcal{T}$ to be empty. Set the iteration counter $k = 0$.

Step 2: Generate neighborhood solutions $\mathbf{J}_i^N$ ($j = 1, \ldots, n^N$) of $\mathbf{J}^{(k)}$ and move to the best solution $\mathbf{J}^*$ among them that is not included in the tabu list $\mathcal{T}$.

Step 3: Add $\mathbf{J}^*$ to $\mathcal{T}$.

Step 4: Remove the oldest solution in $\mathcal{T}$ if the length of the list exceeds the specified value.

Step 5: Let $\mathbf{J}^{(k+1)} = \mathbf{J}^*$ and $k \leftarrow k + 1$. Go to Step 2 if the termination condition is not satisfied; otherwise, output the best solution satisfying the constraints, and terminate the process.

TS is conceived as a deterministic approach if the neighborhood solutions are generated in a deterministic manner. Some *attributes*, including recency, frequency, quality, and influence of the solution, instead of the solution itself, can be stored in the tabu list (Glover and Laguna 1997). For example, the value or the *move* of a specific variable, or a set of variables, can be an attribute.

A.3.2.4 Random search

The simplest probabilistic approach is the random search, which works as follows:

Step 1: Randomly generate the initial solution.

Step 2: Move to the randomly generated neighborhood solution if it satisfies the constraints.

Step 3: Go to Step 2 if the termination condition is not satisfied.

A penalty function approach can also be used for a random search for a constrained problem. Although a random search is not efficient in view of the convergence property to the global optimal solution, it is simple and easy to implement.

An application of random search to truss topology optimization is presented in Sec. 3.8. A *controlled random search* was proposed by Price (1983) for improving convergence properties, and was extended to incorporate the concept of simulated annealing by Mohan and Nguyen (1999), and to incorporate local mutation by Kaelo and Ali (2006). A random move can be incorporated into TS for a single objective problem (Hu 1992) and a multiobjective problem (Baykasoglu, Owen, and Gindy 1999b). An algorithm with jump or restart was presented by Li, Priemer, and Cheng (2004) for problems with real variables. The termination (stopping) rules for random search and local search were investigated by Hart (1998). Surrogate models or metamodels, e.g., response surface approximation (Myers and Montgomery 1995) and kriging (Lee and Jung 2007), can be used for reducing the computational cost (Brigham and Aquino 2007).

A.3.2.5　Simulated annealing

Simulated annealing (SA) was developed to prevent convergence to a local optimal solution by allowing a move to a solution that does not improve the objective function, where the probability of accepting such a solution is defined by the amount of increase (for a minimization problem) of the objective function. The term *simulated annealing* comes from the fact that it simulates the behavior of the metals in an annealing process. The basic algorithm is as follows:

Step 1: Randomly generate the initial solution $\mathbf{J}^{(0)}$, and set the temperature parameter T to a specified initial value T_0. Assign the range parameter δ for scaling the objective function. Set the iteration counter $k = 0$.

Step 2: Randomly generate the neighborhood solution $\mathbf{J}^*$ of $\mathbf{J}^{(k)}$.

Step 3: Define ΔF as

$$\Delta F = F(\mathbf{J}^*) - F(\mathbf{J}^{(k)}) \tag{A.101}$$

If $\Delta F < 0$, or if the random number $0 \le R < 1$ is smaller than P defined by the following equation, accept $\mathbf{J}^*$ and let $\mathbf{J}^{(k+1)} = \mathbf{J}^*$.

$$P = \exp\left(-\frac{\Delta F}{\delta T_k}\right) \tag{A.102}$$

If $\mathbf{J}^*$ is not accepted, let $\mathbf{J}^{(k+1)} = \mathbf{J}^{(k)}$.

Step 4: Decrease T_k to T_{k+1} on the basis of the specified rule.

Step 5: If the termination condition is not satisfied, set $k \leftarrow k + 1$ and go to Step 2; otherwise, output the best solution satisfying the constraints, and terminate the process.

The initial temperature is given so that almost all neighborhood solutions are accepted. The termination condition is defined using the acceptance ratio; e.g., (a) the acceptance ratio among the last several steps is less than the specified value, or (b) the solution does not change within the specified number of cycles. The temperature can be updated in Step 4 using the Metropolis rule:

$$T_{k+1} = cT_k \tag{A.103}$$

with the specified constant c that is slightly less than 1, e.g., 0.95 (Metropolis, Rosenbluth, Rosenbluth, Teller, and Teller 1953). The solutions are usually updated several times at the same temperature to ensure convergence. Alternatively, T_k may be defined as

$$T_k = \frac{K - k}{K} T_0 \tag{A.104}$$

where K is a parameter typically ranging from 15 to 30. Originally, for simulating the annealing process of a metal, the so-called *Bolzman constant b* was used to replace T_k with bT_k; however, b can be incorporated implicitly in the scaling parameter δ of the objective function and/or the definition of T_k itself. Furthermore, our purpose is to solve an optimization problem and not to simulate the physical process of annealing. Therefore, we do not use the Boltzmann constant in this book.

A.4 Multiobjective programming

A.4.1 Definition of multiobjective programming

So far, in the Appendix, we considered optimization problems with a single objective function. However, in the process of structural design, it is natural to consider multiple objective functions to be minimized or maximized.

An optimization problem with multiple objective functions is called a *multiobjective programming* (MOP) problem, and its solution method is called *multiobjective programming* (Cohon 1978; Stadler 1979, 1988; Marler and Arora 2004), which is also called multicriteria optimization and vector optimization. The MOP problem of minimizing the n^F objective functions $\mathbf{F}(\mathbf{x}) = (F_1(\mathbf{x}), \ldots, F_{n^F}(\mathbf{x}))^\top$ is formulated as

$$\text{Minimize} \quad \mathbf{F}(\mathbf{x}) \tag{A.105a}$$

$$\text{subject to} \quad H_i(\mathbf{x}) \leq 0, \quad (i = 1, \ldots, n^I) \tag{A.105b}$$

where only inequality constraints are considered for brevity.

A solution that minimizes all the objective functions simultaneously is called an *absolutely optimal solution*, which does not exist in general. An example of an absolutely optimal solution for $n^F = 2$ is illustrated in Fig. A.10(a) in the *objective function space*, for which the coordinates are defined by the objective functions.

If the absolutely optimal solution does not exist, it is natural to consider a situation in which at least one objective value increases if one of the remaining objective values is decreased. For two feasible solutions $\mathbf{x}^1$ and $\mathbf{x}^2$ satisfying the constraints (A.105b), if $F_i(\mathbf{x}^1) \leq F_i(\mathbf{x}^2)$ for $i = 1, \ldots, n^F$ and $F_j(\mathbf{x}^1) < F_j(\mathbf{x}^2)$ for one of $j \in \{1, \ldots, n^F\}$, then $\mathbf{x}^2$ is said to be dominated by $\mathbf{x}^1$. If there is no solution that dominates a feasible solution $\mathbf{x}^*$, then $\mathbf{x}^*$ is called the *nondominated solution*, the *noninferior solution*, the *compromise solution*, or the *Pareto optimal solution*, which may be simply called *Pareto solution*.

For an MOP problem with continuous variables, generally there exists an infinite number of Pareto optimal solutions, which form a set called the *Pareto optimal set* or the *Pareto front*. The *most preferred solution* is chosen from

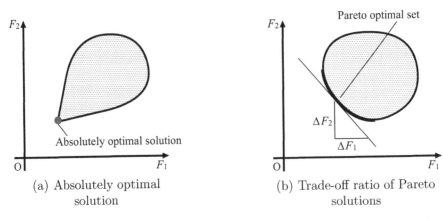

(a) Absolutely optimal
solution

(b) Trade-off ratio of Pareto
solutions

FIGURE A.10: Feasible regions and optimal solutions in the objective function space.

the Pareto set in view of an additional measure of preference. Fig. A.10(b) shows an example of Pareto optimal solutions for $n^F = 2$. The ratio $\Delta F_2/\Delta F_1$ between the two neighboring Pareto solutions defines the trade-off ratio of F_2 to F_1, which serves as one of the *a posteriori* measures of preference.

The process of selecting the most preferred solution is summarized as follows (Marler and Arora 2004):

1. *Approach with* a priori *information*
   The upper bounds, weight coefficients, or ideal values of objective functions are specified, and the single-objective problem is solved using the methods of constraint approach, linear weighted sum approach, goal programming, and so on, which are described in Secs. A.4.2, A.4.3, and A.4.4, respectively. The process is terminated if a single Pareto optimal solution is obtained.

2. *Approach without* a priori *information*

   (a) Enumeration of Pareto optimal solutions:
       Many Pareto optimal solutions are first generated or preferably all the Pareto solutions are enumerated for a problem with integer variables. Then the most preferred solution is selected from the Pareto optimal set in view of additional (*a posteriori*) information.

   (b) Interactive approach:
       A tentative information of preference, e.g., a set of weight coefficients and location of the ideal point in the objective function space, is first specified, and the single Pareto optimal solution is obtained by a similar manner as the approach with *a priori* information. The preference information is next conceived as a parameter that is to be modified interactively to obtain a more preferred solution.

Note that the interactive approach should satisfy the following properties:

1. The relation between the variation of the parameter values and the resulting Pareto optimal solution should be clearly correlated; i.e., the Pareto solution obtained after modification of the parameter should be estimated *a priori* by the decision maker.

2. Pareto optimal solutions can be continuously traced in the objective function space for problems with continuous variables and differentiable functions.

The details of these approaches are explained below.

A.4.2   Constraint approach

Suppose the objective functions are numbered in ascending order with respect to importance; i.e., minimizing $F_1(\mathbf{x})$ is less important than minimizing $F_2(\mathbf{x}), \ldots, F_{n^F}(\mathbf{x})$. Then a single-objective optimization problem is formulated, as follows, to minimize $F_1(\mathbf{x})$ by assigning upper bounds $\overline{F}_2, \ldots, \overline{F}_{n^F}$, respectively, for $F_2(\mathbf{x}), \ldots, F_{n^F}(\mathbf{x})$.

$$\text{Minimize} \quad F_1(\mathbf{x}) \tag{A.106a}$$
$$\text{subject to} \quad F_i(\mathbf{x}) \leq \overline{F}_i, \quad (i = 2, \ldots, n^F) \tag{A.106b}$$
$$H_i(\mathbf{x}) \leq 0, \quad (i = 1, \ldots, n^I) \tag{A.106c}$$

This approach is called the *constraint approach* or the *ε-constraint approach*. Note that the values of the important objective functions are restricted by assigning their upper bounds, whereas the least important objective function is to be minimized; i.e., it is allowed to have an unexpectedly large value. Fig. A.11(a) illustrates the constraint approach for $n^F = 2$, where the gray area is the feasible region. The upper bound can be iteratively modified to find the most preferred solution in an interactive manner. However, the objective value $F_1(\mathbf{x})$ cannot be minimized, if $\overline{F}_i$ $(i = 2, \ldots, n^F)$ are too small and there exists no feasible solution. Therefore, the following procedure is usually used:

Step 1: Assign the objective functions $F_1(\mathbf{x}), \ldots, F_{n^F}(\mathbf{x})$ in ascending order of importance.

Step 2: Solve the single-objective optimization problem for minimizing F_{n^F} under constraints (A.106c) to find only the optimal objective value $\widehat{F}_{n^F}$. Set the iteration counter $k = 1$.

Step 3: Assign a moderately small positive allowable value ε_{n^F-k+1} for relaxing the value of F_{n^F-k+1}, and minimize $F_{n^F-k}(\mathbf{x})$ under constraints (A.106c) and additional constraints

$$F_j(\mathbf{x}) \leq \widehat{F}_j + \varepsilon_j, \quad (j = n^F - k + 1, \ldots, n^F) \tag{A.107}$$

to find the optimal objective value $\widehat{F}_{n^F-k}$.

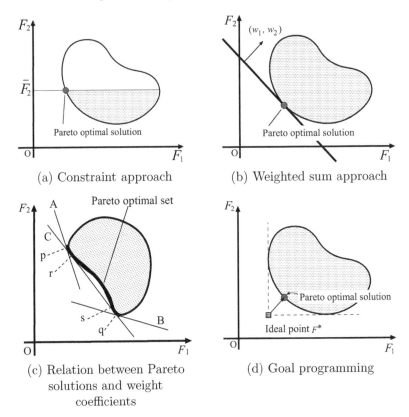

(a) Constraint approach

(b) Weighted sum approach

(c) Relation between Pareto solutions and weight coefficients

(d) Goal programming

FIGURE A.11: Conversion of a multiobjective programming problem to single-objective problem.

Step 4: Stop if enough objective functions are minimized; otherwise let $k \leftarrow k + 1$ and go to Step 3.

A.4.3 Linear weighted sum approach

In the *linear weighted sum approach*, the objective functions $F_i(\mathbf{X})$ are scalarized using the weight coefficients w_i (> 0) in view of the importance of each objective function, and the following optimization problem is solved:

$$\text{Minimize} \quad F(\mathbf{x}) = \sum_{i=1}^{n^F} w_i F_i(\mathbf{x}) \tag{A.108a}$$

$$\text{subject to} \quad H_i(\mathbf{x}) \leq 0, \quad (i = 1, \ldots, n^I) \tag{A.108b}$$

The Pareto optimal solutions are found for each specified set of w_i. This approach is most convenient if appropriate values can easily be determined for

w_i; however, this is generally difficult, because the ranges, units (dimensions), etc., of the objective functions are different.

Consider the simple case of $n^F = 2$, and let $R = w_1/w_2$. Then the slope of the line with constant $F(\mathbf{x})$ at the Pareto solution in the objective function space is $-R$, as illustrated in Fig. A.11(b). Therefore, various Pareto optimal solutions are found by modifying R. However, if the feasible region in the objective function space is not convex, Pareto solutions are found discontinuously even if R is modified continuously, and the complete set of Pareto solutions may not be found using this method.

For example, for the case illustrated in Fig. A.11(c), the line with constant $F(\mathbf{x})$ is indicated by 'A', if R is moderately large, and the Pareto solution 'p' is found. On the other hand, if R is moderately small, the line with constant $F(\mathbf{x})$ is indicated by 'B', and the Pareto optimal solution 'q' is found. Suppose the line with constant $F(\mathbf{x})$ is given by line 'C', and $F(\mathbf{x})$ has the same value at solutions 'r' and 's'. Then, a solution near 'r' is found if R is slightly increased, and a solution near 's' is found if R is slightly decreased. Therefore, the Pareto solutions vary discontinuously with respect to the variation of R; i.e., the solutions between 'r' and 's' cannot be found by this approach. Hence, the linear weighted sum approach is not suitable for interactive search of the most preferred solution, because the decision maker cannot expect a change in the solution as the result of modification of the weight coefficients.

There exist similar approaches, for instance, minimization of the product of $F_1(\mathbf{x}), \ldots, F_{n^F}(\mathbf{x})$, which corresponds to minimization of the sum of $\log[F_i(\mathbf{x})]$. Alternatively, the maximum value of $F_1(\mathbf{x}), \ldots, F_{n^F}(\mathbf{x})$ may be minimized.

A.4.4 Goal programming

The approach for minimizing the distance to the *ideal point* $\mathbf{F}^*$ in the objective function space is called *goal programming*, which is formulated as

$$\text{Minimize} \quad D(\mathbf{F}^*, \mathbf{F}(\mathbf{x})) \tag{A.109a}$$

$$\text{subject to} \quad H_i(\mathbf{x}) \leq 0, \quad (i = 1, \ldots, n^I) \tag{A.109b}$$

where $D(\mathbf{F}^*, \mathbf{F}(\mathbf{x}))$ is the distance between $\mathbf{F}^*$ and $\mathbf{F}(\mathbf{x})$, which may be defined as

$$D(\mathbf{F}^*, \mathbf{F}(\mathbf{x})) = \begin{cases} \displaystyle\sum_{i=1}^{n^F} |F_i^* - F_i(\mathbf{x})| & \text{(Manhattan distance)} \\[2ex] \displaystyle\sqrt{\sum_{i=1}^{n^F} [F_i^* - F_i(\mathbf{x})]^2} & \text{(Euclidean distance)} \\[2ex] \displaystyle\max_i |F_i^* - F_i(\mathbf{x})| & \text{(Chebychev distance)} \end{cases} \tag{A.110}$$

A Pareto solution that is nearest to $\mathbf{F}^*$ is found as the solution of Problem (A.109).

The ideal point can be given using the independently minimized objective values, as illustrated in Fig. A.11(d), or it may be defined simply by the preference of the decision maker. However, it is very difficult to assign the ideal point appropriately. Therefore, the ideal point is modified interactively to search for the most preferred solution among the Pareto optimal solutions. To assist the decision maker in modification of the ideal point, the trade-off ratio, e.g., $\Delta F_2 / \Delta F_1$ in Fig. A.10(b) for $n^F = 2$, can be found analytically utilizing the sensitivity coefficients of the optimal solutions, which is called trade-off analysis. The method based on automatic trade-off analysis is called the *aspiration level approach* (Nakayama 1995).

A.5   Parametric structural optimization problem

A structural optimization problem sometimes depends on a problem parameter that is different from the design variables. For example, the optimal solution of a sizing problem depends on the nodal locations. Therefore, it is important to investigate the sensitivity of the optimal solution with respect to parameters. In this section, we summarize approach called *parametric programming* (Frank 1978; Fiacco 1983; Gal and Greenberg 1997), which was originally developed for postoptimal analysis of linear programming problems. Application of the parametric programming approach is demonstrated for multidisciplinary optimization in Sec. 5.3 and second-order postoptimal analysis of seismic design in Sec. 6.7.

The objective function to be minimized is a function of the design variable vector $\mathbf{x}$, which is denoted by the superscript $(\cdot)^0$ as $F^0(\mathbf{x})$. The equality and inequality constraints are given as $G_i^0(\mathbf{x}, \mathbf{u}(\mathbf{x})) = 0$ $(i = 1, \ldots, n^E)$ and $H_i^0(\mathbf{x}, \mathbf{u}(\mathbf{x})) \leq 0$ $(i = 1, \ldots, n^I)$, respectively, where $\mathbf{u}(\mathbf{x})$ is the vector of nodal displacements, which are conceived as state variables. The optimization problem is formulated as

$$\text{Minimize} \quad F^0(\mathbf{x}) \qquad \qquad \text{(A.111a)}$$

$$\text{subject to} \quad G_i^0(\mathbf{x}, \mathbf{u}(\mathbf{x})) = 0, \quad (i = 1, \ldots, n^E) \qquad \text{(A.111b)}$$

$$\qquad \qquad \quad H_i^0(\mathbf{x}, \mathbf{u}(\mathbf{x})) \leq 0, \quad (i = 1, \ldots, n^I) \qquad \text{(A.111c)}$$

Note that the side constraints for $\mathbf{x}$ are supposed to be included in the general inequality constraints (A.111c).

Let $\mathbf{R}^0(\mathbf{x}, \mathbf{u}(\mathbf{x}))$ denote the vector of constraint functions consisting of the equality constraints (A.111b) and the active inequality constraints in (A.111c) at an optimal solution. The ith component of $\mathbf{R}^0$ is denoted by R_i^0. The function of $\mathbf{x}$ only is defined with a tilde as

$$\widetilde{R}_i^0(\mathbf{x}) = R_i^0(\mathbf{x}, \mathbf{u}(\mathbf{x})) \tag{A.112}$$

The Karush-Kuhn-Tucker conditions for Problem (A.111) are written as

$$\frac{\partial F^0}{\partial x_i} + \sum_{j=1}^{n^A} \lambda_j \frac{\partial \widetilde{R}_j^0}{\partial x_i} = 0, \quad (i = 1, \dots, m) \tag{A.113}$$

where m is the number of variables, n^A is the number of active constraints including the equality constraints, and λ_j (≥ 0) is the Lagrange multiplier. If $\widetilde{R}_j^0$ is related to the active inequality constraint, then $\lambda_j \geq 0$ holds, whereas no restriction in sign exists for λ_j related to an equality constraint.

The partial differentiation of $\widetilde{R}_j^0$ with respect to x_i is obtained as

$$\frac{\partial \widetilde{R}_j^0}{\partial x_i} = \frac{\partial R_j^0}{\partial x_i} + \sum_{k=1}^{n} \frac{\partial R_j^0}{\partial u_k} \frac{\partial u_k}{\partial x_i}, \quad (i = 1, \dots, m; \ j = 1, \dots, n^A) \tag{A.114}$$

where n is the number of components in $\mathbf{u}$. Let $\mathbf{p} = (p_1, \dots, p_{n^P})^\top$ denote a vector of n^P parameters. The parametric form of Problem (A.111) is stated as

$$\text{Minimize} \quad F(\mathbf{x}, \mathbf{p}) \tag{A.115a}$$

$$\text{subject to} \quad G_i(\mathbf{x}, \mathbf{p}) = 0, \quad (i = 1, \dots, n^E) \tag{A.115b}$$

$$H_i(\mathbf{x}, \mathbf{p}) \leq 0, \quad (i = 1, \dots, n^I) \tag{A.115c}$$

where $\mathbf{u}(\mathbf{x})$ is not included in the arguments, for simplicity, because it can be regarded as an implicit function of $\mathbf{x}$.

The design variables, state variables, and the objective value at the optimal solution are functions of $\mathbf{p}$, because they can be obtained for each specified value of $\mathbf{p}$. The vector of active constraints $\mathbf{R}(\mathbf{x}, \mathbf{p})$ is also redefined as

$$\widehat{\mathbf{R}}(\mathbf{p}) = \mathbf{R}(\widehat{\mathbf{x}}(\mathbf{p}), \mathbf{p}) \tag{A.116}$$

In the following, a function of $\mathbf{p}$ only is denoted by a hat as $\widehat{(\cdot)}$. The derivatives of the optimal objective value $\widehat{F}(\mathbf{p}) = F(\widehat{\mathbf{x}}(\mathbf{p}), \mathbf{p})$ with respect to p_k are obtained from

$$\frac{\partial \widehat{F}}{\partial p_k} = \frac{\partial F}{\partial p_k} + \sum_{i=1}^{m} \frac{\partial F}{\partial x_i} \frac{\partial \widehat{x}_i}{\partial p_k} \tag{A.117}$$

Suppose that the active constraints $\widehat{R}_j(\mathbf{p}) = 0$ at an optimal solution remain active for a small variation of p_k; i.e.,

$$\frac{\partial \widehat{R}_j}{\partial p_k} = 0, \quad (j = 1, \dots, n^A) \tag{A.118}$$

Then the following relation holds:

$$\frac{\partial R_j}{\partial p_k} + \sum_{i=1}^{m} \frac{\partial R_j}{\partial x_i} \frac{\partial \hat{x}_i}{\partial p_k} = 0, \quad (j = 1, \ldots, n^{\mathrm{A}}) \tag{A.119}$$

By multiplying $\partial \hat{x}_i / \partial p_k$ on both sides of (A.113) and taking summation over i, we have

$$\sum_{i=1}^{m} \frac{\partial F}{\partial x_i} \frac{\partial \hat{x}_i}{\partial p_k} + \sum_{i=1}^{m} \sum_{j=1}^{n^{\mathrm{A}}} \lambda_j \frac{\partial R_j}{\partial x_i} \frac{\partial \hat{x}_i}{\partial p_k} = 0 \tag{A.120}$$

By multiplying λ_j on both sides of (A.119) and taking summation over j corresponding to the equality constraints and the active inequality constraints, we derive the following relation:

$$\sum_{j=1}^{n^{\mathrm{A}}} \lambda_j \frac{\partial R_j}{\partial p_k} + \sum_{j=1}^{n^{\mathrm{A}}} \sum_{i=1}^{m} \lambda_j \frac{\partial R_j}{\partial x_i} \frac{\partial \hat{x}_i}{\partial p_k} = 0 \tag{A.121}$$

From (A.117), (A.120), and (A.121),

$$\frac{\partial \hat{F}}{\partial p_k} = \frac{\partial F}{\partial p_k} + \sum_{j=1}^{n^{\mathrm{A}}} \lambda_j \frac{\partial R_j}{\partial p_k} \tag{A.122}$$

is derived. We can see from (A.122) that the derivative of the optimal objective value with respect to p_k, which is called the *parametric sensitivity coefficient*, can be obtained without computing the derivatives of the design variables and the Lagrange multipliers.

Next, we compute the second-order derivatives of the optimal objective value $\hat{F}$ with respect to the parameters. Differentiation of (A.113) with respect to p_k leads to

$$\sum_{l=1}^{m} \left(\frac{\partial^2 F}{\partial x_i \partial x_l} + \sum_{j=1}^{n^{\mathrm{A}}} \lambda_j \frac{\partial^2 R_j}{\partial x_i \partial x_l} \right) \frac{\partial \hat{x}_l}{\partial p_k} + \frac{\partial^2 F}{\partial x_i \partial p_k}$$

$$+ \sum_{j=1}^{n^{\mathrm{A}}} \left(\frac{\partial R_j}{\partial x_i} \frac{\partial \hat{\lambda}_j}{\partial p_k} + \lambda_j \frac{\partial^2 R_j}{\partial x_i \partial p_k} \right) = 0 \tag{A.123}$$

The derivatives of $\hat{\lambda}_j$ and $\hat{x}_i$ are computed from a set of $m + n^{\mathrm{A}}$ linear equations (A.120) and (A.123), where the first term on the left-hand side of (A.123) consists of the Hessian of the Lagrangian.

By further differentiating (A.122) with respect to p_r, we obtain

$$\frac{\partial^2 \widehat{F}}{\partial p_k \partial p_r} = \frac{\partial^2 F}{\partial p_k \partial p_r} + \sum_{i=1}^{m} \frac{\partial^2 F}{\partial p_k \partial x_i} \frac{\partial \widehat{x}_i}{\partial p_r} + \sum_{j=1}^{n^A} \sum_{i=1}^{m} \lambda_j \frac{\partial^2 \widetilde{R}_j}{\partial p_k \partial x_i} \frac{\partial \widehat{x}_i}{\partial p_r}$$
$$+ \sum_{j=1}^{n^A} \left(\frac{\partial R_j}{\partial p_k} \frac{\partial \widehat{\lambda}_j}{\partial p_r} + \lambda_j \frac{\partial^2 R_j}{\partial p_k \partial p_r} \right) \tag{A.124}$$

It is observed from (A.124) that the second-order derivative of $\widehat{F}$ can be found without computing the second-order derivatives of $\widehat{\lambda}_j$ and $\widehat{x}_i$, although their first-order derivatives are needed.

A.6 Parametric curves and surfaces

A.6.1 Bézier curve

As is seen in Secs. 6.4, 6.5, and 6.6, parametric curves and surfaces such as a Bézier curve and a Bézier surface can be effectively used for modeling relatively complex curves and surfaces with a small number of variables. Parametric curves are also very effective for generating a smooth optimal shape of a two-dimensional continuum (Braibant and Fleury 1984; Özakca, Hinton, and Rao 1993; Eschenauer, Kobelev, and Schumacher 1994). In this section, we summarize the formulations of Bézier curves and surfaces (Farin 1988; Farin, Hoschek, and Kim 2002). Note that other parametric representations, e.g., B-spline and non-uniform rational B-spline (NURBS) curves/surfaces, are not explained here, because they are not used in this book (see Farin, Hoschek, and Kim (2002) for more details).

The Bézier curve is defined by the function called the *Bernstein basis function* or the *Bernstein polynomial* and the *control points*, as shown in Fig. A.12. The polygon that consists of the control points is called the *control polygon* or the *defining polygon*. In Fig. A.12, the control polygon consists of the control points $\mathbf{R}_0$, $\mathbf{R}_1$, $\mathbf{R}_2$, and $\mathbf{R}_3$.

The Bernstein basis function of order n is given as

$$B_i^n(t) = \binom{n}{i} t^i (1-t)^{n-i}, \quad (i = 0, 1, \ldots, n) \tag{A.125a}$$

$$\binom{n}{i} = \begin{cases} \dfrac{n!}{i!(n-i)!} & \text{for } 0 \le i \le n \\ 0 & \text{for } i < 0 \text{ or } i > n \end{cases} \tag{A.125b}$$

where $0 \le t \le 1$ is the parameter, and $0^0 = 0! = 1$. For example, $B_i^n(t)$ for

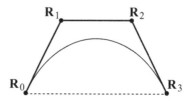

FIGURE A.12: Control points and control polygon of a Bézier curve of order 3.

$n = 3$ are given as

$$B_0^3(t) = (1-t)^3, \quad B_1^3(t) = 3t(1-t)^2,$$
$$B_2^3(t) = 3t^2(1-t), \quad B_3^3(t) = t^3 \tag{A.126}$$

which are plotted in Fig. A.13.

$B_i^n(t)$ can also be defined by the following recursive form:

$$B_i^n(t) = (1-t)B_i^{n-1}(t) + tB_{i-1}^{n-1}(t) \tag{A.127a}$$
$$B_0^0 = 1 \tag{A.127b}$$
$$B_j^n = 0, \quad \text{for } j < 0 \text{ or } j > n \tag{A.127c}$$

For example, the basis functions of order 4 are obtained from $B_i^3(t)$ as

$$B_0^4(t) = (1-t)B_0^3 = (1-t)^4,$$
$$B_1^4(t) = (1-t)B_1^3 + tB_0^3 = 3t(1-t)^3 + t(1-t)^3 = 4t(1-t)^3,$$
$$B_2^4(t) = (1-t)B_2^3 + tB_1^3 = 3t^2(1-t)^2 + t^2(1-t)^2 = 6t^2(1-t)^2, \quad \text{(A.128)}$$
$$B_3^4(t) = (1-t)B_3^3 + tB_2^3 = t^3(1-t) + 3t^3(1-t) = 4t^3(1-t),$$
$$B_4^4(t) = tB_3^3 = t^4$$

By using $B_i^n(t)$, the Bézier curve $\mathbf{P}_b^n(t)$ of order n is given as

$$\mathbf{P}_b^n(t) = \sum_{i=0}^{n} \mathbf{R}_i B_i^n(t) \tag{A.129}$$

where $\mathbf{R}_i$ is the location vector of the ith control point. This way, the Bézier curve of order n is defined with $n+1$ control points, and the order of the basis function is n.

Examples of Bézier curves are illustrated in Fig. A.14 for $n = 3$ with different locations of the control point $\mathbf{R}_3$. As is seen, the shape of the curve can be controlled by moving the control points. Note that these curves exist inside of the *convex hull* $\mathbf{R}_0\mathbf{R}_1\mathbf{R}_3\mathbf{R}_2$, indicated by dotted lines in Fig. A.14(a), formed by the control points.

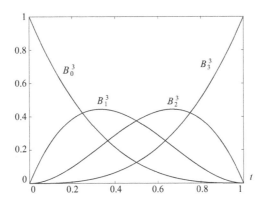

FIGURE A.13: Bernstein basis functions of order 3.

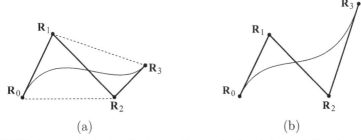

(a)   (b)

FIGURE A.14: Relation between the control points and the shape of the Bézier curve with $\mathbf{R}_0 = (100, 100)$, $\mathbf{R}_1 = (150, 200)$, $\mathbf{R}_2 = (250, 100)$; (a) $\mathbf{R}_3 = (300, 150)$, (b) $\mathbf{R}_3 = (300, 250)$.

It can be easily observed from (A.125) as well as Fig. A.13 that only B_0^n is 1 and the other terms are 0 at $t = 0$. On the other hand, $B_n^n = 1$ and the other terms are 0 at $t = 1$. Therefore, from (A.129), the curve coincides with the points $\mathbf{R}_0$ and $\mathbf{R}_n$ at $t = 0$ and 1, respectively. It can also be derived by differentiating (A.125a) with respect to t that the tangent vectors at both ends with $t = 0$ and 1, respectively, coincide with the direction of the control polygon, as seen in Figs. A.12 and A.14. For example, for $n = 3$, the derivatives of $B_i^3(t)$ with respect to t, denoted by $\dot{B}_i^3(t)$, are obtained as

$$\dot{B}_0^3(t) = -3(1-t)^2, \quad \dot{B}_1^3(t) = 3(1-3t)(1-t),$$
$$\dot{B}_2^3(t) = 3t(2-3t), \quad \dot{B}_3^3(t) = 3t^2 \tag{A.130}$$

Therefore, we have

$$\dot{B}_0^3(0) = -3, \quad \dot{B}_1^3(0) = 3, \quad \dot{B}_1^3(0) = 0, \quad \dot{B}_2^3(0) = 0$$
$$\dot{B}_0^3(1) = 0, \quad \dot{B}_1^3(1) = 0, \quad \dot{B}_1^3(1) = -3, \quad \dot{B}_2^3(1) = 3 \tag{A.131}$$

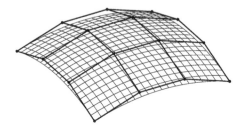

FIGURE A.15:   A tensor product Bézier surface of order 3×3.

which leads to the tangent vector $\dot{\mathbf{P}}_{\mathrm{b}}^3(t)$ at $t = 0$ and 1:

$$\dot{\mathbf{P}}_{\mathrm{b}}^3(0) = 3(\mathbf{R}_1 - \mathbf{R}_0), \quad \dot{\mathbf{P}}_{\mathrm{b}}^3(1) = 3(\mathbf{R}_3 - \mathbf{R}_2) \qquad (A.132)$$

Furthermore, the following equation is derived for any n from (A.125a) and (A.125b):

$$\sum_{i=0}^{n} B_i^n(t) = 1 \qquad (A.133)$$

which leads to the property that the curve exists in the convex hull of the control polygon. These properties are effectively used for controlling the shape of the curve interactively by moving the control points.

A.6.2   Bézier surface

The surface defined as the product of Bézier curves is called the *tensor product Bézier surface*. Let u and v denote the parameters in the two perpendicular directions. The tensor product Bézier surface of order $n \times m$ is defined as

$$\mathbf{P}_{\mathrm{b}}^{n,m}(u, v) = \sum_{i=0}^{n} \sum_{j=0}^{m} \mathbf{R}_{i,j} B_i^n(u) B_j^m(v) \qquad (A.134)$$

where $\mathbf{R}_{i,j}$ $(i = 0, \ldots, n;\ j = 0, \ldots, m)$ are the control points, and $B_i^n(u)$ and $B_j^m(v)$ are the Bernstein basis functions of orders n and m, respectively, in the directions of u and v.

A tensor product Bézier surface of order 3×3 is illustrated in Fig. A.15, where the thick lines are the edges of the control polygon, which is also called the *control net*. Note that the numbers of control points as well as the orders of the Bernstein basis function in the u- and v-directions can be different. As is easily seen from (A.134), the curve of fixed parameter value u or v forms a Bézier curve; therefore, the boundary of the surface is a Bézier curve.

Although the tensor product Bézier surface is very simple, it may not be suitable for modeling a complex shape, including a surface with a circular

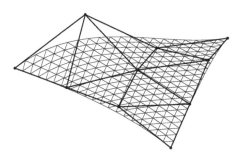

FIGURE A.16: A triangular Bézier patch of order 3.

plan. In this case, the triangular Bézier patch of order n, defined as follows, can be effectively used:

$$\mathbf{P}_t^n(u, v, w) = \sum_{\alpha+\beta+\gamma=n} \mathbf{R}_{\alpha\beta\gamma} B_{\alpha\beta\gamma}^n \qquad (A.135a)$$

$$B_{\alpha\beta\gamma}^n = \frac{n!}{\alpha!\beta!\gamma!} u^\alpha v^\beta w^\gamma \qquad (A.135b)$$

where (u, v, w) are the barycentric coordinates $0 \leq u, \leq 1, 0 \leq v, \leq 1, 0 \leq w, \leq 1$, $u + v + w = 1$, and the summation is made over $\alpha, \beta, \gamma = 0, \ldots, n$ with $\alpha + \beta + \gamma = n$.

Alternatively, the variable w, which is dependent on u and v, is removed as

$$\mathbf{P}_t^n(u, v) = \sum_{\alpha=0}^{n} \sum_{\beta=0}^{n-\alpha} \mathbf{R}_{\alpha\beta} B_{\alpha\beta}^n \qquad (A.136a)$$

$$B_{\alpha\beta}^n = \frac{n!}{\alpha!\beta!(n-\alpha-\beta)!} u^\alpha v^\beta (1-u-v)^{n-\alpha-\beta} \qquad (A.136b)$$

An example of a triangular Bézier patch of order 3 is illustrated in Fig. A.16, where the thick lines show the control net, and the thin lines are the set of points with a constant value of u, v, or $w = 1 - u - v$.

A.6.3 Adjoint curve

Let $\mathbf{P}(t)$ denote a parametric curve, e.g., a Bézier curve, which is regarded as a primary curve or *progenitor curve* (Barnhill 1994; Hoschek 1985; Hoschek and Wissel 1988). A curve defined by parameter t, using the properties of $\mathbf{P}(t)$, is called an *adjoint curve*. For example, the hodograph defined by the velocity vector of a particle in motion, as shown in Fig. A.17, is an adjoint curve (Katoh, Ohsaki, and Tani 2002). In this case, the adjoint curve $\mathbf{P}^a(t)$ to the progenitor curve $\mathbf{P}(t)$ is defined as

$$\mathbf{P}^a(t) = \alpha\dot{\mathbf{P}}(t) \qquad (A.137)$$

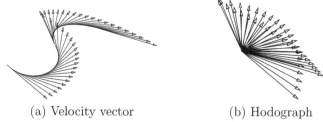

(a) Velocity vector   (b) Hodograph

FIGURE A.17: Velocity vector and hodograph.

TABLE A.1:   Control points of the Bézier curves in Fig. A.18.

Point	Case 1	Case 2	Case 3
0	(100.0, 100.0)	(100.0, 100.0)	(210.0, 400.0)
1	(150.0, 300.0)	(220.0, 300.0)	(290.0, 200.0)
2	(200.0, 100.0)	(200.0, 50.0)	(10.0, 100.0)
3	(250.0, 100.0)	(250.0, 50.0)	(490.0, 100.0)
4	(300.0, 300.0)	(230.0, 300.0)	(210.0, 200.0)
5	(350.0, 100.0)	(350.0, 100.0)	(290.0, 400.0)

where α is a specified scaling parameter.

The *offset curve* that exists at a constant distance from a curve is very important in various fields of engineering, e.g., numerical control machining. The offset curve $\mathbf{D}(t)$ is also an adjoint curve, because it is defined using the unit normal vector $\mathbf{n}^0(t)$ of $\mathbf{P}(t)$ as

$$\mathbf{D}(t) = \mathbf{P}(t) + d\mathbf{n}^0(t) \tag{A.138}$$

where d is the specified offset value. The properties of the offset curve can also be expressed in a parametric form; e.g., the tangent vector of $\mathbf{D}(t)$ is written as

$$\mathbf{D}(t) = \dot{\mathbf{P}}(t) + d\dot{\mathbf{n}}^0(t) \tag{A.139}$$

The offset curves are extensively used for defining the shapes of the arch-type trusses in Secs. 6.4 and 6.6.

Note that the offset curve is not a Bézier curve even if $\mathbf{P}(t)$ is a Bézier curve. A variety of algorithms have been proposed for generating an approximate parametric form of an offset curve or surface. Furthermore, if d is too large or $\mathbf{P}(t)$ has a complex shape, a loop or intersection exists in the offset curve. Examples of offset curves of the progenitor curve defined by the Bézier curve of order 5 are shown in Fig. A.18. The locations of the control points for each case are listed in Table A.1, and the distance d between the progenitor and offset curves is 30. As is seen, if the locations of the control points are not appropriate, there may exist a loop or intersection, as seen in Figs. A.18(b) and (c), respectively.

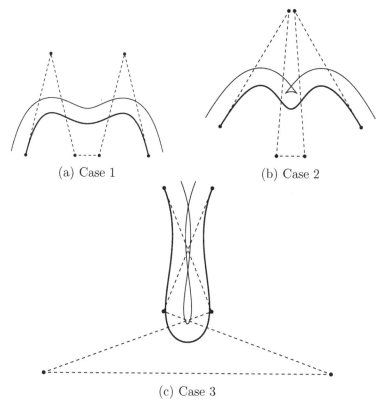

(a) Case 1   (b) Case 2

(c) Case 3

FIGURE A.18: Offset curves of Bézier curves with various locations of control points; thick solid line: progenitor curve, thin solid line: offset curve, dotted line: control polygon.

A.7 Response spectrum approach

A.7.1 SRSS method

The responses of structures under seismic motions can be evaluated using a time-history analysis if the record of input acceleration is given. Although an accurate response can be computed by using time-history analysis for a single deterministic input motion, seismic motions are highly unpredictable in the design process of structures in civil engineering. Therefore, the design response spectrum is usually given for representing the statistical properties of the possible seismic motions at a specific construction site. In this section, we briefly summarize the response spectrum approach for completeness of the book (see textbooks, e.g., Chopra (2001) for details).

Let $\mathbf{K}$ and $\mathbf{M}$ denote the $n \times n$ stiffness matrix and mass matrix, respectively, where n is the number of degrees of freedom. The rth eigenvalue Ω_r and eigenvector $\boldsymbol{\Phi}_r$ are defined by

$$\mathbf{K}\boldsymbol{\Phi}_r = \Omega_r \mathbf{M}\boldsymbol{\Phi}_r, \quad (r = 1, \ldots, n) \tag{A.140}$$

where the eigenvectors are ortho-normalized with respect to the mass matrix as

$$\boldsymbol{\Phi}_r^\top \mathbf{M}\boldsymbol{\Phi}_s = \delta_{rs}, \quad (r, s = 1, \ldots, n) \tag{A.141}$$

where δ_{rs} is the Kronecker delta.

Let $\mathbf{I}$ denote the vector consisting of 1 for the components corresponding to the direction of the input seismic motion and 0 for the remaining components. Then the rth participation factor β_r is given as

$$\beta_r = \boldsymbol{\Phi}_r^\top \mathbf{M}\mathbf{I}, \quad (r = 1, \ldots, n) \tag{A.142}$$

The rth damping ratio h_r may be defined, e.g., using stiffness-proportional damping with h_1^* specified for the first mode; i.e.,

$$h_r = h_1^* \sqrt{\frac{\Omega_r}{\Omega_1}}, \quad (r = 1, \ldots, n) \tag{A.143}$$

In this case, the damping matrix $\mathbf{C}$ is defined as $\mathbf{C} = \alpha_1 \mathbf{K}$ with an appropriate coefficient α_1. If Rayleigh damping is used, the damping ratios for two different modes can be specified, and $\mathbf{C}$ is defined as $\mathbf{C} = \alpha_1 \mathbf{K} + \alpha_2 \mathbf{M}$ with coefficients α_1 and α_2.

The specified displacement response spectrum is denoted by $S_{\mathrm{D}}(\Omega_r, h_r)$, which is a function of Ω_r and h_r. By using the *square-root-of-sum-of-squares* (SRSS) method, we can compute the mean value $U_j^{\max}$ of the maximum response of the displacement component U_j, as follows, under the set of seismic motions that are compatible with $S_{\mathrm{D}}(\Omega_r, h_r)$ denoted simply by S_{D}^r:

$$U_j^{\max} = \sqrt{\sum_{r=1}^{n^{\mathrm{D}}} (S_{\mathrm{D}}^r \beta_r \Phi_{r,j})^2} \tag{A.144}$$

where $\Phi_{r,j}$ is the jth component of $\boldsymbol{\Phi}_r$, and the lowest n^{D} modes are assumed to be used for evaluation of the responses. Note that the higher modes cannot be neglected, especially for long-span domes and structures with some symmetry conditions.

Let $\varepsilon_{r,i}$ denote the strain component of the ith member of a truss or frame corresponding to the mode $\boldsymbol{\Phi}_r$. Then the maximum response strain $\varepsilon_i^{\max}$ of the ith member is also evaluated using the SRSS method:

$$\varepsilon_i^{\max} = \sqrt{\sum_{r=1}^{n^{\mathrm{D}}} (S_{\mathrm{D}}^r \beta_r \varepsilon_{r,i})^2} \tag{A.145}$$

The pseudovelocity response spectrum $S_V(\Omega_r, h_r)$ and the pseudoacceleration response spectrum $S_A(\Omega_r, h_r)$ are defined as

$$S_V^r = S_V(\Omega_r, h_r) = \sqrt{\Omega_r} S_D(\Omega_r, h_r), \quad (r = 1, \ldots, n) \quad \text{(A.146a)}$$
$$S_A^r = S_A(\Omega_r, h_r) = \Omega_r S_D(\Omega_r, h_r), \quad (r = 1, \ldots, n) \quad \text{(A.146b)}$$

Then the maximum nodal velocity $v_i^{\max}$ and acceleration $a_i^{\max}$ corresponding to the ith displacement component are evaluated as

$$v_i^{\max} = \sqrt{\sum_{r=1}^{n_D} (S_V^r \beta_r \Phi_{r,i})^2},$$
$$a_i^{\max} = \sqrt{\sum_{r=1}^{n_D} (S_A^r \beta_r \Phi_{r,i})^2} \qquad \text{(A.147)}$$

Note that the pseudodisplacement response spectrum is to be defined as

$$S_D^r = \frac{1}{\Omega_r} S_A^r \qquad \text{(A.148)}$$

if S_A^r is specified as the design response spectrum.

A.7.2 CQC method

It is well known that the accuracy of the SRSS method deteriorates if the eigenvalues, or frequencies, of the dominant modes are closely spaced, because the SRSS method assumes that the maximum modal responses are not correlated. For this case, the maximum seismic responses may be evaluated using the *complete quadratic combination* (CQC) method (Wilson, Der Kiureghian, and Bayo 1982), which is an extension of the SRSS method.

The CQC method incorporates the correlation between the responses of the dominant modes, where the coefficient τ_{rs} between the rth and sth modes is defined by

$$\tau_{rs} = \frac{8\sqrt{h_r h_s}(h_r + \alpha_{rs} h_s)\alpha_{rs}^{3/2}}{(1 - \alpha_{rs}^2)^2 + 4h_r h_s \alpha_{rs}(1 + \alpha_{rs}^2) + 4(h_r^2 + h_s^2)\alpha_{rs}^2} \qquad \text{(A.149)}$$

where

$$\alpha_{rs} = \sqrt{\frac{\Omega_s}{\Omega_r}}, \quad (r, s = 1, \ldots, n) \qquad \text{(A.150)}$$

Then, for example, the maximum response strain $\varepsilon_i^{\max}$ of the ith member is found from

$$\varepsilon_i^{\max} = \sqrt{\sum_{r=1}^{n_D} \sum_{s=1}^{n_D} [S_D(\Omega_r, h_r)\beta_r \varepsilon_{r,i}]\tau_{rs}[S_D(\Omega_s, h_s)\beta_s \varepsilon_{s,i}]} \qquad \text{(A.151)}$$

As is seen from (A.147) and (A.151), the CQC method reduces to the SRSS method if $\alpha_{rs} = \delta_{rs}$ with δ_{rs} being the Kronecker delta.

The CQC method may also be used if multiple components are considered for input motions. It is assumed here that the three orthogonal directions denoted by directions 1, 2, and 3 correspond to the principal axes of the structure. The participation factor $\beta_r^{(j)}$ of the rth mode for input in the jth direction is defined as

$$\beta_r^{(j)} = \mathbf{\Phi}_r^\top \mathbf{M} \mathbf{I}^{(j)} \tag{A.152}$$

where $\mathbf{I}^{(j)}$ is a vector of which the components corresponding to the displacement in the jth direction are 1, and the remaining components are 0.

Let $S_D^{(j)}(\Omega_r, h_r)$ denote the specified displacement response spectrum for the excitation in the jth direction. Then, the maximum response strain $\varepsilon_i^{\max}$ of the ith member is found from the following equation (Semby and Der Kiureghian 1985):

$$\varepsilon_i^{\max} = \sqrt{\sum_{j=1}^{3}\sum_{r=1}^{n_D}\sum_{s=1}^{n_D}[S_D^{(j)}(\Omega_r, h_r)\beta_r^{(j)}\varepsilon_{r,i}]T_{rs}[S_D^{(j)}(\Omega_s, h_s)\beta_s^{(j)}\varepsilon_{s,i}]} \tag{A.153}$$

A.7.3 Design response spectrum

The response spectrum by Newmark and Hall (1982) is used in most of the examples of seismic optimization in this book. The maximum values for acceleration, velocity, and displacement of the ground motion are denoted by C_A, C_V, and C_D, respectively. The design displacement response spectrum, which is a function of Ω_r, is defined as the minimum values among $S_D^{(1)}, \ldots, S_D^{(5)}$ that are defined as

$$S_D^{(1)}(\Omega_r) = C_A/\Omega_r \tag{A.154a}$$
$$S_D^{(2)}(\Omega_r) = 16.2 C_A A_A \Omega_r^{-1.36} \tag{A.154b}$$
$$S_D^{(3)}(\Omega_r) = C_A A_A/\Omega_r \tag{A.154c}$$
$$S_D^{(4)}(\Omega_r) = C_V A_V/\sqrt{\Omega_r} \tag{A.154d}$$
$$S_D^{(5)}(\Omega_r) = C_D A_D \tag{A.154e}$$

where the amplification factors A_A, A_V, and A_D are given as

$$A_A = 3.21 - 0.68 \log(100 h_r) \tag{A.155a}$$
$$A_V = 2.31 - 0.41 \log(100 h_r) \tag{A.155b}$$
$$A_D = 1.82 - 0.27 \log(100 h_r) \tag{A.155c}$$

The pseudovelocity response spectrum for the standard parameter values $C_A = 2.01 \text{ m/s}^2$, $C_V = 0.25 \text{ m/s}$, and $C_D = 0.1875 \text{ m}$ is plotted in Fig. A.19 with the damping ratio $h_r = 0.02$.

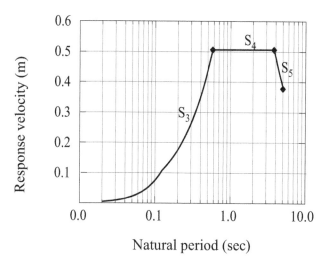

FIGURE A.19: Pseudovelocity response spectrum by Newmark and Hall.

A.7.4 Sensitivity analysis of mean maximum response

Let x denote the design variable representing, e.g., the cross-sectional area of a member of a truss and a nodal coordinate of a frame. The sensitivity coefficients of Ω_r and Φ_r are obtained from (2.39) and (2.42) in Sec. 2.3. Using (A.141) and differentiating (A.142), we obtain the sensitivity coefficient of β_r as

$$\frac{\partial \beta_r}{\partial x} = \Phi_r^{\mathsf{T}} \frac{\partial \mathbf{M}}{\partial x} \mathbf{I} + \left(\frac{\partial \Phi_r}{\partial x}\right)^{\mathsf{T}} \mathbf{M} \mathbf{I} \tag{A.156}$$

Suppose, for simplicity, the responses are evaluated by using the SRSS method. Denoting $S_{\mathrm{D}}^r \beta^r \Phi_{r,j}$ by U_j^r, the following relation is derived by taking the squares of both sides of (A.144) and differentiating them with respect to x:

$$U_j^{\max} \frac{\partial U_j^{\max}}{\partial x} = \sum_{r=1}^{n^{\mathrm{D}}} \left[U_j^r \left(\frac{\partial S_{\mathrm{D}}^r}{\partial x} \beta_r \Phi_{r,j} \right. \right.$$
$$\left. \left. + S_{\mathrm{D}}^r \frac{\partial \beta_r}{\partial x} \Phi_{r,j} + S_{\mathrm{D}}^r \beta_r \frac{\partial \Phi_{r,j}}{\partial x} \right) \right] \tag{A.157}$$

from which the sensitivity coefficient of $U_j^{\max}$ is obtained. The sensitivity coefficient of $S_{\mathrm{D}}^r = S_{\mathrm{D}}(\Omega_r, h_r)$ is easily obtained as

$$\frac{\partial S_{\mathrm{D}}^r}{\partial x} = \frac{\partial S_{\mathrm{D}}^r}{\partial \Omega_r} \frac{\partial \Omega_r}{\partial x} + \frac{\partial S_{\mathrm{D}}^r}{\partial h_r} \frac{\partial h_r}{\partial x} \tag{A.158}$$

Therefore, if h_r is given as a function of Ω_r, then the sensitivity coefficient of the maximum response can be analytically obtained. For the case of stiffness-

TABLE A.2: List of column sections; A ($\times 10^4$ mm^2), I ($\times 10^8$ mm^4), Z ($\times 10^6$ mm^3), Z^{p} ($\times 10^6$ mm^3).

	size	A	I	Z	Z^{p}
1	$\Box - 500 \times 12$	2.268	8.84	3.540	4.100
2	$\Box - 500 \times 16$	2.966	11.30	4.510	5.290
3	$\Box - 500 \times 19$	3.470	13.00	5.180	6.130
4	$\Box - 500 \times 22$	3.957	14.50	5.800	6.920
5	$\Box - 500 \times 25$	4.428	15.90	6.360	7.660
6	$\Box - 500 \times 28$	4.883	17.20	6.870	8.360
7	$\Box - 500 \times 32$	5.463	18.70	7.470	9.210
8	$\Box - 500 \times 36$	6.014	20.00	7.990	9.970

proportional damping, the following relation is derived by taking the squares of both sides of (A.143) and differentiating them with respect to x:

$$2h_r \frac{\partial h_r}{\partial x}\Omega_1 + h_r^2 \frac{\partial \Omega_r}{\partial x} = (h_1^*)^2 \frac{\partial \Omega_r}{\partial x} \tag{A.159}$$

A.8   List of available standard sections of beams and columns

The list of available standard sections of steel members can be found from specifications, e.g., AISC in US and JIS in Japan. The members are classified into the sets, or series, with the same height or width of the section. However, it is not convenient to use the complete list of members of a set or members in different sets for optimization, because the second moment of inertia I, the section modulus Z, and the plastic modulus Z^{p} are not always increasing functions of the cross-sectional area A. Therefore, in structural optimization, several selected sections should be used so that I, Z, and Z^{p} are listed in increasing order of A. The lists in Tables A.2 and A.3 for columns and beams, respectively, are used in the examples in Sec. 5.4.

The values of I, Z, and Z^{p} can be approximated by continuous functions of A (Grierson and Lee 1984; Sadek 1992; Wang and Arora 2006a; Pan, Ohsaki, and Kinoshita 2007). The standard method of least squares may be used for this purpose. However, normalization of A and I, which have different magnitudes and units, may lead to the convergence of the parameter optimization process, as described below. For example, I is approximated as

$$\frac{I(A)}{I_0} = a \left(\frac{A}{A_0} \right)^b \tag{A.160}$$

TABLE A.3:   List of beam sections; A ($\times 10^4$ mm^2), I ($\times 10^8$ mm^4), Z ($\times 10^6$ mm^3), Z^{p} ($\times 10^6$ mm^3).

	size	A	I	Z	Z^{p}
1	H $-$ 500 $\times$ 200 $\times$ 9 $\times$ 12	0.923	3.750	1.500	1.720
2	H $-$ 500 $\times$ 200 $\times$ 9 $\times$ 16	1.076	4.600	1.840	2.080
3	H $-$ 500 $\times$ 200 $\times$ 9 $\times$ 19	1.190	5.210	2.090	2.340
4	H $-$ 500 $\times$ 200 $\times$ 9 $\times$ 22	1.305	5.810	2.330	2.600
5	H $-$ 500 $\times$ 200 $\times$ 12 $\times$ 22	1.442	6.050	2.420	2.760
6	H $-$ 500 $\times$ 250 $\times$ 9 $\times$ 22	1.525	7.070	2.830	3.010
7	H $-$ 500 $\times$ 250 $\times$ 12 $\times$ 22	1.662	7.310	2.920	3.290
8	H $-$ 500 $\times$ 250 $\times$ 12 $\times$ 25	1.804	8.040	3.220	3.610
9	H $-$ 500 $\times$ 250 $\times$ 12 $\times$ 25	1.947	8.750	3.500	3.930

TABLE A.4:   Parameters and approximation errors for a column and a beam.

Column	a	b	R
I	1.030	0.819	2.545×10^{-3}
Z	1.028	0.8189	2.474×10^{-3}
Z^{p}	1.016	0.9017	8.698×10^{-4}
Beam	a	b	R
I	1.038	1.090	1.762×10^{-2}
Z	1.039	1.088	1.840×10^{-2}
Z^{p}	1.019	1.077	3.832×10^{-3}

where A_0 and I_0 are the specified values for normalization. Then the coefficients a and b are obtained by minimizing the error R defined as

$$R = \sum_{i=1}^{n^{\mathrm{L}}} \left[\frac{I(A)}{I_0} - a \left(\frac{A}{A_0} \right)^b \right]^2 \tag{A.161}$$

where n^{L} is the number of sections in the list, and Z and Z^{p} are approximated similarly. Accuracy may be improved if we add the constant term to the right-hand side of (A.160); however, optimization often does not converge because the solution turns out to be non-unique.

Optimization library SNOPT Ver. 7.2 (Gill, Murray, and Saunders 2002) was used for finding the optimal parameter values. The smallest values in the list are used for A_0, I_0, Z_0, and Z_0^{p} for normalization. The minimum errors and the corresponding parameter values for a beam and a column in Tables A.2 and A.3 are listed in Table A.4. The approximate relation for the beam sections is plotted in Fig. A.20, which shows almost linear approximation.

Alternatively, the cross-sectional properties can be approximated as linear functions of the reciprocal of A_i (Grierson and Chan 1993). It is also possible

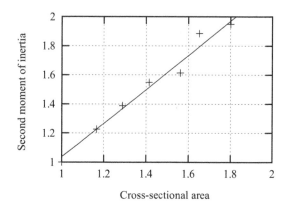

FIGURE A.20: Approximation of second moment of inertia of a beam.

to interpolate I_i and Z_i as piecewise continuously differentiable functions of A_i (Sedaghati and Esmailzadeh 2003), as demonstrated in Example 5.1 in Sec. 5.1.

References

Aarts, E. and J. Korst (1989). *Simulated Annealing and Boltzmann Machines: A Stochastic Approach to Combinatorial Optimization and Neural Computing.* Chichester, UK: Wiley.

Aarts, M. and J. K. Lenstra (Eds.) (1997). *Local Search in Combinatorial Optimization.* Chichester, UK: Wiley.

Abrahamson, N. A., J. F. Schneider, and J. Stepp (1991). Empirical spatial coherency functions for application to soil-structure interaction analysis. *Earthquake Spectra 21*, 1–28.

Achtziger, W. (1996). Truss topology optimization including bar properties different for tension and compression. *Struct. Opt. 12*, 63–74.

Achtziger, W. (1997). Topology optimization of discrete structures. In *Topology Optimization in Structural Mechanics*, CISM Courses and Lectures 374, pp. 57–100. Vienna: Springer-Verlag.

Achtziger, W. (1999a). Local stability of trusses in the context of topology optimization, Part I: Exact modelling. *Struct. Opt. 17*, 235–246.

Achtziger, W. (1999b). Local stability of trusses in the context of topology optimization, Part II: A numerical approach. *Struct. Opt. 17*, 247–258.

Achtziger, W. (2007). On simultaneous optimization of truss geometry and topology. *Struct. Multidisc. Optim. 33*, 285–304.

Achtziger, W., M. P. Bendsøe, A. Ben-Tal, and J. Zowe (1992). Equivalent displacement based formulations for maximum strength truss topology design. *Impact of Computing in Science and Engineering 4*(4), 315–345.

Achtziger, W. and M. Kočvara (2007a). On the maximization of the fundamental eigenvalue in topology optimization. *Struct. Multidisc. Optim. 34*, 181–195.

Achtziger, W. and M. Kočvara (2007b). Structural topology optimization with eigenvalues. *SIAM J. Optim. 18*(4), 1129–1164.

Achtziger, W. and M. Stolpe (2008). Global optimization of truss topology with discrete bar areas, Part I: Theory of relaxed problems. *Comput. Optim. Appl. 40*(2), 247–280.

Achtziger, W. and M. Stolpe (2009). Global optimization of truss topology with discrete bar areas, Part II: Implementation and numerical results. *Comput. Optim. Appl. 44*(2), 315–341.

Adeli, H. (Ed.) (1994). *Advances in Design Optimization*. London: Chapman & Hall.

Adeli, H. and H.-H. Chyou (1987). Microcomputer-aided optimal plastic design of frames. *J. Computing in Civil Eng. 1*(1), 20–34.

Adelman, H. M. and R. T. Haftka (1986). Sensitivity analysis of discrete structural systems. *AIAA J. 24*(5), 823–832.

Agrawal, R., T. Imielinski, and A. Swami (1998). Mining association rules between sets of items in large databases. In *Proc. ACM SIGMOD Conf. on Management of Data (SIGMOD '98)*, New York, pp. 207–216. ACM Press.

AISC (1999). *Load and Resistance Factor Design Specification for Structural Steel Buildings*. Chicago: American Institute of Steel Construction.

Akgün, M. A., J. H. Garcelon, and R. T. Haftka (2001). Fast exact linear and non-linear structural reanalysis and the Sherman–Morrison–Woodbury formulas. *Int. J. Numer. Methods Eng. 50*, 1587–1606.

Allaire, G., F. Jouve, and H. Maliot (2004). Topology optimization for minimum stress design with the homogenization method. *Struct. Multidisc. Optim. 28*, 87–98.

António, C. A. C. (2002). A multilevel genetic algorithm for optimization of geometrically nonlinear stiffened composite structures. *Struct. Multidisc. Optim. 24*, 372–386.

Argyris, J. H. and R. Roy (1972). General treatment of structural modification. *J. Struct. Div. ASCE 98*(ST2), 465–492.

Armentano, V. A. and J. C. Arroyo (2004). An application of a multiobjective tabu search algorithm to a bicriteria flowshop problem. *J. Heuristics 10*, 463–481.

Arora, J. S. (2002). Methods for discrete variable structural optimization. In S. Burns (Ed.), *Recent Advancements in Optimal Structural Design*, pp. 1–40. Raton, VA: ASCE Press.

Arora, J. S. (2004). *Introduction to Optimum Design* (2nd ed.). New York: Academic Press.

Arora, J. S. (Ed.) (2007). *Optimization of Structural and Mechanical Systems*. Singapore: World Scientific.

Arora, J. S., E. J. Haug, and K. Rim (1975). Optimal design of plane frames. *J. Struct. Div. ASCE 101*(ST10), 2063–2078.

Arora, J. S., M. W. Huang, and C. C. Hsieh (1994). Methods for optimization of nonlinear problems with discrete variables. *Struct. Opt. 8*(2–3), 69–85.

Arora, J. S. and C. H. Tseng (1987). IDESIGN User's Manual, Ver. 3.5. ODL-87.1, Technical Report, Optimal Design Laboratory, The University of Iowa.

Arora, J. S. and Q. Wang (2005). Review of formulations for structural and mechanical system optimization. *Struct. Multidisc. Optim. 30*, 251–272.

Atkinson, A. E. (1989). *An Introduction to Numerical Analysis* (2nd ed.). New York: John Wiley & Sons.

Avilés, R., M. B. G. Ajuria, J. Vallejo, and A. Hernández (1997). A procedure for the optimal synthesis of planar mechanisms based on non-linear position problems. *Int. J. Numer. Methods Eng. 40*, 1505–1524.

Avis, D., N. Katoh, M. Ohsaki, I. Streinu, and S. Tanigawa (2007). Enumerating non-crossing minimally rigid frameworks. *Graphs and Combinatorics 23(Suppl)*, 117–134.

Balling, R. J. (1991). Optimal steel frame design by simulated annealing. *J. Struct. Eng. 117*(6), 1780–1795.

Balling, R. J., K. S. Pister, and V. Ciampi (1983). Optimal seismic-resistant design of a planar steel frame. *Earthquake Eng. Struct. Dyn. 11*, 541–556.

Balling, R. J. and J. Sobieszczanski-Sobieski (1996). Optimization of coupled systems: A critical overview of approaches. *AIAA J. 34*(1), 6–17.

Banachiewicz, T. (1937). Zur Berechungung der Determinanten, wie auch der Inversen, und zur darauf basierten Auflösung der Systeme linearer Gleichungen. *Acta Astronomica, Sér. C 3*, 41–67.

Barnhill, R. E. (Ed.) (1994). *Geometry Processing for Design and Manufacturing.* Philadelphia, PA: SIAM.

Barthelemy, J. and J. Sobieszczanski-Sobieski (1983). Optimum sensitivity derivatives of objective functions in nonlinear programming. *AIAA J. 21*, 913–915.

Barthold, F.-J. and N. Gerzen (2009). Application of singular value decomposition in structural optimization. In *Proc. 8th World Congress of Structural and Multidisciplinary Optimization (WCSMO8)*, Paper No. 1392, Lisbon.

Bažant, Z. P. and L. Cedolin (1991). *Stability of Structures.* New York: Oxford University Press.

Baykasoglu, A. (2006). Applying multiple objective tabu search to continuous optimization problems with a simple neighborhood strategy. *Int. J. Numer. Methods Eng. 65*(3), 406–424.

Baykasoglu, A., S. Owen, and N. Gindy (1999a). Solution of goal programming models using a basic taboo search. *J. Operational Research Society 50*(9), 960–973.

Baykasoglu, A., S. Owen, and N. Gindy (1999b). A taboo search based approach to find the Pareto optimal set in multiple objective optimization. *Eng. Opt. 31*, 731–748.

Beckers, M. and C. Fleury (1997). A primal-dual approach in truss topology optimization. *Comput. Struct. 64*, 77–88.

Ben-Israel, A., A. Ben-Tal, and S. Zolbec (1981). *Optimality in Nonlinear Programming: A Feasible Direction Approach.* New York: John Wiley & Sons.

Ben-Tal, A. and M. P. Bendsøe (1993). A new method for optimal truss topology design. *SIAM J. Optim. 3*(2), 322–358.

Ben-Tal, A., F. Jarre, M. Kočvara, A. Nemirovski, and J. Zowe (2000). Optimal design of trusses under a nonconvex global buckling constraint. *Opt. Eng. 1*, 189–213.

Ben-Tal, A., M. Kočvara, and J. Zowe (1993). Two nonsmooth approaches to simultaneous geometry and topology design of trusses. In M. P. Bendsøe and C. M. Mota Soares (Eds.), *Topology Optimization of Structures*, pp. 31–42. Dordrecht, The Netherlands: Kluwer Academic Publishers.

Ben-tal, A. and A. Nemirovski (1994). Potential reduction polynomial time method for truss topology design. *SIAM J. Optim. 4*(3), 596–612.

Ben-Tal, A. and A. Nemirovski (1997). Robust truss topology optimization via semidefinite programming. *SIAM J. Optim. 7*, 991–1016.

Bendsøe, M. P. (1989). Optimal shape design as a material distribution problem. *Struct. Opt. 1*, 193–202.

Bendsøe, M. P., A. Ben-Tal, and J. Zowe (1994). Optimization method for truss geometry and topology design. *Struct. Opt. 7*, 141–159.

Bendsøe, M. P. and O. Sigmund (2003). *Topology Optimization: Theory, Methods and Applications.* Berlin: Springer.

Bennage, W. A. and A. K. Dhingra (1995). Optimization of truss topology using tabu search. *Int. J. Numer. Methods Eng. 38*, 4035–4052.

Bennett, J. A. and M. E. Botkin (Eds.) (1986). *The Optimum Shape: Automated Structural Design.* New York: Plenum Press.

Berke, L. and V. B. Venkayya (1974). Review of optimality criteria approaches to structural optimization. In J. L. A. Schmit (Ed.), *Proc. Structural Optimization Symp., ASME AMD*, Volume 1, pp. 23–34.

Berry, M. J. A. and G. Linoff (1997). *Data Mining Techniques: For Marketing, Sales, and Customer Support.* New York: John Wiley & Sons.

Bersekas, D. P. (1982). *Constrained Optimization and Lagrange Multiplier Methods.* New York: Academic Press.

Bhatti, M. A. and K. S. Pister (1981). A dual criteria approach for optimal design of earthquake resistant structural systems. *Earthquake Eng. Struct. Dyn. 9*, 557–572.

Bloebaum, C. L., P. Hajela, and J. Sobieszczanski-Sobieski (1992). Non-hierarchic system decomposition in structural optimization. *Eng. Opt. 19*, 171–186.

Bochenek, B. and A. Gajewski (1986). Multimodal optimal design of a circular funicular arch with respect to in-plane and out-of-plane buckling. *J. Struct. Mech. 14*, 257–274.

Bojczuk, D. and Z. Mróz (1998a). On optimal design of supports in beam and frame structures. *Struct. Opt. 16*, 47–57.

Bojczuk, D. and Z. Mróz (1998b). Optimal design of trusses with account for topology variation. *Mech. Struct. & Mach. 26*(1), 21–40.

Bojczuk, D. and Z. Mróz (1999). Optimal topology and configuration design of trusses with stress and buckling constraints. *Struct. Multidisc. Optim. 17*, 25–35.

Box, M. J. (1965). A new method of constrained optimization and comparison with other methods. *Computing J. 8*, 42–52.

Boyd, S. and L. Vandenberghe (2004). *Convex Optimization.* Cambridge, UK: Cambridge University Press.

Bradt, M. (Ed.) (1986). *Criteria and Methods of Structural Optimization.* Dordrecht, The Netherlands: Martinus Nijhoff Publishers. (first edition in Polish, 1984).

Braibant, V. and C. Fleury (1984). Shape optimal design using B-splines. *Comput. Methods Appl. Mech. Engrg. 44*, 247–267.

Brigham, J. C. and W. Aquino (2007). Surrogate-model accelerated random search algorithm for global optimization with application to inverse material identification. *Comput. Methods Appl. Mech. Engrg. 196*, 4561–4576.

Bruggi, M. (2008). On an alternative approach to stress constraints relaxation in topology optimization. *Struct. Multidisc. Optim. 36*, 125–141.

Buhl, T., C. B. W. Pedersen, and O. Sigmund (2000). Stiffness design of geometrically nonlinear structures using topology optimization. *Struct. Multidisc. Optim. 19*, 93–104.

Burns, S. (Ed.) (2002). *Recent Advancements in Optimal Structural Design.* Raton, VA: ASCE Press.

Calladine, C. R. and S. Pellegrino (1991). First-order infinitesimal mechanisms. *Int. J. Solids Struct. 27*(4), 505–515.

Canyurt, O. E. and P. Hajela (2005). A cellular framework for structural analysis and optimization. *Comput. Methods Appl. Mech. Engrg. 194*, 3516–3534.

Cerny, V. (1985). Thermodynamical approach to the travelling salesman problem: An efficient simulation algorithm. *J. Optimization Theory Appl. 45*, 41–51.

Chan, C.-M., D. E. Grierson, and A. N. Sherbourne (1995). Automatic optimal design of tall steel building frameworks. *J. Struct. Eng. 121*(5), 838–847.

Chan, H. S. Y. (1969). On Foulkes mechanism in portal frame design for alternative loads. *J. Appl. Mech. 36*, 73–75.

Chen, S. H., X. M. Wu, and Z. J. Yang (2006). Eigensolution reanalysis of modified structures using epsilon-algorithm. *Int. J. Numer. Methods Eng. 66*, 2115–2130.

Chen, S. H., X. W. Yang, and H. D. Lian (2000). Comparison of several eigenvalue reanalysis methods for modified structures. *Struct. Multidisc. Optim. 20*, 253–259.

Chen, T.-Y. and C.-J. Chen (1997). Improvements of simple genetic algorithms in structural design. *Comput. Methods Appl. Mech. Engrg. 40*, 1323–1334.

Chen, W.-F. (1991). *Stability Design of Steel Frames*. Boca Raton, FL: CRC Press.

Chen, W. F. and S. E. Kim (1997). *LRFD Steel Design using Advanced Analysis*. Boca Raton, FL: CRC Press.

Cheng, F. Y. (2002). Multiobjective optimum design of seismic-resistant structures. In S. Burns (Ed.), *Recent Advancements in Optimal Structural Design*, pp. 241–255. Raton, VA: ASCE Press.

Cheng, G. (1995). Some aspects of truss topology optimization. *Struct. Opt. 10*, 173–179.

Cheng, G. and X. Guo (1997). ε-relaxed approach in structural topology optimization. *Struct. Opt. 13*, 258–266.

Cheng, G. and Z. Jiang (1992). Study on topology optimization with stress constraints. *Eng. Opt. 20*, 129–148.

Chern, J.-M. and W. Prager (1972). Optimal design of trusses for alternative loads. *Ing.-Arch. 41*, 225–231.

Choi, K. K. and N.-H. Kim (2004). *Structural Sensitivity Analysis and Optimization 1, Linear Systems*. New York: Springer.

Choi, K. K. and S.-L. Twu (1988). On equivalence of continuum and discrete methods of shape sensitivity analysis. *AIAA J. 27*(10), 1418–1424.

Chopra, A. K. (2001). *Dynamics of Structures: Theory and Applications to Earthquake Engineering*. Englewood Cliffs, NJ: Prentice-Hall.

Cilly, F. H. (1900). The exact design of statically indeterminate frameworks, an exposition of its possibility but futility. *Trans. ASCE 43*, 353–407.

Clough, R. W. and J. Penzien (1975). *Dynamics of Structures*. New York: McGraw-Hill.

Coello Coello, C. A. and G. T. Pulido (2005). Multiobjective structural optimization using a microgenetic algorithm. *Struct. Multidisc. Optim. 30*, 388–403.

Coello Coello, C. A., G. T. Pulido, and M. S. Lechuga (2004). Handling multiple objectives with particle swarm optimization. *IEEE Trans. Evolutionary Computation 8*(3), 256–279.

Cohon, J. L. (1978). *Multiobjective Programming and Planning*. Mathematics in Science and Engineering, 140. New York: Academic Press.

Cramer, E. J., J. E. Dennis, Jr., P. D. Frank, R. M. Lewis, and G. R. Shubin (1994). Problem formulation for multidisciplinary optimization. *SIAM J. Optim. 4*(4), 754–776.

Crisfield, M. A. (1991). *Non-Linear Finite Element Analysis of Solids and Structures, I: Essentials*. New York: Wiley.

Culmann, K. (1875). *Die graphische Statik*. Zürich: Meyer & Zeller.

Czyzak, P. and A. Jaszkiewicz (1998). Pareto simulated annealing: A metaheuristic technique for multi-objective combinatorial optimization. *J. Multi-Criteria Decision Analysis 7*, 34–47.

Dantzig, G. B. and P. Wolfe (1960). Decomposition principle for linear programs. *Operational Research 8*(1), 101–111.

Dasgupta, D. and Z. Michalewicz (Eds.) (1997). *Evolutionary Algorithms in Engineering Applications*. Berlin: Springer.

de Boer, H. and F. van Keulen (2000). Refined semi-analytical design sensitivities. *Int. J. Solids Struct. 37*, 6961–6980.

de Klerk, E., C. Roos, and T. Terlaky (1995). Semi-definite problems in truss topology optimization. Technical Report No. 95-128, Delft University of Technology, Delft, The Netherlands.

DebChaudhury, A. and G. D. Gazis (1988). Response of MDOF systems to multiple support seismic excitation. *J. Eng. Mech. 114*(4), 583–603.

Dems, K. and W. Gatkowski (1995). Optimal design of a truss configuration under multiloading conditions. *Struct. Opt. 9*, 262–265.

Deng, L. and M. Ghosn (2001). Pseudoforce method for nonlinear analysis and reanalysis of structural systems. *J. Struct. Eng. 127*(5), 570–578.

Der Kiureghian, A. (1980). Structural response to stationary excitation. *J. Eng. Mech. Div. ASCE 106*(6), 1195–1213.

Der Kiureghian, A. (1996). A coherency model for spatially varying ground motions. *Earthquake Eng. Struct. Dyn. 25*, 99–111.

Der Kiureghian, A. and A. Neuenhofer (1991). A response spectrum method for multiple-support seismic excitations. Technical Report No. UCB/EERC-91/08, Earthquake Engineering Research Center, University of California, Berkeley, CA.

374 *Optimization of Finite Dimensional Structures*

Der Kiureghian, A. and A. Neuenhofer (1992). Response spectrum method for multiple-support seismic excitation. *Earthquake Eng. Struct. Dyn. 21*, 713–740.

Dey, A. and V. Gupta (1998). Response of multiply supported secondary systems to earthquakes in frequency domain. *Earthquake Eng. Struct. Dyn. 27*, 187–201.

Dhingra, A. K. and W. A. Bennage (1995). Discrete and continuous variable structural optimization using tabu search. *Eng. Opt. 24*, 177–196.

Ding, Y. and B. J. D. Esping (1991). An improved multilevel optimization approach. *Comput. Struct. 38*(5–6), 557–567.

Dobbs, M. W. and R. B. Nelson (1975). Application of optimality criteria to automated structural design. *AIAA J. 14*(10), 1436–1443.

Dobbs, W. and L. P. Felton (1969). Optimization of truss geometry. *J. Struct. Div. ASCE 95*(ST10), 2105–2119.

Doi, K., M. Yoshimura, S. Nishiwaki, and K. Izui (2009). Optimization method for lifecycle designs of products having different lifecycle characteristics. In *Proc. 8th World Congress of Structural and Multidisciplinary Optimization (WCSMO8)*, Paper No. 1071, Lisbon.

Dorn, W., R. Gomory, and H. Greenberg (1964). Automatic design of optimal structures. *J. de Mecanique 3*, 25–52.

Doğan, E. and M. P. Saka (2009). Particle swarm design optimization of moment resisting steel frames with semi-rigid connections to LRFD-AISC. In *Proc. 8th World Congress of Structural and Multidisciplinary Optimization (WCSMO8)*, Paper No. 1109, Lisbon.

Drucker, D. C. and R. T. Shield (1961). Bounds on minimum weight design. *Quart. Appl. Math. 15*, 269–281.

Dubois, D. and P. Fortemps (1999). Computing improved optimal solutions to max–min flexible constraint satisfaction problems. *European J. Operational Research 118*, 95–126.

Duysinx, P. and M. P. Bendsøe (1998). Topology optimization of continuum structures with local stress constraints. *Int. J. Numer. Methods Eng. 43*, 1453–1478.

Duysinx, P. and O. Sigmund (1998). New developments in handling stress constraints in optimal material distribution. In *Proc. 7th Symp. on Multidisciplinary Analysis and Optimization*, Paper AIAA-98-4906, pp. 1501–1509.

Ebenau, C., J. Rottschäfer, and G. Thierauf (2005). An advanced evolutionary strategy with an adaptive penalty function for mixed-discrete structural optimization. *Advances in Engineering Software 36*, 29–38.

Ekeland, I. and R. Témam (1999). *Convex Analysis and Variational Problems*. Philadelphia, PA: SIAM.

Elishakoff, I., R. T. Haftka, and J. Fang (1994). Structural design under bounded uncertainty–optimization with anti-optimization. *Comput. Struct. 53*(6), 1401–1405.

Elishakoff, I. and M. Ohsaki (2010). *Optimization and Anti-Optimization of Structures*. London: Imperial College Press.

Erdal, F. and M. P. Saka (2009). Optimum design of castellated beams using harmony search method. In *Proc. 8th World Congress of Structural and Multidisciplinary Optimization (WCSMO8)*, Paper No. 1102, Lisbon.

Eriksson, A. (2008). Optimization in target movement simulations. *Comput. Methods Appl. Mech. Engrg. 197*, 4207–4215.

Eschenauer, H. A., V. V. Kobelev, and A. Schumacher (1994). Bubble method for topology and shape optimization of structures. *Struct. Opt. 8*, 42–51.

Evgrafov, A. (2005). On globally stable singular truss topologies. *Struct. Multidisc. Optim. 29*, 170–177.

Evgrafov, A. (2006). Simultaneous optimization of topology and geometry of flow networks. *Struct. Multidisc. Optim. 32*, 99–109.

Evgrafov, A. and M. Patriksson (2003). Stable relaxations of stochastic stress-constrained weight minimization problems. *Struct. Multidisc. Optim. 25*, 189–198.

Farin, G. (1988). *Curves and Surfaces for Computer Aided Geometric Design*. Boston: Academic Press.

Farin, G., J. Hoschek, and M.-S. Kim (2002). *Handbook of Computer Aided Geometric Design*. Amsterdam: North-Holland.

Faux, I. D. and M. J. Pratt (1979). *Computational Geometry for Design and Manufacture*. Chichester, UK: Ellis Horwood.

Fiacco, A. V. (1983). *Introduction to Sensitivity and Stability Analysis in Nonlinear Programming*. New York: Academic Press.

Fiacco, A. V. and G. P. M. Cormic (1968). *Nonlinear Programming: Sequential Unconstrained Minimization Techniques*. New York: John Wiley & Sons.

Fleury, C. (1979). A unified approach to structural weight minimization. *Comput. Methods Appl. Mech. Engrg. 20*, 17–38.

Fleury, C. (1980). An efficient optimality criteria approach to the minimum weight design of elastic structures. *Comput. Struct. 11*, 163–173.

Fleury, C. (1989a). CONLIN: An efficient dual optimizer based on convex approximation. *Struct. Opt. 1*(2), 81–89.

Fleury, C. (1989b). Efficient approximation concepts using second order information. *Int. J. Numer. Methods Eng. 28*, 2041–2058.

Floudas, C. A. (1995). *Nonlinear and Mixed-Integer Optimization*. New York: Oxford University Press.

Foley, C. M. (2002). Optimized performance-based design of buildings. In S. Burns (Ed.), *Recent Advancements in Optimal Structural Design*, pp. 169–240. Raton, VA: ASCE Press.

Forsythe, G. E. and W. R. Wasow (1960). *Finite-Difference Methods for Partial Differential Equations*. New York: Wiley.

Foulkes, J. (1954). Minimum weight design of structural frames. *Proc. Royal Soc., London, Ser. A 82*, 492–494.

Fox, R. L. and M. P. Kapoor (1968). Rates of change of eigenvalues and eigenvectors. *AIAA J. 6*, 2426–2429.

Frank, P. M. (1978). *Introduction to System Sensitivity Theory*. New York: Academic Press.

Fredricson, H. (2005). Topology optimization of frame structures – joint penalty and material selection. *Struct. Multidisc. Optim. 30*, 193–200.

Fredricson, H., T. Johansen, A. Klarbring, and J. Petersson (2003). Topology optimization of frame structures with flexible joints. *Struct. Multidisc. Optim. 25*, 199–214.

Freeman, S. A. (2004). Review of the development of the capacity spectrum method. *ISEL J. Earthquake Technology 41*(1), 1–13.

Friedman, Z. and M. B. Fuchs (1987). Multilevel optimal design of thin-walled continuous beams. *Comput. Struct. 25*, 405–414.

Frye, M. J. and G. A. Morris (1975). Analysis of frames with flexibly connected steel frames. *Canadian J. Civil Eng. 2*(3), 280–291.

Fuchs, M. B. (1993). *N*th-order stiffness sensitivities in structural analysis. *Struct. Opt. 5*, 207–212.

Fujisawa, K., M. Kojima, and K. Nakata (1997). SDPA (semidefinite programming algorithm). In *Proc. 2nd Workshop on High Performance Optimization Techniques*, Rotterdam, The Netherlands.

Fujita, S. and M. Ohsaki (2009). Shape optimization of shells considering strain energy and algebraic invariants of parametric surface. In *Proc. 9th Asian Pacific Conf. on Shell and Spatial Struct. (APCS2009)*, Paper No. P0012, Nagoya, Japan.

Fukuda, T., K. Mori, and M. Tsukiyama (1999). Parallel search for multimodal function optimization with diversity and learning of immune algorithm. In *Artificial Immune Systems and Their Applications*. Berlin: Springer.

Gal, T. (1979). *Postoptimal Analyses, Parametric Programming, and Related Topics*. New York: Mc-Graw Hill.

Gal, T. and H. J. Greenberg (Eds.) (1997). *Advances in Sensitivity Analysis and Parametric Programming*. Norwell, MA: Kluwer Academic Publishers.

Galileo Galilei (1638). *Discorsi e Dimonstrazioni Matematiche, Interno, a Due Nuove Scienze*. Leida.

Ganzerli, S., C. P. Pantelides, and L. D. Reaveley (2000). Performance-based design using structural optimization. *Earthquake Eng. Struct. Dyn. 29*, 1677–1690.

Garcea, G., G. Formica, and R. Casciaro (2005). A numerical analysis of infinitesimal mechanisms. *Int. J. Numer. Methods Eng. 62*, 979–1012.

Gavarini, C. and D. Veneziano (1972). Minimum weight limit design under uncertainty. *Meccanica 7*(2), 98–104.

Geem, Z. W., J.-H. Kim, and G. V. Loganathan (2001). A new heuristic optimization algorithm: Harmony search. *Simulation 76*(2), 60–68.

Ghasemi, M. R. and M. Farshchin (2009). Multi-objective weight and eigen-period optimization of steel moment frames under seismic conditions, using ant colony method. In *Proc. 8th World Congress of Structural and Multidisciplinary Optimization (WCSMO8)*, Paper No. 1362, Lisbon.

Giger, M. and P. Ermanni (2006). Evolutionary truss topology optimization using a graph-based parameterization concept. *Struct. Multidisc. Optim. 32*, 313–326.

Gill, P. E., W. Murray, and M. A. Saunders (2002). SNOPT: An SQP algorithm for large-scale constrained optimization. *SIAM J. Optim. 12*, 979–1006.

Glover, F. (1975). Improved linear integer programming formulations of nonlinear integer problems. *Management Science 22*(4), 455–460.

Glover, F. (1989). Tabu search: Part I. *ORSA J. Computing 1*(3), 190–206.

Glover, F. and M. Laguna (1997). *Tabu Search*. Boston: Kluwer Academic Publishers.

Goldberg, D. E. (1989). *Genetic Algorithms in Search, Optimization, and Machine Learning*. Reading, MA: Addison-Wesley.

Goldberg, D. E. and J. Richerdson (1987). Genetic algorithms with sharing for multimodal function optimization. In *Proc. 2nd Int. Conf. Genetic Algorithms on Genetic Algorithms and Their Application*, pp. 41–49. Hillsdale, NJ: L. Erlbaum Associates Inc.

Gondzio, J. (1995). HOPDM – a fast LP solver based on a primal-dual interior point method. *European J. Operational Research 85*, 221–225.

Graver, J., B. Servatius, and H. Servatius (1993). *Combinatorial Rigidity*. Graduate Studies in Mathematics 2. Providence: American Mathematical Society.

Greiner, D., J. M. Emperador, and G. Winter (2004). Single and multi-objective frame optimization by evolutionary algorithms and the auto-adaptive rebirth operator. *Comput. Methods Appl. Mech. Engrg. 193*, 3711–3743.

Grierson, D. E. and C.-M. Chan (1993). An optimality criteria design method for tall steel buildings. *Advances in Engineering Software 16*, 119–125.

Grierson, D. E. and C. W. Chiu (1984). Optimal synthesis of frameworks under multilevel performance constraints. *Comput. Struct. 18*(5), 889–898.

Grierson, D. E. and W. H. Lee (1984). Optimal synthesis of steel frameworks using standard sections. *J. Struct. Mech. 12*(3), 335–370.

Grierson, D. E. and W. H. Pak (1993). Optimal sizing, geometrical and topological design using a genetic algorithm. *Struct. Opt. 6*, 151–159.

Grierson, D. E. and L. A. Schmit (1982). Synthesis under service and ultimate performance constraints. *Comput. Struct. 15*(4), 405–417.

Guillet, S., F. Noël, and J. C. Léon (1996). Structural shape optimization of parts bounded by free-form surfaces. *Struct. Opt. 11*, 159–169.

Gunnlaugsson, G. A. and J. B. Martin (1973). Optimality conditions for fully stressed designs. *SIAM J. Appl. Math. 25*(3), 474–482.

Guo, X. and G. Cheng (2000). An extrapolation approach for the solution of singular optima. *Struct. Multidisc. Optim. 19*, 255–262.

Guo, X., G. Cheng, and K. Yamazaki (2001). A new approach for the solution of singular optima in truss topology optimization with stress and local buckling constraints. *Struct. Multidisc. Optim. 22*, 364–372.

Gupta, V. (1997). Acceleration transfer function of secondary systems. *J. Eng. Mech. 123*(7), 678–675.

Gurav, S. P., M. Langhaar, J. F. L. Goosen, and F. van Keulen (2005). Uncertainty-based design optimization of MEMS structures using combined cycle-based alternating anti-optimization. In *Proc. 6th World Congress of Structural and Multidisciplinary Optimization (WCSMO6)*, Rio de Janeiro.

Haber, R. B. and J. F. Abel (1982). Initial equilibrium solution methods for cable reinforced membranes, Part I: Formulations. *Comput. Methods Appl. Mech. Engrg. 30*, 263–284.

Haftka, R. T. (1985). Simultaneous analysis and design. *AIAA J. 23*(7), 1099–1103.

Haftka, R. T. and H. M. Adelman (1989). Recent developments in structural sensitivity analysis. *Struct. Opt. 1*, 137–151.

Haftka, R. T., Z. Gürdal, and M. P. Kamat (1990). *Elements of Structural Optimization*. Dordrecht, The Netherlands: Kluwer Academic Publishers.

Haftka, R. T. and M. P. Kamat (1989). Simultaneous nonlinear structural analysis and design. *Comp. Mech. 4*, 409–416.

Haftka, R. T., J. A. Nachlas, L. T. Watson, T. Rizzo, and R. Desai (1987). Two-point constraint approximation in structural optimization. *Comput. Methods Appl. Mech. Engrg. 60*, 289–301.

Haftka, R. T. and J. H. Starnes, Jr. (1976). Application of a quadratic extended interior penalty function for structural optimization. *AIAA J. 14*, 718–724.

Haftka, R. T. and L. T. Watson (2005). Multidisciplinary design optimization with quasiseparable subsystems. *Optim. Eng. 6*, 9–20.

Hagishita, T. and M. Ohsaki (2007). Optimization of placement of braces for a frame with semi-rigid joints by scatter search. In *Proc. 7th World Congress of Structural and Multidisciplinary Optimization (WCSMO7)*, Seoul, pp. 692–701.

Hagishita, T. and M. Ohsaki (2008a). Optimal placement of braces for steel frames with semi-rigid joints by scatter search. *Comput. Struct. 86*, 1983–1993.

Hagishita, T. and M. Ohsaki (2008b). Topology mining for optimization of framed structures. *J. Advanced Mechanical Design, Systems, and Manufacturing, JSME 2*(3), 417–428.

Hagishita, T. and M. Ohsaki (2009). Topology optimization of trusses by growing ground structure approach. *Struct. Multidisc. Optim. 37*(4), 377–393.

Hajela, P. (1997). Stochastic search in discrete structural optimization simulated annealing, genetic algorithm and neural networks. In W. Gutkowski (Ed.), *Discrete Structural Optimization*, CISM Courses and Lectures 373. Vienna: Springer-Verlag.

Hajela, P. and E. Lee (1995). Genetic algorithms in truss topological optimization. *Int. J. Solids Struct. 32*(22), 3341–3357.

Hajela, P. and C.-Y. Lin (1991). Optimal design of viscoelastically damped beam structures. *Appl. Mech. Rev. 44*, 96–102.

Hajela, P. and C.-Y. Lin (1992). Genetic search strategies in multicriterion optimal design. *Struct. Opt. 4*, 99–107.

Hand, D., H. Mannila, and P. Smyth (2001). *Principles of Data Mining*. Cambridge, MA: MIT Press.

Hansen, M. P. (1997). Tabu search for multiobjective optimization: MOTS. In *Proc. 13th Int. Conf. on MCDM*, Cape Town, pp. 6–10.

Hao, H. and X. N. Duan (1995). Seismic response of asymmetric structures to multiple ground motions. *J. Struct. Eng. 121*(11), 1557–1564.

Haque, M. I. (1996). Optimal frame design with discrete members using the complex method. *Comput. Struct. 59*(5), 847–858.

Harichandran, R. S., A. Hawwari, and B. N. Swedian (1996). Response of long-span bridges to spatially varying ground motion. *J. Struct. Eng. 122*(5), 476–484.

Hart, W. (1998). Sequential stopping rules for random optimization methods with applications to multistart local search. *SIAM J. Optim. 9*(1), 270–290.

Harville, D. (1976). Extension of the Gauss–Markov theorem to include the estimation of random effects. *Ann. Statist. 4*(2), 384–395.

Hasan, R., L. Xu, and D. E. Grierson (2002). Push-over analysis for performance-based seismic design. *Comput. Struct. 80*, 2483–2493.

Haug, E. J. and J. Cea (Eds.) (1981). *Optimization of Distributed Parameter Structures 1, 2*. Alphen aan den Rijn, The Netherlands: Noordhoff.

Haug, E. J. and K. K. Choi (1982). Systematic occurrence of repeated eigenvalues in structural optimization. *J. Optimization Theory Appl. 38*, 251–274.

Haug, E. J., K. K. Choi, and V. Komkov (1986). *Design Sensitivity Analysis of Structural Systems*. New York: Academic Press.

Hayalioglu, M. S. (2000). Optimum design of geometrically non-linear elastic-plastic steel frames via genetic algorithm. *Comput. Struct. 77*, 527–538.

Hayalioglu, M. S. and S. O. Degertekin (2005). Minimum cost design of steel frames with semi-rigid connections and column bases via genetic optimization. *Comput. Struct. 83*, 1849–1863.

Hegemier, G. A. and W. Prager (1969). On Michell trusses. *Int. J. Mech. Sci. 11*, 209–215.

Hemp, W. S. (1973). *Optimum Structures*. Oxford, UK: Clarendon Press.

Henderson, C. R., O. Kempthorne, S. R. Searle, and C. M. von Krosigk (1959). The estimation of environmental and genetic trends from records subject to culling. *Biometrics 15*(2), 192–218.

Henderson, H. V. and R. Searle (1981). On deriving the inverse of a sum of matrices. *SIAM Review 23*, 53–60.

Heyman, J. (1959). On the absolute minimum weight design of framed structures. *Quart. J. Mech. Appl. Math. 12*(3), 314–324.

Ho, S. L., S. Yang, H. C. Wong, and G. Ni (2003). A simulated annealing algorithm for multiobjective optimizations of electromagnetic devices. *IEEE Transactions on Magnetics 39*(3), 1285–1288.

Holland, J. H. (1975). *Adaptation in Natural and Artificial Systems.* Ann Arbor, MI: The University of Michigan Press.

Horn, R. A. and C. R. Johnson (1990). *Matrix Analysis* (Reprint ed.). Cambridge, MA: Cambridge University Press.

Horst, R., P. M. Pardalos, and N. V. Thoai (1995). *Introduction to Global Optimization.* Dordrecht, The Netherlands: Kluwer Academic Publishers.

Horst, R. and H. Tuy (1985). *Global Optimization.* Berlin: Springer-Verlag.

Hoschek, J. (1985). Offset curves in the plane. *Comput.-Aided Design 17,* 77–82.

Hoschek, J. and N. Wissel (1988). Optimal approximate conversion of spline curves and spline approximation of offset curves. *Comput.-Aided Design 20,* 475–483.

Hotelling, H. (1943). Some new methods in matrix calculation. *Ann. Math. Statist. 14,* 1–34.

Hsieh, C. C. and J. S. Arora (1984). Design sensitivity analysis and optimization of dynamic response. *Comput. Methods Appl. Mech. Engrg. 43,* 195–219.

Hu, N. (1992). Tabu search method with random moves for globally optimal design. *Int. J. Numer. Methods Eng. 35,* 1055–1070.

Hu, T. C. and R. T. Shield (1961). Uniqueness in the optimum design of structures. *J. Appl. Mech. 83,* 1–4.

ILOG (2007). *ILOG CPLEX 10.2 User's Manual.* ILOG Inc.

Imai, K. and L. A. Schmit (1982). Configuration optimization of trusses. *J. Struct. Div. ASCE 107*(ST5), 745–756.

Izmailov, A. F. and M. V. Solodov (2008). Mathematical programs with vanishing constraints: Optimality conditions, sensitivity, and a relaxation method. *J. Optimization Theory Appl. 142*(3), 69–99.

Jármai, K., J. Farkas, and Y. Kurobane (2006). Optimum design of a multistorey steel frame. *Engineering Structures 28,* 1038–1048.

Jarre, F., M. Kočvara, and J. Zowe (1998). Optimal truss design by interior-point methods. *SIAM J. Optim. 8*(4), 1084–1107.

Jenkins, W. M. (1991). Towards structural optimization via the genetic algorithm. *Comput. Struct. 40,* 1321–1327.

Jilla, C. D. and D. W. Miller (2001). Assessing the performance of a heuristic simulated annealing algorithm for the design of distributed satellite systems. *Acta Astronautica 48*(5), 529–543.

Jog, C. S. and R. B. Haber (1996). Stability of finite element models for distributed-parameter optimization and topology design. *Comput. Methods Appl. Mech. Engrg. 130,* 203–226.

Jones, D. F., S. K. Mirrazavi, and M. Tamiz (2002). Multi-objective meta-heuristics: An overview of the current state-of-the-art. *European J. Operational Research 137*, 1–9.

Jutte, C. V. and S. Kota (2008). Design of nonlinear springs for prescribed load-displacement functions. *J. Mech. Design 130*(8), Paper–081403.

Kaelo, P. and M. M. Ali (2006). Some variants of the controlled random search algorithm for global optimization. *J. Optimization Theory Appl. 130*(2), 253–264.

Kameshki, E. S. and M. P. Saka (2001a). Genetic algorithm based optimum bracing design of non-sway tall plane frames. *J. Const. Steel Res. 57*, 1081–1097.

Kameshki, E. S. and M. P. Saka (2001b). Optimum design of nonlinear steel frames with semi-rigid connections using genetic algorithm. *Comput. Struct. 79*, 1593–1604.

Kameshki, E. S. and M. P. Saka (2003). Genetic algorithm based optimum design of nonlinear planar steel frames with various semi-rigid connections. *J. Const. Steel Res. 59*, 109–134.

Kaneko, I. and C. D. Ha (1983). A decomposition procedure for large-scale optimum plastic design problems. *Int. J. Numer. Methods Eng. 19*, 873–889.

Kanno, Y. and M. Ohsaki (2007). Maximization of minimal eigenvalue of structures by using sequential semidefinite programming. In *Proc. 7th World Congress of Structural and Multidisciplinary Optimization (WCSMO7)*, Seoul, pp. 1121–1130.

Kanno, Y., M. Ohsaki, and N. Katoh (2001). Sequential semidefinite programming for optimization of framed structures under multimodal buckling constraints. *Int. J. Structural Stability and Dynamics 1*(4), 585–602.

Kanno, Y., M. Ohsaki, and N. Katoh (2002). Symmetricity of the solution of semidefinite programming. *Struct. Multidisc. Optim. 24*, 225–232.

Kanno, Y., M. Ohsaki, K. Murota, and N. Katoh (2001). Group symmetry in interior-point methods for semi-definite program. *Optim. Eng. 2*, 293–320.

Kargahi, M., J. C. Andersen, and M. M. Dessouky (2007). Structural weight optimization of frames using tabu search, Part I: optimization procedure. *J. Struct. Eng. 132*(12), 1858–1868.

Karmarkar, N. (1984). A new polynomial-time algorithm for linear programming. *Combinatorica 4*, 373–395.

Katoh, N., M. Ohsaki, and A. Tani (2002). *Introduction to Architectural Systems*. Creators Library 3. Tokyo: Kyoritsu Shuppan (in Japanese).

Kaveh, A. (1986). Statical bases for an efficient flexibility analysis of planar trusses. *J. Struct. Mech. 14*(4), 475–488.

Kaveh, A. (1991). Graph and structures. *Comput. Struct. 40*(4), 893–901.

Kaveh, A. (2004). *Structural Mechanics: Graph and Matrix Methods* (3rd ed.). Somerset, UK: Research Studies Press.

Kaveh, A., B. F. Azar, and S. Talatahari (2008). Ant colony optimization for design of space trusses. *Int. J. Space Structures 23*(3), 167–181.

Kavlie, D., H. Graham, and G. H. Powell (1971). Efficient reanalysis of modified structures. *J. Struct. Div. ASCE 97*(ST1), 377–392.

Kawamoto, A. (2005). Path-generation of articulated mechanisms by shape and topology variations in non-linear truss presentation. *Int. J. Numer. Methods Eng. 64*, 1557–1574.

Kawamoto, A., M. P. Bendsøe, and O. Sigmund (2004a). Articulated mechanism design with a degree of freedom constraint. *Int. J. Numer. Methods Eng. 61*, 1520–1545.

Kawamoto, A., M. P. Bendsøe, and O. Sigmund (2004b). Planar articulated mechanism design by graph theoretical enumeration. *Struct. Opt. 27*, 295–299.

Kawamura, H., H. Ohmori, and N. Koto (2002). Truss topology optimization by a modified genetic algorithm. *Struct. Multidisc. Optim. 23*, 467–472.

Kennedy, J. (1997). The particle swarm: Social adaptation of knowledge. In *Proc. Int. Conf. Evolutionary Computation*, Piscataway, NJ, pp. 303–308. IEEE.

Khot, N. S., L. Berke, and V. B. Venkayya (1978). Comparison of optimality criteria algorithms for minimum weight design of structures. *AIAA J. 17*(2), 182–190.

Khot, N. S. and M. P. Kamat (1985). Minimum weight design of truss structures with geometric nonlinear behavior. *AIAA J. 23*, 139–144.

Kicher, T. P. (1966). Optimum design-minimum weight versus fully stressed. *J. Struct. Div. ASCE 92*(ST6), 265–279.

Kicinger, R., T. Arciszewski, and K. de Jong (2005). Evolutionary computation and structural design: A survey of the state-of-the-art. *Comput. Struct. 83*, 1943–1978.

Kim, H., R. T. Haftka, W. H. Mason, L. T. Watson, and B. Grossman (2002). Probabilistic modeling of errors from structural optimization based on multiple starting points. *Optim. Eng. 3*, 415–430.

Kim, H. M., N. F. Michelena, P. Y. Papalambos, and T. Jiang (2003). Target cascading in optimal structural design. *J. Mech. Design 125*, 474–480.

Kim, M.-J., G.-W. Jang, and Y. Y. Kim (2008). Application of a ground beam-joint topology optimization method for multi-piece frame structure design. *J. Mechanical Design 130*(8), Paper–081401.

Kim, Y. Y., G. W. Jang, J. H. Park, J. S. Hyun, and S. J. Nam (2005). Configuration design of rigid link mechanisms by an optimization method: A first step. In *Proc. IUTAM Symp. on Topological Design Optimization of Structures, Machines and Materials*, Dordrecht, The Netherlands, pp. 251–260. Springer.

Kimura, T. and H. Ohmori (2008). Computational morphogenesis of free form shells. *J. Int. Assoc. Shells and Spatial Struct. 49*(3), 175–180.

Kirkpatrick, S., C. D. Gelatt, Jr., and M. P. Vecchi (1983). Optimization by simulated annealing. *Science 220*(4598), 671–680.

Kirsch, U. (1972). Optimum design by partitioning into substructures. *J. Struct. Div. ASCE 98*(ST1), 249–267.

Kirsch, U. (1975). Multilevel approach to optimum structural design. *J. Struct. Div. ASCE 101*(ST4), 957–974.

Kirsch, U. (1989a). Optimal topologies of truss structures. *Appl. Mech. Rev. 42*, 223–239.

Kirsch, U. (1989b). Optimal topologies of truss structures. *Comput. Methods Appl. Mech. Engrg. 72*, 15–28.

Kirsch, U. (1990). On singular topologies in optimum structural design. *Struct. Opt. 2*, 133–142.

Kirsch, U. (1993). Efficient reanalysis for topological optimization. *Struct. Opt. 6*, 143–150.

Kirsch, U. (1994). Efficient sensitivity analysis for structural optimization. *Comput. Methods Appl. Mech. Engrg. 117*, 143–156.

Kirsch, U. (1995). Improved stiffness-based first-order approximations for structural optimization. *AIAA J. 33*(1), 143–150.

Kirsch, U. (1996). Integration of reduction and expansion process in layout optimization. *Struct. Opt. 11*, 13–18.

Kirsch, U. (2000). Combined approximation: A general reanalysis approach for structural optimization. *Struct. Multidisc. Optim. 20*, 97–106.

Kirsch, U. and S. Liu (1995). Exact structural reanalysis by a first-order reduced basis approach. *Struct. Opt. 10*, 153–158.

Kirsch, U. and G. I. N. Rozvany (1994). Alternative formulations of structural optimization. *Struct. Opt. 7*, 32–41.

Kirsch, U. and M. F. Rubinstein (1972). Structural reanalysis by iteration. *Comput. Struct. 2*, 497–510.

Kishi, N. and W.-F. Chen (1990). Moment-rotation relations of semi-rigid connections with angles. *J. Struct. Eng. 116*(7), 1813–1834.

Klarbring, A., J. Petersson, B. Torstenfelt, and M. Karlsson (2003). Topology optimization of flow networks. *Comput. Methods Appl. Mech. Engrg. 192*, 3909–3932.

Kleiber, M. (1997). *Parameter Sensitivity in Nonlinear Mechanics*. Chichester, UK: John Wiley & Sons.

Kocer, F. Y. and J. S. Arora (1999). Optimal design of H-frame transmission poles for earthquake loading. *J. Struct. Eng. 125*(11), 1299–1309.

Kogiso, N., L. T. Watson, Z. Gürdal, and R. T. Haftka (1994). Genetic algorithm with local improvement for composite laminate design. *Struct. Opt. 7*, 207–218.

Kojima, M., S. Shindoh, and S. Hara (1997). Interior-point methods for the monotone semidefinite linear complementarity problems. *SIAM J. Optim. 7*, 86–125.

Kollár, L. (Ed.) (1999). *Structural Stability in Engineering Practice*. London: E & FN Spon.

Koski, J. and R. Silvennoisen (1987). Norm methods and partial weighting in multicriterion optimization of structures. *Int. J. Numer. Methods Eng. 24*, 1101–1121.

Kočvara, M. (1997). Topology optimization with displacement constraints: A bilevel programming approach. *Struct. Opt. 14*, 256–263.

Kočvara, M. and J. V. Outrata (2006). Effective reformulation of the truss topology design problem. *Optim. Eng. 7*, 201–219.

Kočvara, M. and J. Zowe (1995). How to optimize mechanical structures simultaneously with respect to topology and geometry. In *Proc. 1st World Congress of Structural and Multidisciplinary Optimization (WCSMO1)*, Goslar, Germany, pp. 135–140.

Kravanja, S., Z. Kravanja, and B. S. Bedenik (1998). The MINLP optimization approach to structural optimization, Part I: General view on simultaneous topology and parameter optimization. *Int. J. Numer. Methods Eng. 43*, 263–292.

Kreisselmeier, G. and R. Steinhauser (1983). Application of vector performance optimization to a robust control loop design for a fighter aircraft. *Int. J. Control 37*, 251–284.

Krishna, P. (1979). *Cable Suspended Roofs*. New York: Mc-Graw Hill.

Krishnamoorthy, C. S., P. P. Venkatesh, and R. Sudarshan (2002). Object-oriented framework for genetic algorithms with application to space truss optimization. *J. Computing Civil Eng. 16*(1), 66–75.

Kutylowski, K. (2002). On nonunique solutions in topology optimization. *Struct. Multidisc. Optim. 23*, 398–403.

Kwan, A. S. (1998). An evolutionary approach for layout optimization of truss structures. *Int. J. Space Struct. 13*(3), 145–155.

Lagaros, N., M. Papadrakakis, and G. Kokossalakis (2002). Structural optimization using evolutionary algorithm. *Comput. Struct. 80*, 571–598.

Lagaros, N., L. D. Psarras, M. Papadrakakis, and G. Kokossalakis (2008). Optimum design of steel structures with web opening. *Eng. Struct. 30*, 2528–2537.

Laguna, M. and R. Marti (2003). *Scatter-Search Methodology and Implementation in C.* Dordrecht, The Netherlands: Kluwer Academic Publishers.

Lamberti, L. (2008). An efficient simulated annealing algorithm for design optimization of truss structures. *Comput. Struct. 86*, 1936–1953.

Lee, K. S. and Z. W. Geem (2004). A new structural optimization method based on the harmony search. *Comput. Struct. 82*, 781–798.

Lee, T. H. and J. J. Jung (2007). Kriging metamodel based optimization. In J. S. Arora (Ed.), *Optimization of Structural and Mechanical Systems.* Singapore: World Scientific.

Leelataviwat, S., S. C. Goel, and B. Stojadinović (2002). Energy-based seismic design of structures using yield mechanism and target drift. *J. Struct. Eng. 128*(8), 1046–1054.

Lemonge, A. C. C., H. J. C. Barbosa, and L. G. Fonseca (2009). A genetic algorithm for configuration and sizing optimization of dome structures including cardinality constraints. In *Proc. 8th World Congress of Structural and Multidisciplinary Optimization (WCSMO8)*, Paper No. 1409, Lisbon.

Lewiński, T. and G. I. N. Rozvany (2007). Exact analytical solutions for some popular benchmark problems in topology optimization, II: Three-sided polygonal supports. *Struct. Multidisc. Optim. 33*, 337–349.

Li, C., R. Priemer, and K.-H. Cheng (2004). Optimization by random search with jumps. *Int. J. Numer. Methods Eng. 60*, 1301–1315.

Li, G., R.-G. Zhou, L. Duan, and W.-F. Chen (1999). Multiobjective and multilevel optimization for steel frames. *Eng. Struct. 21*, 519–529.

Li, L. J., Z. B. Huang, F. Liu, and Q. H. Wu (2007). A heuristic particle swarm optimizer for optimization of pin connected structures. *Comput. Struct. 85*, 340–349.

Liew, J. Y. R., D. W. White, W. F. Chen, and S. Toma (1993a). Second-order refined plastic hinge analysis for frame design, Part I. *J. Struct. Eng. 119*(3), 3196–3216.

Liew, J. Y. R., D. W. White, W. F. Chen, and S. Toma (1993b). Second-order refined plastic hinge analysis for frame design, Part II. *J. Struct. Eng. 119*(11), 3217–3237.

Lin, J. H., W. Y. Che, and Y. S. Yu (1982). Structural optimization on geometrical configuration and element sizing with statical and dynamical constraints. *Comput. Struct. 15*, 507–515.

Lio, M. D., V. Cossalter, and R. Lot (2000). On the use of natural coordinates in optimal synthesis of mechanisms. *Mechanism and Machine Theory 35*, 1367–1389.

Liu, M. and S. A. Burns (2003). Multiple fully stressed designs of steel frame structures with semi-rigid connections. *Int. J. Numer. Methods Eng. 58*, 821–838.

Liu, M., S. A. Burns, and Y. K. Wen (2006). Genetic algorithm based construction-conscious minimum weight design of seismic steel moment-resisting frames. *J. Struct. Eng. 132*(1), 50–58.

Liu, M., Y. K. Wen, and S. A. Burns (2004). Life cycle cost oriented seismic design optimization of steel moment frame structures with risk-taking preference. *Eng. Struct. 26*, 1407–1421.

Liu, M., Y. K. Wen, and S. A. Burns (2005). Multiobjective optimization for performance-based seismic design of steel moment-resisting frames structures. *Earthquake Eng. Struct. Dyn. 34*, 289–306.

Liu, S. and H. Qiao (2009). Topology optimization of continuum structures with different tensile and compressive properties in bridge layout design. In *Proc. 8th World Congress of Structural and Multidisciplinary Optimization (WCSMO8)*, Paper No. 1328, Lisbon.

Luco, J. E. and H. L. Wong (1986). Response of a rigid foundation to a spatially random ground motion. *Earthquake Eng. Struct. Dyn. 14*, 891–908.

Luenberger, D. G. (2003). *Linear and Nonlinear Programming*. Boston: Kluwer Academic Publishers.

Luh, G.-C. and C.-H. Chueh (2004). Multi-objective optimal design of truss structure with immune algorithm. *Comput. Struct. 82*, 829–844.

Luh, G.-C. and C.-Y. Lin (2008). Optimal design of truss structures using ant algorithm. *Struct. Multidisc. Optim. 36*, 365–379.

Luo, Z.-Q., J.-S. Pang, and D. Ralph (1996). *Mathematical Programs with Equilibrium Constraints*. Cambridge, UK: Cambridge University Press.

Maar, B. and V. Schulz (2000). Interior point multigrid methods for topology optimization. *Struct. Multidisc. Optim. 19*, 214–224.

Machaly, E. S. B. (1986). Optimum weight analysis of steel frames with semirigid connections. *Comput. Struct. 23*(4), 461–474.

Mahin, S., J. Malley, and R. Hamburger (2002). Overview of the FEMA/SAC program for reduction of earthquake hazards in steel moment frame structures. *J. Const. Steel Res. 58*, 511–528.

Makode, P. V., R. B. Corotis, and M. R. Ramirez (1999). Nonlinear analysis of frame structures by pseudodistortions. *J. Struct. Eng. 125*(11), 1309–1317.

Mangasarian, O. L. (1969). *Nonlinear Programming*. New York: McGraw-Hill.

Marcelin, J. L., P. Trompette, and R. Dornberger (1995). Optimization of composite beam structures using a genetic algorithm. *Struct. Opt. 9*, 236–244.

Marler, T. and J. S. Arora (2004). Survey of multi-objective optimization methods for engineering. *Struct. Multidisc. Optim. 26*(6), 369–395.

Martínez, P., P. Martí, and O. M. Querin (2007). Growth method for size, topology, and geometry optimization of truss structures. *Struct. Multidisc. Optim. 33*, 13–26.

Massa, F., B. Lallemanda, and T. Tison (2009). Fuzzy multiobjective optimization of mechanical structures. *Comput. Methods Appl. Mech. Engrg. 198*, 631–643.

Masur, E. F. (1984). Optimal structural design under multiple eigenvalue constraints. *Int. J. Solids Struct. 20*, 211–231.

Maxwell, J. C. (1890). On reciprocal figures, frames, and diagrams of forces. *Scientific Papers 2*, 161–207.

McKeown, J. J. (1998). Growing optimal pin-jointed frames. *Struct. Opt. 15*, 92–100.

McNeil, W. A. (1971). Structural weight minimization using necessary and sufficient conditions. *J. Optimization Theory Appl. 8*(6), 454–466.

Mehrotra, S. (1992). On the implementation of a primal-dual interior point method. *SIAM J. Optim. 2*, 575–601.

Meirovitchand, L. and A. L. Hale (1981). On the substructure synthesis method. *AIAA J. 19*(7), 940–947.

Melosh, R. J. and R. Luik (1968). Multiple configuration analysis of structures. *J. Struct. Div. ASCE 94*(ST11), 2581–2596.

Mesarović, M. D., D. Macko, and Y. Takahara (1970). *Theory of Hierarchical, Multilevel, Systems*. New York: Academic Press.

Metropolis, N., A. Rosenbluth, M. Rosenbluth, A. Teller, and E. Teller (1953). Equation of state calculations by fast computing machines. *J. Chem. Phys 21*, 1087–1092.

Michell, A. G. M. (1904). The limits of economy in frame structures. *Philosophical Magazine Sect. 6, 8*(47), 589–597.

Mijar, A. R., C. Swan, J. S. Arora, and I. Kosaka (1998). Continuum topology optimization for concept design of frame bracing system. *J. Struct. Eng. 124*(5), 541–550.

Mittelmann, H. D. and D. Roose (Eds.) (1990). *Continuation Techniques and Bifurcation Problems*. Basel, Germany: Birkhäuser Verlag.

Mohammadi, R. K., M. H. El Naggar, and H. Moghaddam (2004). Optimum strength distribution for seismic resistant shear buildings. *Int. J. Solids Struct. 41*, 6597–6612.

Mohan, C. and H. T. Nguyen (1999). A controlled random search technique incorporating the simulated annealing concept for solving integer and mixed integer global optimization problems. *Comp. Opt. Appl. 14*, 103–132.

Mróz, Z. and D. Bojczuk (2003). Finite topology variations in optimal design of structures. *Struct. Multidisc. Optim. 25*, 153–173.

MSC Software (2005). *ADAMS 2005 User's Manual*. Santa Ana, CA: MSC Software Corp.

Mueller, K. M., M. Liu, and S. A. Burns (2002). Fully stressed design of frame structures and multiple load paths. *J. Struct. Eng. 128*(6), 806–814.

Munro, J. and P.-H. Chuang (1986). Optimal plastic design with imprecise data. *J. Eng. Mech. Div. ASCE 112*(9), 888–903.

Muralidhar, R. and J. Rao (1997). New models for optimal truss topology in limit design based on unified elastic/plastic analysis. *Comput. Methods Appl. Mech. Engrg. 140*, 109–138.

Myers, R. H. and D. C. Montgomery (1995). *Response Surface Methodology: Process and Product Optimization using Design Experiments*. New York: Wiley.

Nagtegaal, J. C. (1973). A new approach to optimal design of elastic structures. *Comput. Methods Appl. Mech. Engrg. 2*, 255–264.

Nakamura, T. and M. Ohsaki (1988). Sequential optimal truss generator for frequency ranges. *Comput. Methods Appl. Mech. Engrg. 67*(2), 189–209.

Nakamura, T. and M. Ohsaki (1992). A natural generator of optimum topology of plane trusses for specified fundamental frequency. *Comput. Methods Appl. Mech. Engrg. 94*(1), 113–129.

Nakanishi, Y. and S. Nakagiri (1996a). Optimization of frame topology using boundary cycle and genetic algorithm. *JSME Int. J., Series A. 39*(2), 279–285.

Nakanishi, Y. and S. Nakagiri (1996b). Optimization of truss topology using boundary cycle: Derivation of design variables to avoid inexpedient structure. *JSME Int. J., Series A. 39*(3), 415–421.

Nakanishi, Y. and S. Nakagiri (1997). Structural optimization under topological constraint represented by homology groups: Topological constraint on one-dimensional complex by use of zero- and one-dimensional groups. *JSME Int. J., Series A. 40*(3), 219–227.

Nakayama, H. (1995). Aspiration level approach to interactive multi-objective programming and its applications. In P. M. Pardalos, Y. Siskos, and C. Zopounidis (Eds.), *Advances in Multicriteria Analysis*, Dordrecht, The Netherlands, pp. 147–174. Kluwer Academic Publishers.

Narayanan, S. (2006). *Space Structures: Principles and Practice, Vol. 1,2.* Brentwood, UK: Multi-Science Publishing.

Newmark, N. M. and W. J. Hall (1982). Earthquake spectra and design. Technical Report, Earthquake Engineering Research Institute, Berkeley, CA.

Nguyen, D. T. (1987). Multilevel substructuring sensitivity analysis. *Comput. Struct. 25*(2), 191–202.

Nguyen, V., J. Strodiot, and C. Fleury (1987). A mathematical convergence analysis of the convex linearization method for engineering design. *Eng. Opt. 11*, 195–216.

Nishiwaki, S., S. Min, J. Yoo, and N. Kikuchi (2001). Optimal structural design considering flexibility. *Comput. Methods Appl. Mech. Engrg. 190*, 4457–4504.

Noor, A. K. and A. E. Lowder (1975). Structural reanalysis via a mixed method. *Comput. Struct. 5*, 9–12.

Oberndorfer, J. M., W. Achtziger, and H. R. E. M. Hörnlein (1996). Two approaches for truss topology optimization: A comparison for practical use. *Struct. Opt. 11*, 137–144.

Ogawa, K. and M. Tada (1994). Computer program for static and dynamic analysis of steel frames considering the deformation of joint panel. In *Proc. 17th Symp. on Computational Technology of Information, System and Applications*, pp. 79–84 (in Japanese).

Ohsaki, M. (1995). Genetic algorithm for topology optimization of trusses. *Comput. Struct. 57*(2), 219–225.

Ohsaki, M. (1997a). Optimization of building structural systems using parametric optimization method. *J. Struct. Eng., Architectural Inst. Japan 43B*, 79–88 (in Japanese).

Ohsaki, M. (1997b). Simultaneous optimization of topology and geometry of a regular plane truss. *Comput. Struct. 66*(1), 69–77.

Ohsaki, M. (2001a). Random search method based on exact reanalysis for topology optimization of trusses with discrete cross-sectional areas. *Comput. Struct. 79*(6), 673–679.

Ohsaki, M. (2001b). Sensitivity of optimum designs for spatially varying ground motions. *J. Struct. Eng. 127*(11), 1324–1329.

Ohsaki, M. (2003a). Application of optimization methods to architectural structures. In G. Yagawa (Ed.), *Handbook of Structural Engineering*, pp. 647–649. Tokyo: Maruzen (in Japanese).

Ohsaki, M. (2003b). Earthquake response analysis and optimum design of arch-type trusses considering stiffness of supporting structure. *J. Struct. Constr. Eng., Architectural Inst. Japan* (566), 53–58 (in Japanese).

Ohsaki, M. (2003c). System engineering. In J. Kanda (Ed.), *Architecture and Engineering, Visual Introduction of Architecture 9*, pp. 204–211. Tokyo: Shokokusha (in Japanese).

Ohsaki, M. (2005a). Design sensitivity analysis and optimization for nonlinear buckling of finite-dimensional elastic conservative structures. *Comput. Methods Appl. Mech. Engrg. 194*, 3331–3358.

Ohsaki, M. (2005b). Spatial structures: Optimization of structures with discrete cross-sectional areas using heuristic approaches. In *Intelligible Soft Computing for Architecture, City and Environment*, pp. 77–82. Architectural Institute of Japan (in Japanese).

Ohsaki, M. (2006a). Local and global searches of approximate optimal designs of regular frames. *Int. J. Numer. Methods Eng. 67*, 132–147.

Ohsaki, M. (2006b). Local and global searches of approximate optimal designs of regular frames. In *Proc. 3rd European Conference on Computational Mechanics (ECCM 2006)*, Paper No. 1174, Lisbon.

Ohsaki, M. (2008). Local search for multiobjective optimization of steel frames. In *Proc. 5th China-Japan-Korea Joint Symp. on Optimization of Structural and Mechanical Systems (CJK-OSM5)*, Jeju, Korea.

Ohsaki, M. and J. S. Arora (1993). A direct application of higher order parametric programming techniques to structural optimization. *Int. J. Numer. Methods Eng. 36*, 2683–2702.

Ohsaki, M. and J. S. Arora (1994). Design sensitivity analysis of elastoplastic structures. *Int. J. Numer. Methods Eng. 37*, 737–762.

Ohsaki, M., K. Fujisawa, N. Katoh, and Y. Kanno (1999). Semi-definite programming for topology optimization of trusses under multiple eigenvalue constraints. *Comput. Methods Appl. Mech. Engrg. 180*, 203–217.

Ohsaki, M. and M. Hayashi (2000). Fairness metrics for shape optimization of ribbed shells. *J. Int. Assoc. Shells and Spatial Struct. 41*(1), 31–39.

Ohsaki, M. and K. Ikeda (2007). *Stability and Optimization of Structures – Generalized Sensitivity Analysis*. Mechanical Engineering Series. New York: Springer.

Ohsaki, M. and Y. Kanno (2007). Semidefinite programming for engineering applications. In J. S. Arora (Ed.), *Optimization of Structural and Mechanical Systems*, pp. 541–567. Singapore: World Scientific.

Ohsaki, M. and Y. Kato (1997). Simultaneous optimization of topology and nodal locations of a plane truss associated with a Bézier curve. In *Proc. Annual Meeting, Architectural Inst. Japan*, Volume B-1, pp. 447–448 (in Japanese).

Ohsaki, M. and Y. Kato (1999). Simultaneous optimization of topology and nodal locations of a plane truss associated with a Bézier curve. In *Structural Engineering in the 21st Century, Proc. Structures Congress*, pp. 582–585. ASCE.

Ohsaki, M. and N. Katoh (2005). Topology optimization of trusses with stress and local constraints on nodal stability and member intersection. *Struct. Multidisc. Optim. 29*, 190–197.

Ohsaki, M., N. Katoh, T. Kinoshita, S. Tanigawa, D. Avis, and I. Streinu (2009). Enumeration of optimal pin-jointed bistable compliant mechanisms with non-crossing members. *Struct. Multidisc. Optim. 37*, 645–651.

Ohsaki, M., Y. Nagano, and K. Wakamatsu (2000). A two-level optimization method for seismic design of elastic three-dimensional frames. In *Proc. 12th World Conference on Earthquake Engineering*, Paper #0686, Auckland, New Zealand.

Ohsaki, M. and T. Nakamura (1993). A natural generator of optimum topology of a space frame for specified fundamental frequency. In G. Parke and C. Howard (Eds.), *Space Structures 4, Proc. 4th Int. Conf. on Space Structures*, pp. 1221–1229. Guildford, UK: Thomas Telford.

Ohsaki, M. and T. Nakamura (1994). Optimum design with imperfection sensitivity coefficients for limit point loads. *Struct. Opt. 8*, 131–137.

Ohsaki, M. and T. Nakamura (1996). Minimum constraint perturbation method for topology optimization of systems. *Eng. Opt. 26*, 171–186.

Ohsaki, M., T. Nakamura, and Y. Isshiki (1998). Shape-size optimization of plane trusses with designer's preference. *J. Struct. Eng. 124*(11), 1323–1330.

Ohsaki, M., T. Nakamura, and M. Kohiyama (1997). Shape optimization of a double-layer space truss described by a parametric surface. *Int. J. Space Struct. 12*(2), 109–119.

Ohsaki, M. and S. Nishiwaki (2004). Shape design of bistable compliant mechanism by utilizing snapthrough behavior. *Trans. Japan Soc. of Mech. Eng., Ser. A 70*(700), 23–28 (in Japanese).

Ohsaki, M. and S. Nishiwaki (2005). Shape design of pin-jointed multistable compliant mechanism using snapthrough behavior. *Struct. Opt. 30*, 327–334.

Ohsaki, M. and S. Nishiwaki (2007a). Generation of link mechanism by shape-topology optimization of trusses considering geometrical nonlinearity. In *Proc. 7th World Congress of Structural and Multidisciplinary Optimization (WCSMO7)*, Seoul. (also available in *J. Computational Science and Technology, JSME 3* (1), pp. 46–53, 2009).

Ohsaki, M. and S. Nishiwaki (2007b). Generation of link mechanism by shape-topology optimization of trusses considering geometrical nonlinearity. *Trans. Japan Soc. of Mech. Eng., Ser. A* 73(729), 659–665 (in Japanese).

Ohsaki, M., T. Ogawa, and R. Tateishi (2003). Shape optimization of curves and surfaces considering fairness metrics and elastic stiffness. *Struct. Multidisc. Optim.* 24, 449–456. Erratum: 27, pp. 250–258, 2004.

Ohsaki, M., H. Tagawa, and Y. Kato (2000). Optimum design of structures subjected to spatially varying ground motion. *J. Struct. Eng., Architectural Inst. Japan* 46B, 9–18 (in Japanese).

Ohsaki, M., H. Tagawa, and P. Pan (2009). Shape optimization of reduced beam section for maximum plastic energy dissipation under cyclic loads. *J. Const. Steel Res.* 65, 1511–1519.

Ohsaki, M. and R. Watada (2008). Linear mixed integer programming for topology optimization of trusses and plates. In *Proc. 6th Int. Conf. on Computation of Shell and Spatial Structures, IASS-IACM*, Ithaca, NY.

Ohsaki, M., J. Y. Zhang, and S. Kimura (2005). An optimization approach to design of geometry and forces of tensegrities. In *Proc. IASS Symp. 2005*, Bucharest, pp. 603–610. Int. Assoc. Shell and Spatial Struct.

Ohsaki, M., J. Y. Zhang, and Y. Ohishi (2008). Force design of tensegrity structures by enumeration of vertices of feasible region. *Int. J. Space Struct.* 23(2), 117–126.

Olhoff, N. (1980). Optimal design with respect to structural eigenvalues. In *Proc. 15th IUTAM Congress*, Toronto, Canada, pp. 133–149.

Olhoff, N. and S. H. Rasmussen (1977). On single and bimodal optimum buckling loads of clamped columns. *Int. J. Solids Struct.* 13, 605–614.

Olhoff, N. and J. E. Taylor (1979). On optimal structural remodelling. *J. Optimization Theory Appl.* 27(4), 571–582.

Orozco, C. and O. Ghattas (1997). A reduced SAND method for optimal design of non-linear structures. *Int. J. Numer. Methods Eng.* 40, 2759–2774.

Otto, F. (1967). *Tensile Structures: Volume 1, Pneumatic Structures.* Cambridge, MA: MIT Press.

Otto, F. (1969). *Tensile Structures: Volume 2, Cable Structures.* Cambridge, MA: MIT Press.

Özakca, M., E. Hinton, and N. V. R. Rao (1993). Shape optimization of axisymmetric structures with adaptive finite element procedures. *Struct. Opt.* 5, 256–264.

Paeng, J. K. and J. S. Arora (1989). Dynamic response optimization of mechanical systems with multiplier methods. *AIAA J.* 111, 73–80.

Pan, P., M. Ohsaki, and T. Kinoshita (2007). Constraint approach to performance-based design of steel moment-resisting frames. *Eng. Struct. 29*, 186–194.

Pan, P., M. Ohsaki, and H. Tagawa (2007). Shape optimization of H-beam flange for maximum plastic energy dissipation. *J. Struct. Eng. 133*(8), 1176–1179.

Papalambos, P. Y. and D. J. Wilde (2000). *Principles of Optimal Design: Modeling and Computation* (2nd ed.). Cambridge, UK: Cambridge University Press.

Patnaik, S. N. and P. Dayaratnam (1970). Behavior and design of pin connected structures. *Int. J. Numer. Methods Eng. 2*, 579–595.

Patnaik, S. N. and A. Hopkins (1998). Optimality of a fully stressed design. *Comput. Methods Appl. Mech. Engrg. 165*, 215–221.

Pavlovčič, L., A. Krajnc, and D. Beg (2004). Cost function analysis in the structural optimization of steel frames. *Struct. Multidisc. Optim. 28*, 286–295.

Pedersen, P. (1972). On the optimal layout of multi-purpose trusses. *Comput. Struct. 2*, 695–712.

Pedersen, P. (1973). Optimal joint positions for space trusses. *J. Struct. Div. ASCE 99*(ST12), 2459–2476.

Pedersen, P. and N. L. Pedersen (2009). A discussion on the application of optimality criteria for compliance with forced support displacement. In *Proc. 8th World Congress of Structural and Multidisciplinary Optimization (WCSMO8)*, Paper No. 1047, Lisbon.

Pellegrino, S. (1993). Structural computation with the singular value decomposition of the equilibrium matrix. *Int. J. Solids Struct. 30*(21), 3025–3035.

Pellegrino, S. and C. R. Calladine (1986). Matrix analysis of statically and kinematically indeterminate frameworks. *Int. J. Solids Struct. 22*(4), 409–428.

Peressini, A. L., F. E. Sullivan, and J. J. Uhl (1988). *The Mathematics of Nonlinear Programming*. New York: Springer.

Pereyra, V., D. Lawver, and J. Isenberg (2003). An algorithm for optimal design of steel frame structures. *Applied Numerical Mathematics 47*, 503–514.

Petersson, J. (1999). A finite element analysis of optimal variable thickness sheets. *SIAM J. Numer. Anal. 36*(6), 1759–1778.

Petersson, J. (2001). On continuity of the design-to-state mappings for trusses with variable topology. *J. Eng. Sci. 39*, 1119–1141.

Pierre, D. A. and M. J. Lowe (1975). *Mathematical Programming via Augmented Lagrangians.* Reading, MA: Addison-Wesley.

Pólik, I. (2005). *Addendum to the SeDuMi User Guide, Ver. 1.1.* Advanced Optimization Laboratory, McMaster University, Ontario, Canada. Available at: http://sedumi.ie.lehigh.edu/.

Prager, W. (1967). Optimum plastic design of a portal frame for alternative loads. *J. Appl. Mech. 34*, 772–773.

Prager, W. (1971). Foulkes mechanism in optimal plastic design for alternative loads. *Int. J. Mech. Sci. 13*, 971–973.

Prager, W. (1972). Conditions for structural optimality. *Comput. Struct. 2*, 833–840.

Prager, W. (1974a). *Introduction to Structural Optimization.* Vienna: Springer.

Prager, W. (1974b). A note on discretized Michell structure. *Comput. Methods Appl. Mech. Engrg. 3*, 349–355.

Prager, W. (1976). Geometric discussion of the optimal design of a simple truss. *J. Struct. Mech. 4*(1), 57–63.

Prager, W. and J. E. Taylor (1968). Problem of optimal structural design. *J. Appl. Mech. 35*(1), 102–106.

Price, T. E. and M. O. Eberhard (1998). Effects of spatially varying ground motions on short bridges. *J. Struct. Eng. 124*(8), 948–955.

Price, W. L. (1983). Global optimization by controlled random search. *J. Optimization Theory Appl. 40*, 333–348.

Pugnale, A. and M. Sassone (2007). Morphogenesis and structural optimization of shell structures with the aid of a genetic algorithm. *J. Int. Assoc. Shells and Spatial Struct. 48*(3), 161–166.

Putresza, J. T. and P. Kolakowski (2001). Sensitivity analysis of frame structures: Virtual distortion method approach. *Int. J. Numer. Methods Eng. 50*, 1307–1329.

Qiu, G. Y. and X. S. Li (2010). A note on the derivation of global stress constraints. *Struct. Multidisc. Optim. 40*, 625–628.

Rahmatalla, S. and C. Swan (2003). Form finding of sparse structures with continuum topology optimization. *J. Struct. Eng. 129*(12), 1707–1716.

Rajasekaran, S., V. S. Mohan, and O. Khamis (2004). The optimization of space structures using evolution strategies with functional networks. *Engineering with Computers 20*, 75–87.

Rajeev, S. and C. S. Krishnamoorthy (1997). Genetic algorithms-based methodologies for design optimization of trusses. *J. Struct. Eng. 123*, 350–358.

Ramm, E., K.-U. Bletzinger, and R. Reitinger (1993). Shape optimization of shell structures. *Bulletin of Int. Assoc. for Shell and Spatial Struct 34*(2), 103–121.

Ramrakhyami, D. S., M. I. Frecker, and G. A. Lesieutre (2009). Hinged beam elements for the topology design of compliant mechanisms using the ground structure approach. *Struct. Multidisc. Optim. 37*, 557–567.

Rando, T. and J. A. Roulier (1991). Designing faired parametric surfaces. *Comput.-Aided Des. 23*, 492–497.

Rao, S. S. (1987). Multi-objective optimization of fuzzy structural systems. *Int. J. Numer. Methods Eng. 24*, 1157–1171.

Ray, T. and K. M. Liew (2002). A swarm metaphor for multiobjective design optimization. *Eng. Opt. 34*(2), 141–153.

Razani, R. (1965). Behavior of fully stressed design of structures and its relationship to minimum weight design. *AIAA J. 3*, 2262–2268.

Rechenberg, I. (1965). Cybernetic solution path of an experimental problem. Library Translation 1122, Royal Aircraft Establishment, Franborough, UK.

Reddy, G. M. and J. Cagan (1995a). An improved shape annealing algorithm for truss topology generation. *J. Mech. Eng. 117*, 315–321.

Reddy, G. M. and J. Cagan (1995b). Optimally directed truss topology generation using shape annealing. *J. Mech. Eng. 117*, 206–209.

Reemtsen, R. and J.-J. Rückmann (Eds.) (1998). *Semi-Infinite Programming*. Norwell, MA: Kluwer Academic Publishers.

Reeves, C. (1995). *Modern Heuristic Techniques for Combinatorial Problems*. New York: McGraw-Hill.

Richtmyer, R. D. and K. W. Morton (1967). *Difference Methods for Initial Value Problems*. New York: Interscience.

Ringertz, U. T. (1985). On topology optimization of trusses. *Eng. Opt. 9*, 209–218.

Ringertz, U. T. (1986). A branch and bound algorithm for topology optimization of truss structures. *Eng. Opt. 10*, 111–124.

Rocakfellar, R. T. (1970). *Convex Analysis*. Princeton, NJ: Princeton University Press.

Rodorigues, H. C., J. M. Guedes, and M. P. Bendsøe (1995). Necessary conditions for optimal design of structures with a nonsmooth eigenvalue based criterion. *Struct. Opt. 9*, 52–56.

Rogers, D. F. and J. A. Adams (1990). *Mathematical Elements for Computer Graphics*. New York: McGraw-Hill.

Roulier, J. A. and T. Rando (1994). Measures of fairness for curves and surfaces. In N. S. Spadis (Ed.), *Designing Fair Curves and Surfaces*, pp. 75–122. Philadelphia, PA: SIAM.

Rozvany, G. I. N. (1976). *Optimal Design of Flexural Systems*. Oxford, UK: Pergamon Press.

Rozvany, G. I. N. (1989). *Structural Design via Optimality Criteria*. Dordrecht, The Netherlands: Kluwer Academic Publishers.

Rozvany, G. I. N. (1996). Difficulties in truss topology optimization with stress, local buckling and system stability constraints. *Struct. Opt. 11*, 213–217.

Rozvany, G. I. N. (Ed.) (1997). *Topology Optimization in Structural Mechanics*. CISM Courses and Lectures 374. Vienna: Springer-Verlag.

Rozvany, G. I. N. (2001). On design-dependent constraints and singular topologies. *Struct. Multidisc. Optim. 21*(2), 164–172.

Rozvany, G. I. N. and M. Zhou (1994). Optimality criteria methods for large discretized systems. In H. Adeli (Ed.), *Advances in Design Optimization*, pp. 41–108. London: Chapman & Hall.

Rozvany, G. I. N., M. Zhou, and T. Birker (1992). Generalized shape optimization without homogenization. *Struct. Optim. 4*, 250–252.

Rule, W. K. (1994). Automatic truss design by optimized growth. *J. Struct. Eng. 120*(10), 3063–3070.

Rychter, Z. and A. Musiuk (2007). Topological sensitivity to diagonal member flips of two-layered statically determinate trusses under worst loading. *Int. J. Solids Struct. 44*, 4942–4957.

Sadek, E. A. (1986). Dynamic optimization of framed structures with variable layout. *Int. J. Numer. Methods Eng. 23*, 1273–1294.

Sadek, E. A. (1989). An optimality criterion method for dynamic optimization of structures. *Int. J. Numer. Methods Eng. 28*, 579–592.

Sadek, E. A. (1992). Optimization of structures having general cross-sectional relationships using an optimality criterion method. *Comput. Struct. 43*(5), 959–969.

Saka, M. P. (1980). Shape optimization of trusses. *J. Struct. Div. ASCE 106*(ST5), 1155–1174.

Saka, M. P. (2007). Optimum topological design of geometrically nonlinear single layer latticed domes using coupled genetic algorithm. *Comput. Struct. 85*, 1635–1646.

Saka, M. P. and F. Erdal (2009). Harmony search based algorithm for the optimum design of grillage systems to LRFD-AISC. *Struct. Multidisc. Optim. 38*, 25–41.

Saka, M. P. and M. Ulker (1991). Optimum design of geometrically nonlinear space trusses. *Comput. Struct. 42*(3), 289–299.

Salajegheh, E. (1996). Approximate discrete variable optimization of frame structures with dual method. *Int. J. Numer. Methods Eng. 39*, 1607–1617.

Salerno, G. (1992). How to recognize the order of infinitesimal mechanisms: A numerical approach. *Int. J. Numer. Methods Eng. 35*, 1351–1395.

Sankaranarayanan, S., R. T. Haftka, and R. K. Kapania (1994). Truss topology optimization with simultaneous analysis and design. *AIAA J. 32*(2), 420–424.

Sarma, K. and H. Adeli (2002). Life-cycle cost optimization of steel structures. *Int. J. Numer. Methods Eng. 55*, 1451–1462.

Save, M. A. (1983). Remarks on minimum-volume designs of a three-bar truss. *J. Struct. Mech. 11*(1), 101–110.

Sawada, K. and A. Matsuo (2003). A revised enumeration algorithm for elastic plastic discrete optimization of steel building frames. *J. Struct. Constr. Eng., Architectural Inst. Japan 574*, 93–98 (in Japanese).

Saxena, A. (2005). Synthesis of compliant mechanisms for path generation using genetic algorithm. *J. Mech. Design 127*, 745–752.

Saxena, A. and G. K. Ananthasuresh (2000). On an optimal property of compliant topologies. *Struct. Multidisc. Optim. 19*, 36–49.

Schmidt, L. C. (1962). Minimum weight layouts of elastic, statically determinate, triangulated frames under alternative load systems. *J. Mech. Phys. Solids 10*, 139–149.

Schmit, L. A. and K. J. Chang (1984). Optimum design sensitivity based on approximation concepts and dual method. *Int. J. Numer. Methods Eng. 20*, 39–75.

Schmit, L. A. and B. Farshi (1974). Some approximation concepts for structural synthesis. *AIAA J. 12*, 692–699.

Schmit, L. A. and R. L. Fox (1965). An integrated approach to structural synthesis and analysis. *AIAA J. 3*, 1104–1112.

Schur, J. (1917). über Potenzreihen, die im Innern des Einheitskreises beschränkt sind. *J. Reine Angew Math. 147*, 205–232.

Schutte, J. F. and A. A. Groenwold (2003). Sizing design of truss structures using particle swarm. *Struct. Multidisc. Optim. 25*, 261–269.

Sedaghati, R. and E. Esmailzadeh (2003). Optimum design of structures with stress and displacement constraints using the force method. *Int. J. Mech. Sci. 45*, 1369–1389.

Sekimoto, T. and H. Noguchi (2001). Homologous topology optimization in large displacement and buckling problems. *JSME Int. J., Series A 44*, 610–615.

Semby, W. and A. Der Kiureghian (1985). Modal combination rules for multicomponent earthquake excitation. *Earthquake Eng. Struct. Dyn. 13*, 1–12.

Sergeyev, O. and Z. Mróz (2000). Sensitivity analysis and optimal design of 3D frame structures for stress and frequency constraints. *Comput. Struct. 75*, 167–185.

Sewell, M. J. (1987). *Maximum and Minimum Principles*. Cambridge, UK: Cambridge University Press.

Seyranian, A. P. (1993). Sensitivity analysis of multiple eigenvalues. *Mech. Struct. & Mach. 21*, 261–284.

Seyranian, A. P., E. Lund, and N. Olhoff (1994). Multiple eigenvalues in structural optimization problem. *Struct. Opt. 8*, 207–227.

Shames, I. H. and F. A. Cozzarelli (1997). *Elastic and Inelastic Stress Analysis*. Washington, DC: Taylor & Francis.

Shea, K., J. Cagan, and S. J. Fenves (1997). A shape annealing approach to optimal truss design with dynamic grouping of members. *J. Mech. Eng. 118*, 388–394.

Shea, K. and F. C. Smith (2006). Improving full-scale transmission tower design through topology and shape optimization. *J. Struct. Eng. 132*(5), 781–790.

Sherman, J. and W. J. Morrison (1950). Adjustment of an inverse matrix corresponding to changes in one element of a given matrix. *Ann. Math. Statist. 21*, 124–127.

Sheu, C. Y. and L. A. Schmit (1972). Minimum weight design of elastic redundant trusses under multiple static loading conditions. *AIAA J. 10*(2), 155–162.

Shield, R. T. (1960). On the optimum design of shells. *J. Appl. Mech. 27*, 316–322.

Shimizu, K., Y. Ishizuka, and J. F. Bard (1997). *Nondifferentiable and Two-Level Mathematical Programming*. Norwell, MA: Kluwer Academic Publishers.

Shin, Y. S., R. T. Haftka, and R. H. Plaut (1987). Simultaneous analysis and design or eigenvalue maximization. *AIAA J. 26*(6), 738–744.

Shin, Y. S., R. T. Haftka, L. T. Watson, and R. H. Plaut (1988). Tracing structural optima as a function of available resources by a homotopy method. *Comput. Methods Appl. Mech. Engrg. 70*, 151–164.

Simões, L. M. C. (1989). Isolated global optimality in truss sizing problems. *Comput. Struct. 33*(2), 375–384.

Smith, B., P. Bjørstad, and W. Gropp (1996). *Domain Decomposition: Parallel Multilevel Methods for Elliptic Partial Differential Equations.* New York: Cambridge University Press.

Sobieszczanski-Sobieski, J. (1992). A technique for locating function roots and for satisfying equality constraints in optimization. *Struct. Opt. 4*(3–4), 241–243.

Sobieszczanski-Sobieski, J. and R. T. Haftka (1996). Multidisciplinary aerospace design optimization. In *Proc. 34th Aerospace Science Meeting and Exhibit*, Paper AIAA-96-0711, Reno, NV. AIAA.

Sobieszczanski-Sobieski, J., B. James, and A. Dovi (1985). Structural optimization by multilevel decomposition. *AIAA J. 23*(11), 1775–1782.

Stadler, W. (1979). A survey of multicriteria optimization of the vector maximum problem, Part I: 1776–1960. *J. Optimization Theory Appl. 29*(1), 1–52.

Stadler, W. (Ed.) (1988). *Multicriteria Optimization in Engineering and in the Sciences.* New York: Plenum Press.

Stadler, W. (1999). Discrete geometry, the Steiner problem, and topology optimization. In *Proc. 3rd World Congress of Structural and Multidisciplinary Optimization (WCSMO3).*

Stolpe, M. (2004). Global optimization of minimum weight truss topology problems with stress, displacement, and local buckling constraints using branch-and-bound. *Int. J. Numer. Methods Eng. 61*(8), 1270–1309.

Stolpe, M. and K. Svanberg (2001a). An alternative interpolation scheme for minimum compliance topology optimization. *Struct. Multidisc. Optim. 22*, 116–124.

Stolpe, M. and K. Svanberg (2001b). On the trajectories of the epsilon-relaxation approach for stress-constrained truss topology optimization. *Struct. Multidisc. Optim. 21*, 140–151.

Stolpe, M. and K. Svanberg (2003a). Modeling topology optimization problems as linear mixed 0-1 programs. *Int. J. Numer. Methods Eng. 57*, 723–739.

Stolpe, M. and K. Svanberg (2003b). A note on stress-constrained truss topology optimization. *Struct. Multidisc. Optim. 25*, 62–64.

Storaasli, O. O. and J. Sobieszczanski-Sobieski (1983). On the accuracy of the Taylor approximation for structural resizing. *AIAA J 21*, 1571–1580.

Sturm, J. F. (1999). Using SeDuMi 1.02, a Matlab toolbox for optimization over symmetric cones. *Optimization Methods and Software 11-12*, 625–653.

Sui, Y. K., J. Du, and Y. Guo (2006). Independent continuum mapping for topological optimization of frame structures. *Struct. Multidisc. Optim. 22*, 611–619.

Sui, Y. K. and X. C. Wang (1997). Second-order method of generalized geometric programming for spatial frame optimization. *Comput. Methods Appl. Mech. Engrg. 141*, 117–123.

Svanberg, K. (1981). Optimization of geometry in truss design. *Comput. Methods Appl. Mech. Engrg. 28*, 63–80.

Svanberg, K. (1984). On local and global minima in structural optimization. In E. Atrek, R. H. Gallagher, K. M. Ragsdel, and O. C. Zienkiewicz (Eds.), *New Directions in Structural Design 34*, pp. 327–341. Chichester, UK: John Wiley & Sons.

Svanberg, K. (1987). The method of moving asymptotes – A new method for structural optimization. *Int. J. Numer. Methods Eng. 24*, 359–373.

Svanberg, K. (1994). On the convexity and concavity of compliances. *Struct. Opt. 7*, 42–46.

Svanberg, K. and M. Werme (2009). On the validity of using small positive lower bounds on design variables in discrete topology optimization. *Struct. Multidisc. Optim. 37*, 325–334.

Sved, G. (1954). The minimum weight of certain redundant structures. *Austral. J. Appl. Sci. 5*, 1–8.

Sved, G. and Z. Ginos (1968). Structural optimization under multiple loading. *Int. J. Mech. Sci. 10*, 803–805.

Svensson, B. (1987). A substructuring approach to optimum structural design. *Comput. Struct. 25*, 251–258.

Tagawa, H. and M. Ohsaki (1999). A continuous topology transition model for shape optimization of plane trusses with uniform cross-sectional area. In *Proc. 3rd World Congress of Structural and Multidisciplinary Optimization (WCSMO3)*, pp. 254–256.

Takagi, J. and M. Ohsaki (2004). Design of lateral braces for columns considering critical imperfection of buckling. *Int. J. Structural Stability and Dynamics 4*(1), 69–88.

Taleb-Agha, G. and R. B. Nelson (1975). Method for the optimum design of truss-type structures. *AIAA J. 14*(4), 436–445.

Tam, T. K. H. and A. Jennings (1989). Classification and comparison of LP formulations for the plastic design of frames. *Eng. Struct. 11*, 163–178.

Taylor, J. E. (1967). Minimum mass bar for axial vibration at specified natural frequency. *AIAA J. 5*(10), 1911–1913.

Thompson, B. S. and C. K. Sung (1986). A survey of finite element techniques for mechanism design. *Mechanism and Machine Theory 21*(4), 3517–359.

Tonon, F. (1999). Multiobjective optimization of uncertain structures through fuzzy set and random theory. *Computer-Aided Civil and Infrastructure Eng. 14*, 119–140.

Topping, B. H. V. (1984). Shape optimization of skeletal structures: A review. *J. Struct. Eng. 109*(8), 1933–1951.

Topping, B. H. V. (1992). Mathematical programming techniques for shape optimization of skeletal structures. In G. I. N. Rozvany (Ed.), *Shape and Layout Optimization of Structural Systems and Optimality Criteria Methods*, pp. 349–375. Vienna: Springer.

Topping, B. V. H., A. I. Khan, and J. P. Leite (1996). Topological design of truss structures using simulated annealing. *Struct. Engng. Rev. 8*, 301–314.

Tosserams, S., L. F. P. Etman, P. Y. Papalambos, and J. E. Rooda (2006). An augmented Lagrangian relaxation for analytical target cascading using the alternating direction method of multipliers. *Struct. Multidisc. Optim. 31*, 176–189.

Tuttle, E. R., S. W. Peterson, and J. E. Titus (1989). Enumeration of basic kinematic chains using the theory of finite groups. *J. Mechanisms, Transmissions, and Automation in Design, ASME 111*, 498–503.

Twu, S.-L. and K. K. Choi (1992). Configuration design sensitivity analysis of built-up structures, Part I: Theory. *Int. J. Numer. Methods Eng. 35*, 1127–1150.

van Keulen, F., R. T. Haftka, and N. H. Kim (2005). Review of options for structural design sensitivity analysis, Part 1: Linear systems. *Comput. Methods Appl. Mech. Engrg. 194*(30-33), 3213–3243.

Vanderplaats, G. N. (1988). Multidiscipline design optimization. *Appl. Mech. Rev. 41*, 257–262.

Vanderplaats, G. N. (1999). *Numerical Optimization Techniques for Engineering Design*. Colorado Springs, CO: Vanderplaats Research & Development Inc.

Vanderplaats, G. N. and N. Yoshida (1985). Efficient calculation of optimum design sensitivity. *AIAA J. 23*, 1798–1803.

Vassart, N., R. Laporte, and R. Motro (2000). Determination of mechanism's order for kinematically and statically indeterminate systems. *Int. J. Solids Struct. 37*, 3807–3839.

Venkayya, V. B. (1978). Structural optimization: A review and some recommendations. *Int. J. Numer. Methods Eng. 13*, 203–228.

Venkayya, V. B., N. S. Khot, and L. Berke (1973). Application of optimality criteria approaches on automated design of large practical structures. In *Proc. 2nd Symp. on Structural Optimization, AGARD-CP-123*, Milan, Italy, pp. 3.1–3.19.

Venkayya, V. B. and V. A. Tishler (1983). Optimization of structures with frequency constraints. In *Computer Methods in Nonlinear Solids Structural Mechanics, ASME-AMD-54*, pp. 239–259. New York: ASME.

Viana, A. and J. P. de Sousa (2000). Using metaheuristics in multiobjective resource constrained project. *European J. Operational Research 120*, 359–374.

Villaverde, R. (1997). Seismic design of secondary structures: State of the art. *J. Struct. Eng. 123*(8), 1011–1019.

Visual Numerics Inc. (1997). *IMSL Math/Library Ver. 4.01*. Houston: Visual Numerics Inc.

VR&D (1999). *DOT User's Manual, Ver 5.0*. Colorado Springs, CO.

Wang, Q. and J. S. Arora (2006a). Alternative formulation for structural optimization: An evaluation using frames. *J. Struct. Eng. 132*(12), 1880–1889.

Wang, Q. and J. S. Arora (2006b). Optimization of large-scale truss structures using sparse SAND formulations. *Int. J. Numer. Methods Eng. 69*, 390–407.

Watada, R. and M. Ohsaki (2009a). Continuation approach for investigation of non-uniqueness of optimal topology for minimum compliance. In *Proc. 8th World Congress of Structural and Multidisciplinary Optimization (WCSMO8)*, Paper No. 1073, Lisbon.

Watada, R. and M. Ohsaki (2009b). Topology optimization of trusses consisting of traditional layouts. In *Proc. 9th Asian Pacific Conf. on Shell and Spatial Struct. (APCS2009)*, Paper No. P0026, Nagoya, Japan.

Watada, R. and M. Ohsaki (2009c). Topology optimization of trusses consisting of traditional layouts. *J. Struct. Constr. Eng., Architectural Inst. Japan 74*(639), 857–863 (in Japanese).

Watson, L. T. and R. T. Haftka (1989). Modern homotopy methods in optimization. *Comput. Methods Appl. Mech. Engrg. 74*(3), 289–305.

Wen, Y. K. and Y. J. Kang (2001). Minimum building life-cycle cost design criteria, I: Methodology. *J. Struct. Eng. 127*(3), 330–337.

Whidborne, J. F., D. W. Gu, and I. Postlethwaite (1997). Simulated annealing for multiobjective control system design. *IEE Proc. Control Theory Appl. 144*(6), 582–588.

Whittaker, A., M. Constantinou, and P. Tsopelas (1998). Displacement estimates for performance-based seismic design. *J. Struct. Eng., ASCE 124*(8), 905–912.

Wilson, E. L., A. Der Kiureghian, and E. P. Bayo (1982). A replacement of the SRSS method in seismic analysis. *Earthquake Engineering and Structural Dynamics 13*, 1–12.

Witten, I. H. and E. Frank (2000). *WEKA, Machine Learning Algorithms in Java*. San Francisco, CA: Morgan Kaufmann Publishers.

Wolkowicz, H., R. Saigal, and L. Vandenberghe (Eds.) (2000). *Handbook of Semidefinite Programming: Theory, Algorithms, and Applications*. Boston, MA: Kluwer Academic Publishers.

Woo, T. H. and L. A. Schmit (1981). Decomposition in optimal plastic design of structures. *Int. J. Solids Struct. 17*, 39–56.

Woodbury, M. A. (1950). Inverting modified matrices. *Memorandom Report 42*, Statistical Research Group, Princeton, NJ.

Wrenn, G. A. (1998). An indirect method for numerical optimization using the Kreisselmeier-Steinhause function. NASA Contractor Report 4220, NASA Langley Research Center.

Wu, C. C. and J. S. Arora (1987). Simultaneous analysis and design optimization of nonlinear response. *Engineering with Computers 2*, 53–63.

Xie, Y. M. and G. P. Steven (1993). A simple evolutionary procedure for structural optimization. *Comput. Struct. 49*, 885–896.

Xu, L. (2002). Design and optimization of semi-rigid frames structures. In S. Burns (Ed.), *Recent Advancements in Optimal Structural Design*, pp. 147–168. Raton, VA: ASCE Press.

Xu, L., Y. Gong, and D. E. Grierson (2006). Seismic design optimization of steel building frameworks. *J. Struct. Eng. 132*(2), 277–286.

Xu, L. and D. E. Grierson (1993). Computer-automated design of semirigid steel frameworks. *J. Struct. Eng. 199*(6), 1740–1759.

Xue, Q. and C.-C. Chen (2003). Performance-based seismic design of structures: a direct displacement-based approach. *Eng. Struct. 25*, 1803–1813.

Yang, X. Y., Y. M. Xie, G. P. Steven, and O. M. Querin (1999a). Bidirectional evolutionary method for stiffness optimization. *AIAA J. 37*(11), 1483–1488.

Yang, X. Y., Y. M. Xie, G. P. Steven, and O. M. Querin (1999b). Topology optimization for frequencies using an evolutionary method. *J. Struct. Eng. 125*(12), 1432–1438.

Yang, Y. and C. K. Soh (2002). Automated optimum design of structures using genetic programming. *Comput. Struct. 80*, 1537–1546.

Yoshikawa, N., I. Elishakoff, and S. Nakagiri (1998). Worst case estimation of homology design by convex analysis. *Comput. Struct. 67*, 191–196.

Yu, M., Z.-S. Liu, and D.-J. Wang (1996). Comparison of several approximate modal methods for computing mode shape derivatives. *Comput. Struct. 62*(2), 381–393.

Yun, Y. M. and B. H. Kim (2005). Optimum design of plane steel frame structures using second-order inelastic analysis and a genetic algorithm. *J. Struct. Eng. 131*(12), 1820–1831.

Zembaty, Z. (1996). Spatial seismic coefficients, some sensitivity results. *J. Eng. Mech. 122*, 379–382.

Zerva, A. (1990). Response of multi-span beams to spatially incoherent seismic ground motions. *Earthquake Eng. Struct. Dyn. 19*, 819–832.

Zhou, M. (1996). Difficulties in truss topology optimization with stress and local buckling constraints. *Struct. Opt. 11*, 134–136.

Zhou, M. and R. T. Haftka (1995). A comparison of optimality criteria methods for stress and displacement constraints. *Comput. Methods Appl. Mech. Engrg. 124*, 253–271.

Zhou, M. and G. I. N. Rozvany (1991). The COC algorithm, Part II: Topological, geometrical and generalized shape optimization. *Comput. Methods Appl. Mech. Engrg. 89*, 309–336.

Zhou, M. and G. I. N. Rozvany (1992). DCOC: A new optimality criteria method for large systems, Part I: Theory. *Struct. Opt. 5*, 12–25.

Zhou, M. and G. I. N. Rozvany (1993). DCOC: A new optimality criteria method for large systems, Part II: Algorithm. *Struct. Opt. 6*, 250–262.

Zou, X.-K. and C.-M. Chan (2005). An optimal resizing technique for seismic drift design of concrete buildings subjected to response spectrum and time history loadings. *Comput. Struct. 83*, 1689–1704.

Index

positive definite, 41, 42, 77, 315, 332

positive semidefinite, 41, 42, 46, 140, 143, 315, 337

rank, 70
  full, 70
  full row, 70, 73
  singular, 72

melting node, 160

merit function, 332

method of feasible directions (MFD), 332
  pushoff factor, 333

method of moving asymptote (MMA), 333
  move limit, 334

Michell truss, 4, 85, 87

minimal surface, 4, 260

mixed integer nonlinear programming (MINLP), 10, 89, 106, 113, 338

mixed integer programming (MIP), 89, 113, 118, 338

multibody dynamics, 176

multidisciplinary optimization (MDO), 215
  discipline, 215

multilevel optimization, 215

multiobjective programming (MOP), 50, 150, 235, 260, 275, 283, 345
  ε-constraint approach, 347
  a posteriori measures of preference, 346
  a priori information of preference, 234, 346
  aspiration level approach, 51, 350
  combinatorial, 235
  compromise solution, 345
  constraint approach, 253, 276, 347
  dominated solution, 345
  goal programming, 349
  ideal point, 349

interactive approach, 346
most preferred solution, 346
noninferior solution, 345
Pareto front, 243, 345
Pareto optimal set, 345
Pareto optimal solution, 52, 276, 285, 345
trade-off analysis, 350
trade-off design, 275
trade-off ratio, 346
weighted sum approach, 348

multiple load sets, 5

multiple loading conditions, 5, 15, 23, 94

mutual energy, 43

natural coordinate, 177
natural frequency, 43, 69
natural period, 43, 69, 311
neighborhood solution, 326
nested analysis and design, 55
nodal cost, 87, 123, 167
nonlinear programming (NLP), 2, 5, 7, 89, 162, 189, 321
  constrained, 323
  multistart, 253
  unconstrained, 321
normal distribution, 243
numerical instability, 90

objective function, 320
  lower bound, 102, 108
  optimal value, 206, 320
  space, 150, 241, 345
offset curve, 288
offset vector
  normal, 274
  unit normal, 274
  vertical, 274
open angle, 262
optimal control, 185
optimal plastic design, 4, 11
optimal solution
  absolute, 345
  approximate, 52, 206

Author Index

417